ENERGY POLICY: ECONOMIC EFFECTS, SECURITY ASPECTS AND ENVIRONMENTAL ISSUES

Energy Policy: Economic Effects, Security Aspects and Environmental Issues

Noah B. Jacobs
Editor

Nova Science Publishers, Inc.
New York

For permission to use material from this book please contact us:
Telephone 631-231-7269; Fax 631-231-8175
Web Site: http://www.novapublishers.com

LIBRARY OF CONGRESS CATALOGING-IN-PUBLICATION DATA

Available Upon Request

ISBN 978-1-60692-894-3

Published by Nova Science Publishers, Inc. ✢ New York

CONTENTS

PREFACE

Energy policy is the manner in which a given entity (often governmental) has decided to address issues of energy development including energy production, distribution and consumption. The attributes of energy policy may include legislation, international treaties, incentives to investment, guidelines for energy conservation, taxation and other public policy techniques.A national energy policy comprises a set of measures involving that country's laws, treaties and agency directives. This new book presents the latest research on the economic effects, security aspects and environmental issues connected with energy policy.

Chapter 1 - In the context of the Kyoto protocol and the European Union's energy strategies, different policies focusing on renewable energies has been developed in many European regions. Navarre (Spain) has been considered as one of the leading regions in renewables worldwide over the last ten years. Thus, the authors are going to analyze in this paper the reasons for this thriving development having the purpose of studying the main policies carried out during the period 1984-2006 in Navarre to guide the industrial policy to its crest. The success of renewable energies in this region has been the result of the conjoint impact of some characteristic traits: decisive institutional support, significant industrial initiatives and the social consensus. Since the implementation of the first windmill as energy generator in Navarre in 1995, this enhancement of renewables has had a significant economic impact in terms of employment and R&D in that region.

Nevertheless, this positive situation could change in the future if the economic and energy growth rates follow the same pattern than in previous years. Consequently, the authors are going to study the future of Navarre in terms of the renewable energies contribution into the energy consumption. Particularly, they carefully consider the role played by the substitution process between non-renewable and renewable energies. Concerning sustainability, the implementation of renewable energy in Navarre has involved some effects of lower increase of the pollutant gases emissions, but not in the expected quantity. Furthermore, the renewable energies have not been revealed as substitutive energies to the traditional and non-sustainable ones, albeit they have increased the awareness of doing so in the nearby future. Therefore, the conclusion for this forecast is that the impact of the renewable development is mainly constrained by the economic growth rates. Finally, another important achievement in the energy policy in renewables in Navarre, is the easiness of spreading the policies and results described here to other regions worldwide.

Chapter 2 - Since the start of utilizing the electric energy in its basic shape, until today, the way in delivering energy to customers have gone through various stages according to the needs, capability, knowledge, technology, and rules. Starting without regulators until it is found necessary to have regulators govern the small utilities before more than a century ago. Then, vertically integrated utilities became the most popular practice all over the world with little differences in detailed practices. Despite those differences, typically, such utilities own three major business lines namely generation, transmission and distribution covering a specific geographic area. Such utility is self dependent in all aspects from technical support to auxiliary services and billings. However, some of utilities are getting financial support (subsidies) from their governments to avoid high bills on customers.

Chapter 3 - Many researchers argue that renewable energy (RE) will eventually have to replace the present fossil fuel dominated global energy supply system, for two reasons. First, resource depletion will increasingly limit annual availability of oil, and later natural gas and even coal. Second, the European Union is calling for a 2 °C cap on allowable temperature increases compared with the pre-industrial era, in order to avert dangerous anthropogenic climate change. Keeping within this 2 °C limit will mean cutting present world emissions of greenhouse gases to as little as 15 % of their 2000 value. Some researchers also argue that other solutions, such as nuclear energy and carbon sequestration and storage, will merely solve some problems but introduce others. This leaves RE, which can be conveniently divided into non-intermittent and intermittent sources. The first category includes hydro, biomass, and geothermal. Using an earth system science approach, we analyse each of these in turn, and demonstrate that their likely future potential is small, because of limited availability, competition for scarce inputs (especially biomass), or because their potential will be adversely affected by unavoidable climate change. The second category includes solar and wind energy, each with much higher global technical potential, and the various forms of ocean energy. However, the technical potential for both wind and solar—particularly solar thermal electricity with its need for direct insolation—are unevenly distributed among the earth's regions. For this reason, large-scale very long-distance energy transmission will be increasingly needed as RE penetration rises. Further, these intermittent sources will require conversion of electricity to some other energy carrier, possibly hydrogen, and storage. The need for international transmission of energy, for example from the North African Sahara to the countries of Western Europe, raises security issues similar to those for oil. The energy losses and capital expenditure associated with energy conversion, storage and transmission will both raise costs of delivered energy and delay their implementation. Overall, energy security, together with economic and environmental cost considerations, point to a lower energy future than most energy analysts forecast.

Chapter 4 - This chapter aims at giving a brief overview of the current knowledge about global CO_2 emissions and related effects on our climate. Additionally, some of the contributors of CO_2 emissions are quantified. The chapter is divided into four separate parts that aim at investigating topics related to CO_2 emissions and climate change. All topics presented here directly impact on both short and long term future of humankind. The article combines an overview over the current state of affairs concerning our oil reserves, the potential future of a hydrogen economy, explains how mass media try to create a controversy in the discussion about global warming that does not exist in the scientific community, and finally calculates different sector contributions to green house gas emissions, outlining how largely varying these numbers are between different sources.

Chapter 5 - Latin American countries have increasingly turned to fossil fuel-generated electricity to fill demand growth in recent years as the hydro applications that comprise 73%

of the region's energy generation portfolio have proven unreliable in years of drought. These countries, which experienced a wave of restructuring in the 1990s, cannot attract private investors that are willing to construct hydro applications because these sources of new generation necessitate huge capital expenditures. Recent fossil fuel prices have caused some countries to consider how to steer new capacity additions away from fuel oil generation with high operating costs. Many of these countries have implemented new or strengthened renewable energy legislation to promote generation with stable operating costs. Countries have pursued diverse renewable energy policy incentives like production tax credits in Argentina, feed-in tariffs in Ecuador, and portfolio mandates in Chile. The impact this legislation will have on future capacity additions will depend on how it is incorporated in the existing energy sectors and the longevity of the support. Countries like Brazil that have strong renewable energy laws may neglect this new legislation as domestic future fuel prices drop in response to new fossil fuel discoveries like the large off-shore oil deposit found in late 2007. The decision made by these countries on how to fulfill future demand growth is critical for climate change mitigation. However, an international Mechanism to involve developing countries in the effort to mitigate climate change ironically may be rendered useless if a Latin American country has taken domestic steps to promote renewable energy applications.

Chapter 6 - This chapter offers an assessment of projects to reduce greenhouse gas emissions, in particular projects involving biomass utilization, under the Clean Development Mechanism (CDM) of the Kyoto Protocol, as well as an analysis of price setting strategies for investing firms and host country governments involved in CDM projects. Drawing on a published database of registered CDM projects and project design documents, the assessment presented here shows that biomass utilization projects are particularly prevalent—and are of great significance—in Asia. To analyze price setting strategies, we employ a simple microeconomic model in which a foreign firm sets the price at which it purchases certified emission reduction credits obtained in a CDM project, and a host country government owning or regulating a utility or electricity distribution company sets the price at which it purchases the electricity the project produces. The model shows that the valuation of the social benefit of projects by host country governments plays a critical role in price setting strategies.

Chapter 7 - The development of electricity policy in Hong Kong has illustrated the arduous task of incorporating environmental principles into a traditionally growth-led model. Attempts to reform the obsolete policy framework prove ineffective due to some path-dependent institutional constraints which are inherited from its development model and intensified by the recent political contests. These have precluded a timely transformation of the relations between the power supply and demand. The state has failed to tackle these structural impediments and tends to focus more on technical solutions than truly collaborative initiatives. The minimal progress in local democratic development and the planned extension of the monopolist status of the existing utilities have also created public distrust against the state and the utilities. Environmental initiatives then become technocratic or rhetoric. The analysis reveals that the ecological rationalization of the electricity policy has been confined by the inherent institutional dilemma. Capacities for developing a sustainable electricity policy have diminished as some existing institutions resist changing, despite preliminary forms of responsive institutions are established.

Chapter 8 - This article discusses the emergency management strategies of municipal authorities for securing the electricity supply, according to a networked, or "governance", control and direction structure, and how this influences the relationship between electricity companies and Swedish municipalities. The article also discusses the role of the households during outages, and how they perceive their responsibility during an outage, but also how municipalities and grid companies understand the households' responsibility. Since electricity market deregulation, the responsibility for electricity supply security has become unclear; field studies of Swedish municipalities indicate that all actors still seem to be seeking to find their proper roles in the deregulated market. Municipalities still expect to exercise influence over private energy company decisions regarding prioritization of emergency power deliveries. Energy companies vacillate between emphasizing their need to regard economic factors and their sense of responsibility for providing a secure electricity supply to vital municipal functions (even though municipalities may lack contracts specifying this). The households' responsibility for being prepared for an outage was fuzzy for the actors. Both municipalities and grid companies expected the customers to have some preparedness. The household meant that they lacked any responsibility to be prepared for an outage even if they *need* to be prepared to survive.

Chapter 9 - The world energy scenario in the late XX century was characterized by a widespread trend towards liberalization and privatization in the area of domestic policies. In the domain of international energy-related transactions, a naive belief in the omnipotence of technology and free market mechanisms contributed to consolidate uneven trade and geopolitical relations of forces strongly favoring advanced Western countries, and historically low prices prevailed in world energy markets.

This scenario underwent a sea-change around the year 2000, leading to a very different situation in the early XXI century. New large and middle-sized economic and geopolitical powers emerged in the developing world, from China to Venezuela. The South as a whole is less financially, economically, and politically and dependent on the North than it was the case in the past. Direct public control on energy resources is emerging again as a key factor of economic and geopolitical strength, with nation-states and SOEs increasingly dominating world markets. Multi-billion long term bilateral and multilateral agreements regulate an increasing share of world energy trade. The relative weight and role of material, non-reproducible resources is re-establishing itself vis-à-vis that of non-material, reproducible assets such as knowledge and technology.

Taking into account the profound change in geopolitical relations of strength worldwide, and the inadequate development of alternative energy sources, and notwithstanding the serious impact of the present international economic crisis, the persistent dependency of advanced and developing market economies alike on the availability of ultimately finite fuel reserves is bound to lead to an era of historically high energy prices in the medium-long term, and is not sustainable in the very long term.

Chapter 10 - In the post-war period, energy (and especially electricity) provision was heavily regulated by the state which guaranteed long-term security of supply through a centrally planned predict-and-provide model. In the early 1990s, this was abandoned in favour of a neoliberal model which saw whole-sale privatisation of the sector. This resulted in lower electricity and gas prices for consumers in the short term, and the abandonment of investment and R&D in the energy sector. A few years later the political taste for intervention in the energy sector increased again because of environmental and (more recently) security-of-supply concerns. But despite a range of state incentives, including the setting of national targets and the provision of substantial subsidies, the private sector has been unable to deliver many of the renewable energy solutions which have been proven in other countries. The UK government has belatedly recognised the limitations of reliance on such large-scale private sector initiatives in the power sector and it has started to embrace the idea of supporting more community based initiatives. The authors examine the contrasting and ambiguous role of the state in the development of renewable energy provision in Denmark and the UK. Danish policy, based on the creation of an enabling environment for local initiatives –many of which are cooperatives-, has been highly successful (20% of national electricity supply). By contrast, UK policy for local-level initiatives has mainly sponsored social enterprise initiatives that are embedded in existing non-profit organisations such as schools and municipalities. The potential role of electrification cooperatives in developing the renewable energy sector in the UK (currently stuck at 3-4% of national electricity supply) is highlighted through a comparison with North American electricity co-operatives who are leading the US electricity sector with regards to renewable energy.

Chapter 11 - The main goal of this paper is to show that in the Brazilian Electricity Sector the Regulatory Authority (ANEEL) faces a significant challenge. The authors show that due to the physical features of the Brazilian Electricity Sector coordination is more important than competition, so bundled transactions are more efficient than transactions through verticalized firms. They concluded that firms strategies are to be organized as holding companies, and this lead them to have a high market power, which demands from ANEEL a strong position against the abuse of market power.

Chapter 12 - Energy consumption in the transport sector is significantly growing at the global level, with negative socioeconomic and environmental implications, such as concerns over security of energy supply and CO_2 emissions. According to the Intergovernmental Panel of Climate Change (IPCC) energy consumption in this sector has doubled between 1971 and 2000 (from almost 1000 to 2000 million tonnes of oil equivalent), despite the significant energy efficient improvements in automobiles. This trend is expected to continue in the short, medium and long terms. According to the same source, energy consumption is expected to more than double between 2000 and 2050. Although developing countries are expected to account for the largest increase in energy consumption in this sector, some developed countries will probably increase their energy consumption in the transport sector significantly, such as Spain. This is so because scale effects are likely to offset technological effects and other factors, leading to a net increase in energy consumption and CO_2 emissions. This paper aims to provide a diagnostic on the main factors influencing energy consumption and CO_2 emissions in the passenger transport sector in Spain. The paper will also identify those policy measures which are likely to be more effective to reduce this energy consumption.

Chapter 13 - There are two important challenges: new oil shock and climate change where energy efficiency improvement is considered as a measure of them. In fact, IPCC and EEA suggested that energy efficiency improvement policy can be the most effective and economic measure in the short term. According to these suggestions, organizations and governments are planning for efficiency improvement in the international and national levels. But this argument has been refuted in the point where goal of this policy will not be accomplished if the market mechanism is not considered seriously. The main issue is who will take responsibility for this policy. Will it be the government or the market? This paper checks the roles of the market and government in respect to energy efficiency improvement. This paper also reviews the controversies of market failure, government failure and rebound effect theoretically and empirically. In conclusion, it shows the policy implication of energy efficiency improvement in the context of climate change. Also it suggests that smart government, which hopes to reflect market mechanism in this policy, has to reform traditional tax system ecologically.

Short Communication - Since the early 1990s, industrial energy end-use policy activity in Sweden has been scarce, in particular towards small- and medium-sized industrial enterprises (SMEs) and non-energy-intensive manufacturers. While the other Nordic states have taken considerable actions towards the industrial sector, Sweden has not. Rising electricity prices in Sweden and the Energy End-use Efficiency and Energy Services Directive (ESD), among other things, address the need for the Swedish authorities to design and adopt energy policies towards the non-energy-intensive and SME industrial sectors. This chapter aims to explore this by discussing plausible policy solutions, based on current research on barriers to energy efficiency and industrial energy programs. Initially, results from the three available Swedish studies on barriers to energy efficiency for these sectors are outlined. Then, a few examples of successful and less successful actions towards the industry and in particular SMEs are presented and finally, a discussion is held regarding plausible energy end-use polices, based on the presented research. In conclusion, a highly useful approach towards non-energy-intensive manufacturers and SMEs in particular is to provide energy audits free of charge and involve the local authority energy consultants.

RESEARCH AND REVIEW STUDIES

In: Energy Policy: Economic Effects, Security Aspects… ISBN 978-1-60692-894-3
Editor: Noah B. Jacobs

Chapter 1

ENERGY POLICY IN RENEWABLES AND ITS ECONOMIC AND ENVIRONMENTAL CONSEQUENCES AT REGIONAL LEVEL: THE CASE OF NAVARRE (SPAIN)

Javier Faulin*

Department of Statistics and Operations Research, Los Magnolios Building, first floor, Public University of Navarre, Campus Arrosadia, s/n, Pamplona, 31006, Spain

Fernando Lera-López, Alejandro Arizkun†

Department of Economics, Los Madroños Building, second floor, Public University of Navarre, Campus Arrosadia, s/n, Pamplona, 31006, Spain

Jesús M. Pintor

Department of Mechanical Engineering. Los Pinos Building, ground floor Public University of Navarre, Campus Arrosadia, s/n, Pamplona, 31006, Spain

ABSTRACT

In the context of the Kyoto protocol and the European Union's energy strategies, different policies focusing on renewable energies has been developed in many European regions. Navarre (Spain) has been considered as one of the leading regions in renewables worldwide over the last ten years. Thus, we are going to analyze in this paper the reasons for this thriving development having the purpose of studying the main policies carried out during the period 1984-2006 in Navarre to guide the industrial policy to its crest. The success of renewable energies in this region has been the result of the conjoint impact of some characteristic traits: decisive institutional support, significant industrial initiatives and the social consensus. Since the implementation of the first windmill as energy

* Corresponding author. Department of Statistics and Operations Research, Los Magnolios Building, first floor, Public University of Navarre, Campus Arrosadia, s/n, Pamplona, 31006, Spain. Email: javier.faulin@unavarra.es. Ph: +34-948169211. Fax: +34-948169204

† Member of the association "Scientifics for the Environment-CiMa".

generator in Navarre in 1995, this enhancement of renewables has had a significant economic impact in terms of employment and R&D in that region.

Nevertheless, this positive situation could change in the future if the economic and energy growth rates follow the same pattern than in previous years. Consequently, we are going to study the future of Navarre in terms of the renewable energies contribution into the energy consumption. Particularly, we carefully consider the role played by the substitution process between non-renewable and renewable energies. Concerning sustainability, the implementation of renewable energy in Navarre has involved some effects of lower increase of the pollutant gases emissions, but not in the expected quantity. Furthermore, the renewable energies have not been revealed as substitutive energies to the traditional and non-sustainable ones, albeit they have increased the awareness of doing so in the nearby future. Therefore, the conclusion for this forecast is that the impact of the renewable development is mainly constrained by the economic growth rates. Finally, another important achievement in the energy policy in renewables in Navarre, is the easiness of spreading the policies and results described here to other regions worldwide.

1. Introduction

The development of the renewable energies involves a big economic effort and a strong political will. The justification for that effort and will is different according to the outlook of the society and its agents. Nevertheless, the idea of reaching the sustainability is spreading in the modern societies (The Economist, 2008). Then, the main reasons of performing that effort are the environmental impacts associated to the production of energy and the climatic change (e.g. the reduction of local and global pollutants). Similarly, socioeconomic impacts of renewable energies expansion should be considered. This includes the diversification of energy supply, regional and rural economic development and the domestic industry creation. Thus, it is possible to face new scenarios, different to current ones, in which the hydrocarbons are not cheap and abundant. Keeping in mind this new scenario worldwide, we are going to focus our attention on the energy policy in a small region in Spain called Navarre.

Therefore, we will center the purpose of this paper on the study of the renewables development in Navarre, as a way of explaining a success energy policy in enhancing renewable energies at regional level. Accordingly, we indirectly seek that the Navarre case serves as a blueprint to encourage other energy experiences worldwide. Concerning the paper organization, it can be described as follows. We begin with an introductory section pinpointing Navarre in Spain and in the world with special emphasis in energy issues. The second section analyzes the energy situation in Navarre and its main difficulties. The third section is devoted to the historical analysis of the Navarrese case in relation to renewables. It is particularly interesting the renewables driving factors subsection. Section four explains the relationship between renewable energies and the sustainability problem. We have painstakingly described this section trying to connect those popular topics in energetic terms. Later on, section five describes the economic consequences of the thriving development of renewables, striving to precise the main characteristics of the energy policy carried out in Navarre. Conclusions section and references conclude the paper.

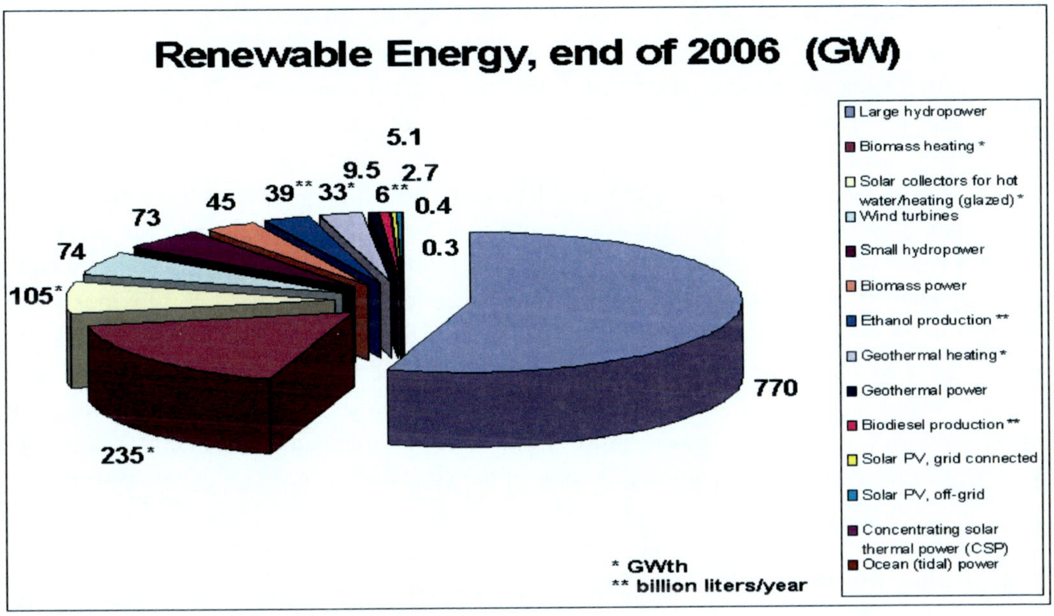

Source: http://en.wikipedia.org/wiki/Image:Ren2006.png

Figure 1. Weight of each type of renewable energies worldwide.

1.1. What are the Renewable Energies?

Renewable energies can be defined as the energies generated from natural resources, such as sunlight, wind, rain, tides and geothermal heat, among others, which are renewable (naturally replenished). Renewable energy technologies include solar power, wind power, hydroelectricity, micro hydro, biomass and biofuels. In 2006, worldwide speaking, about 18% of global final energy consumption came from renewables, with 13% coming from traditional biomass, such as wood-burning. Hydropower was the next largest renewable source, providing 3%, followed by hot water/heating, which contributed 1.3%. Modern technologies, such as geothermal, wind, solar, and ocean energy together provided some 0.8% of final energy consumption. The technical potential for their use is very large, exceeding all other readily available sources. Renewable energy technologies are sometimes criticized for being intermittent or unsightly, yet the market is growing for many forms of renewable energy. Wind power is growing at the rate of 30 percent annually, with a worldwide installed capacity of over 100 GW, and it is widely used in several European countries and the United States (REN21, 2007). The relative importance in the world of each type of renewable energy can be seen in Figure 1.

1.2. The Primary and Final Energies Concepts

We are going to explain the concepts of *primary and final energies* which are indispensable to understand the next paragraphs. It is widely known that the productive system needs from different types of energy and from specific consumption units (inputs). We call this kind of energies, *final energies*. Furthermore, in order to make use from those energies adapted to the final technologies, it has been necessary to follow a path of energy transformation since their direct obtaining from nature. That transformation path could have been more or less long in phases and in time, and during it, energy has been used and spread. Consequently, the amount of initial energy obtained from nature is bigger in volume than the effective energy employed in its direct use. Moreover, the extraction of energy sources, their transformation and the energy transportation has involved the use of energy. Similarly, the previous process is losing energy because of leakages or dispersion in the way of heat. Thus, the energy sources directly obtained from nature are called *primary energies*. Obviously, the primary energy volume is greater than the final energy volume. In sustainability terms, the limits of the used energies are close to the primary energies consumed, and it is considered as an upper bound to measure the maximum amount of energy to be spent. Therefore, the primary energies analysis is much more interesting to manage sustainability concepts. Likewise, this analysis will be a good predictor of the emissions caused by the energy use.

1.3. Defining the Scenario: The Main Energy Policy Tenets

We are going to describe the main energy policies in the European Union, and specifically in Spain and Navarre, to introduce in this paper the consequences of renewables policy at regional level (Club Español de la Energía, 2007). The main references to reach that objective are Moreno and López (2008) and MEC *et al.* (2005). The key problems related to energy policies were modeled some decades ago (Hogan, 2002), but now they are more closely associated to the regional policies of the local governments. Thus, Moreno and López (2008) explained the local energy policy in Asturias (Spain) and Faulin *et al.* (2006) did a similar study in Navarre (Spain). Those references point out the foremost aspects we are going to discuss in the current paper.

In contrast, the most important objectives defined by the European Union in the context of energy policy by 2020 are explained in EREC (2004). Traditionally, the European Union has been a legislative power in the area of energy policy for most of its existence (Reiche, 2002): this has its roots in the original European Coal and Steel Community. The introduction of a mandatory and comprehensive European energy policy was approved at the meeting of the European Council in October 2005, and the first draft policy was published in January 2007. The Commission has five key points in its energy policy: increase competition in the internal market, encourage investment and boost interconnections between electricity grids; diversify energy resources with better systems to respond to a crisis; establish a new treaty framework for energy co-operation with Russia while improving relations with energy-rich states in Central Asia and North Africa; use existing energy supplies more efficiently while increasing use of renewable energy; and finally increase funding for new energy technologies. These key points have to be developed according to the specific instructions given by the Research Community Programs in the European Union such as Intelligent Energy-Europe Program (IEE), FP7, LIFE+ and Marco Polo and also the Structural Funds such as the European Regional Development Fund (ERDF) and the Cohesion Fund (European Union, 2008). Renewable energy sources can also help improve industrial competitiveness and have a positive impact on regional development and employment. The European renewable energy industry has already reached a turnover of €10 billion and employs 200,000 people. Europe is now the global leader of renewable energies.

Currently, the EU imports 82% of its oil and 57% of its gas, making it the world's leading importer of these fuels. There are concerns that the EU is largely dependent on other countries, primarily Russia, for its energy. This concern has grown following a series of clashes between Russia and its neighbors, threatening the flow of gas. The last Russian crisis related to the gas delivery took place in January 2009. As a result the EU, is attempting to diversify its energy supply. Thus, the main energetic measures in the European Union are described in the next goals:

a. The White Paper target of 12% of energy consumption provided by renewable energy by 2010 (15% in 2015 and 20% in 2020).
b. The electricity target of 22.1% of electricity production provided by renewable energy by 2010 (33% in 2020).
c. The biofuel target of 5.75% of the total fuel consumption by 2010 (8% in 2015 and 10% in 2020).

Furthermore, the benefits of this greater contribution by renewables are estimated as:

i) A CO_2 reduction of 728 Mt/year in 2020, representing a 17.6% decrease of the total EU GHG emissions in 1990.
ii) Investments of €443 billion made in the period 2001-2020.
iii) An estimated €115.8 billion saved in avoided fuel costs and from €126.7 and €323.9 billion in avoided external costs.
iv) An additional 2,023,000 people in full time employment in the renewable sector in the period 2001-2020, what means at least 0.8% of the total employment in Europe.

Similarly, the objectives fixed for 2010 to Spain are the following:

a. 12.2% of the energy consumption provided by renewable energy by 2010.
b. 30.3% of production of electric energy from renewable energy by 2010.
c. Use of biofuel of 5.83% on the total use of fuel by 2010.

These objectives involve an annual increase of 22.4% of the produced energy by renewable sources. To reach this objective, €23,600M in investments have been estimated (the public investment is the 3%). The increase in renewable energy has estimated to produce a CO_2 reduction from 547 to 796 tCO_2/year (76,983,254 tn in the period 2005-2010). Also, additional 94,925 people in full time employment in the Spanish renewable sector have been estimated (Ministerio de Industria, Turismo y Comercio, 2005). The European Commission, following the methodology developed by the TERES II Project estimated 44,970 additional jobs expected in Spain in 2010 and 84,397 in 2020 (European Commission, 1996). By 2020, ISTAS (2008) has estimated from 228,435 to 270,788 jobs the total employment in the renewable sector in Spain. See Sánchez-López (2006) for a review of the literature about the estimated employment related to the renewable energy development.

1.4. Navarre in Spain and in Europe

Navarre is one of the seventeen regions (autonomous communities) which constitute Spain. Navarre is situated in the North of Spain, between the Basque Country, Aragon and France. Its surface is 10,391 sq km (4,012 sq mi) showing a great range in climates and landscapes. For instance, Navarre stretches from the Atlantic environment at the Pyrenees and the continental climate in the central area to the desert plains in the South. On the other hand, Navarre enjoys a strategic position in the Spanish communications network because she is the key gate in the passage between Spain and France (with a 100 km border), holding a preeminent position in the Ebro valley (one of the most important focus of development of the continental Spain). Similarly, Navarre benefits from being next to the Atlantic Axis (European regions in the Atlantic shore) in the trade relationships between region (Figure 2).

The population in Navarre in 2007 was 605,876 inhabitants, living in the capital of the region (Pamplona) the 30%. The population density is just 58.31 inhabitants per sq km, which is clearly below the Spanish average (90 inhabitants/sq km). Furthermore, the population growth rate is around 1.67 and the natural population increase is almost nought, taking into account that the percentage of the immigrant population in Navarre is circa 8.5% (Navarrese Institute of Statistics, 2008; Sanz, 2009).

Historically speaking, Navarre was chiefly a rural and hermetic region in the period 1900-1960, having a strong presence of the public sector and with a hardly diversified economy. Nonetheless, the Navarrese society underwent a great transformation during the second half of the twentieth century: its economic development was very important, much greater than the average of the Spanish economy. That is, perhaps, the reason of the prosperity of the Navarrese economy, making it closer to the European economic standards than the majority of the regions of Spain.

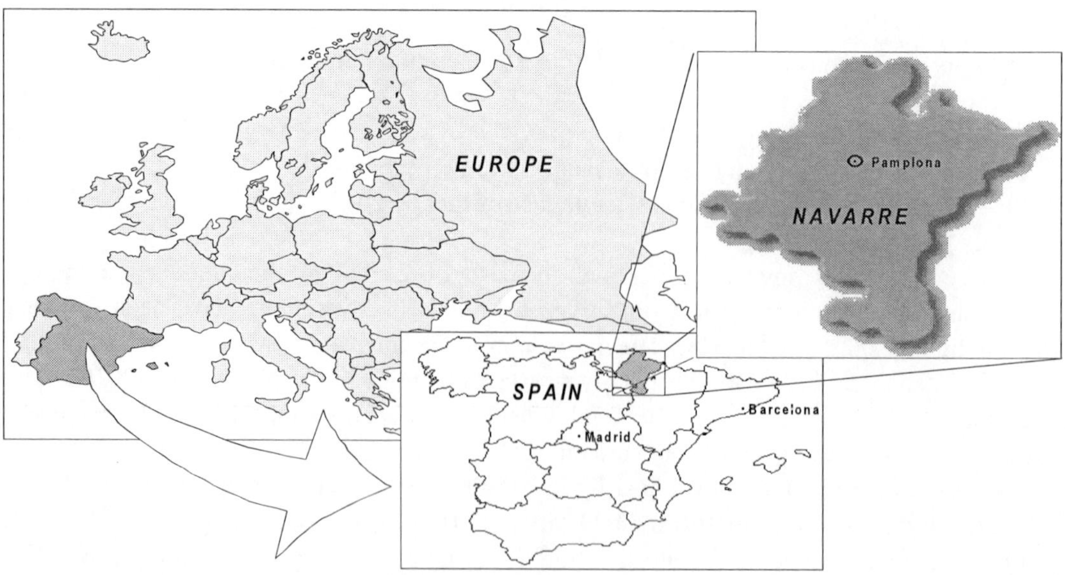

Figure 2. Pinpointing the Autonomous Community of Navarre in Spain.

Next, we are going to summarize some of the most characteristic traits of the economy of Navarre, taking into account the statistical data provided by Spanish National Statistics Institute (2008), Navarrese Institute of Statistics (2008) and Eurostat (2006):

1. Navarre shows a small size within the Spanish context, not only in surface terms (2% of the total Spanish surface), but also in demographic or economic points of view (1.35% in population, 1.68% in Gross Domestic Product (GDP) and 2.96% in industrial GDP in 2007).
2. Navarre enjoys an outstanding place inside the Spanish economy according to its income per person, labor productivity and employment rate. For instance, taking into account the Eurostat 2006 report, Navarre during 2005 is the third region in Spain and the 36th region in Europe in regional GDP (PPS per inhabitant in percentage of the EU-27 average).
3. Likewise, Navarre has an important specialization in industrial activities, but with a smaller development of the service sector. The set of Navarrese companies is chiefly formed by small and medium-sized enterprises (SME), but with a significant presence of the multinational firms. This kind of companies supplies around the 33% of the industrial employment.
4. On the other hand, Navarre shows a high standard in the training of human resources, and in the education level (Navarre is the second one in Spain in population percentage having a university degree (Bancaja, 2003)). Moreover, the development of infrastructures and the institutional policy of supporting innovation have attracted the foreign investment, funding companies of high technological standards.

Similarly, Navarre presents some unique historical characteristics (it was the last kingdom in the Iberian Peninsula which was incorporated to Spain, because of the development of the national states in the fifteenth century) and some legal privileges (the well-known *Fueros*) which makes that the specific Navarrese tax system has been different to the Spanish system. The Navarrese tax system has a complete autonomy in budget and in law in many arenas. Table 1 summarizes some of the previous data of Navarre in relation to Spain and the European Union-27.

Table 1. Data of Navarre in Spain and in Europe, Year 2007

Variables	**Navarre**	**Spain**	**European Union-27**
Surface (Km^2)	10,391	504,790	4,324,782
Population (in thousands)	605	44,630	499,021
Unemployment rate	4.8	8.3	7.3
Employment distribution by sectors			
Agriculture	4.5	4.0	2.3
Industry and construction	38.7	28.4	30.0
Services	55.8	67.6	67.7
GDP per capita	129	103	100

Source: Own design taking information from the Eurostat, the Spanish National Statistics Institute (INE) and the Spanish Ministry of Economy.

2. The Energy Sector in Navarre. An Overview from 1984 to 2006

We are going to present a historical outlook of the energy sector in Navarre during the interval 1984-2006, which is the period of disposal of energy balances given by the Government of Navarre. Specifically, we will explain the main characteristics of final energies in Navarre. Thus, the final energy is defined as the set of energies which are put at disposal of the final consumer (Sánchez-López, 2008). Other important references to describe this historical review are Moreno and López (2008) and ISTAS (2008) about energy demand and energy consumption.

2.1. Transforming Primary Energies into Final Energies

Historically speaking, the energetic transformation to obtain other final energy sources has been rather small. Hence, the production[1] of final energies was constrained, up to 1993, to the electricity generation with minihydraulic centrals and the use of wood as biomass to be burnt. The remainder of the used energy should be imported. Nevertheless, from 1994 a huge increase in the production of energy was generated due to the wind energy. The extensive use of natural gas to produce energy is carried out at the beginning of the twenty-first century at biomass' expense. Moreover, the new final energies which appear in the last years, such as biodiesel or solar-thermal, have no real influence in the global energy balance and its main interest is based on their future development (Figure 3).

The production of final energies in Navarre has been increased in the period 1984-2006 by the factor 2.2. This involved an annual increase of 3.85%, although it is necessary to distinguish into two different periods: 1984-1995 and 1996-2006. Up to 1995, the energy production decrease is -3.12%. Nevertheless, from 1996 it has been an important increase in a 9.15% per year. However, this augment seems to have been stopped in the year 2006. This increase in the electric production has been based on the popularity of wind energy since 1994 and, mainly, on the natural gas combined cycle centrals since 2002.

Thus, the evolution of the final energies production in Navarre can be summarized with the next milestones (Government of Navarre, 2007a):

- 1991. Setup of a biomass plant to generate electric energy in Sangüesa (Navarre).
- 1994. Setup of the first wind park in the Perdón hills in the surrounding of Pamplona.
- 2000. Setup of the first "solar garden" to generate photovoltaic energy in Southern Navarre.
- 2002. Setup of two centrals of combined cycle to produce electricity from gas in Castejón (Navarre).

[1] Scientifically speaking, energy is neither created nor destroyed, but transformed. We used in the text the words "generation" or "production" for brevity reasons, but they should be understood in the sense explained here.

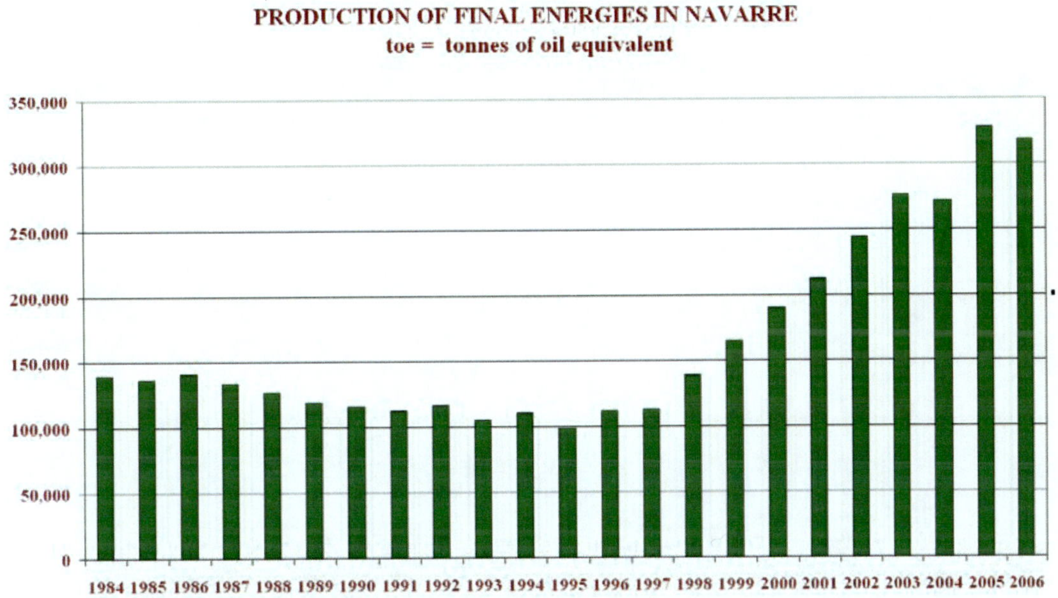

Source: Own elaboration from the Energy Balances of Navarre in 1984-2006 (Government of Navarre, 1984-2006; Government of Navarre, 1994-2006).

Figure 3. Profile of final energies in Navarre in the period 1984-2006.

- 2004. Starting up of the heat production with thermal-solar shells.
- 2005. Beginning of the commercial production of biodiesel from vegetal oils in Caparroso (Navarre).
- 2006. Beginning of the urban waste processing plant to obtain biogas to produce energy in Tudela (Navarre).

2.2. The Use of Final Energies

In spite of the energy production increase in Navarre, the energy consumption has soared, being multiplied by 2.6 in a continuous augment in the period 1994-2006. All final energy sources have experimented an unremitting rising in use, except the biomass whose exploitation fell in a 50%. Conversely, the natural gas two-folded its use reaching the second position in the energy sources ranking. Nonetheless, it is necessary to point out the peak of the oil derivatives in the period 1984-2006, reaching an increase percentage in that period of time between the 49% and 59% of the total used final energy (Figure 4).

Furthermore, the use of final energies by economic sector has tremendously increased since 1993, highlighting the energy consumption in the transport sector in contrast to a light reduction in energy use in the industrial sector during the last years of the period 1984-2006 (Figure 5). In the study by sectors, we have organized the economic sectors according to the next criteria: a) "Primary" Sector includes Agriculture and Farming activities b) Sector "Industry" contains public works, manufacturing and construction activities and c) Sector "Transportation" contains, among others, the use of gasoline and diesel by consumers d) Sector "Other economic activities" includes commerce, and public services, among others.

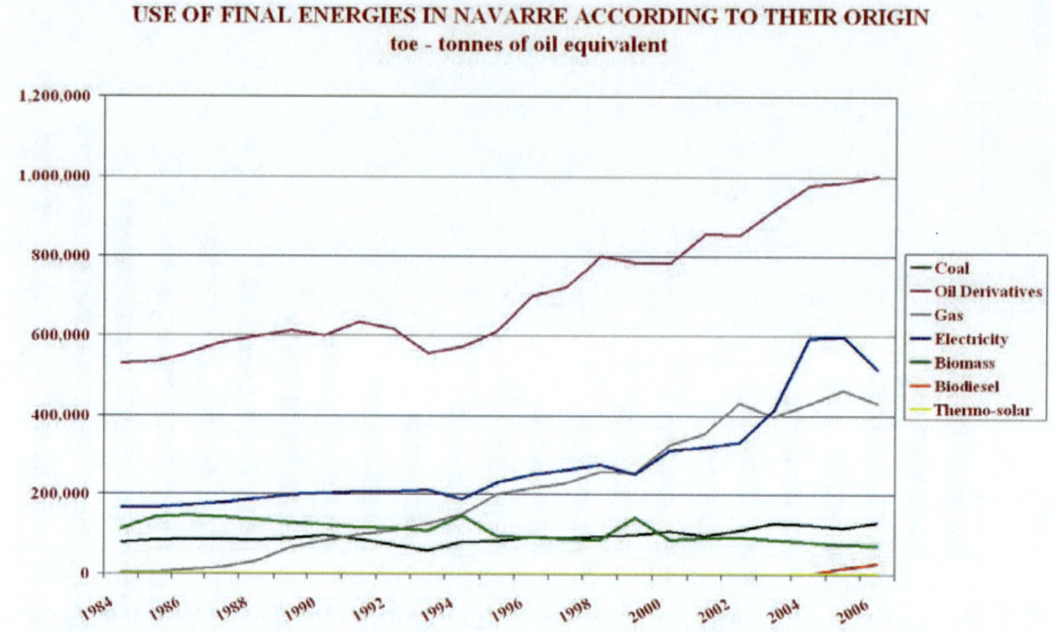

Source: Own elaboration from the Energy Balances of Navarre in 1984-2006 (Government of Navarre, 1984-2006; Government of Navarre, 1994-2006).

Figure 4. Use of final energies in Navarre in the period 1984-2006 according to their origin.

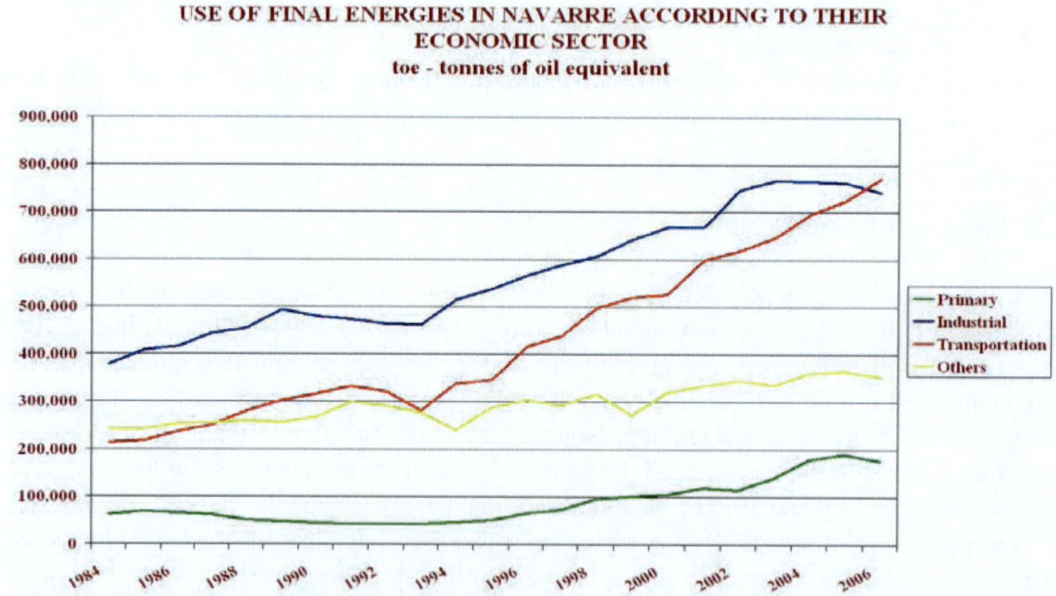

Source: Own elaboration from the Energy Balances of Navarre in 1984-2006 (Government of Navarre, 1984-2006; Government of Navarre, 1994-2006).

Figure 5. Use of final energies in Navarre in the period 1984-2006 according to their economic sector.

Apart from the previous data, some meaningful pie charts (Figure 6) can help to understand the Copernican change associated to the final energy sources during the period 1984-2006. Making a comparison between the final energy sources in the years 1984 and 2006, it is clear that oil derivatives have reduced their use in percentage during that period of time. Undoubtedly, this change has been due to the promotion of the renewable energies in Navarre with appropriate policies.

Consequently, it is required to analyze in detail the strategic policy followed in Navarre. The next section explains the driving factors behind the renewable boom in Navarre.

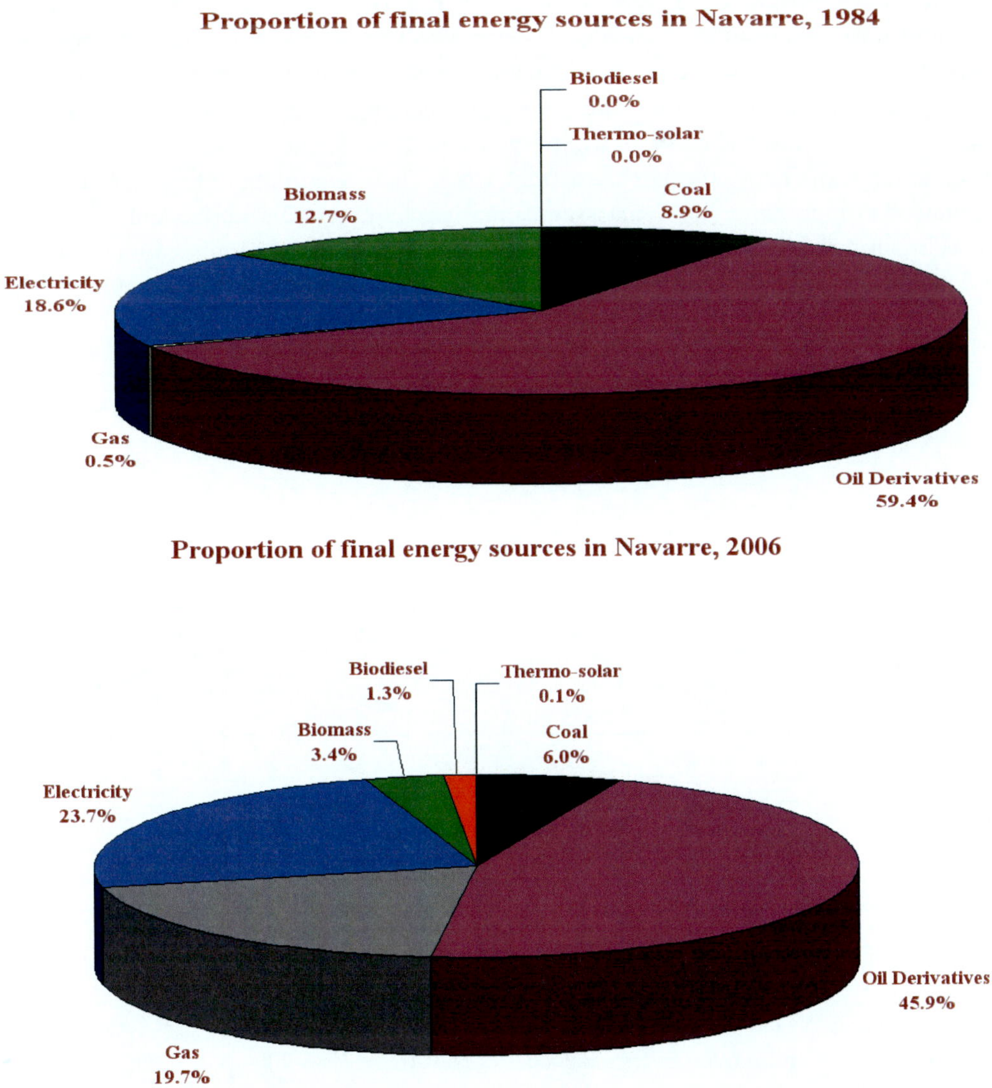

Source: Own elaboration from the Energy Balances of Navarre in 1984-2006 (Government of Navarre, 1984-2006; Government of Navarre, 1994-2006).

Figure 6. Percentages of use of final energy sources in Navarre in the border years 1984 and 2006.

3. Development of Renewable Energy in Navarre in the Period 1984-2006

3.1. Initial Stage: 1984-1995

The use and development of energy sources are traditionally constrained by the natural resources in a region or in a country and rarely can they be only explained by the political success of a regional or national government. Nevertheless, different situations can enhance the particular expansion of a specific kind of energy source: the renewable energies. Historically speaking, from 1950 to 2003 Navarre was characterized as being heavily energy-dependent on other regions or countries. Navarre has no source of primary energy such as oil, gas or coal. Navarre's only original energy resources are hydroelectric power stations and the biomass, as it has been shown in the previous section. Figure 7 summarizes Navarre's situation in terms of energy sources from 1984-1995. Then, according to Figure 7, it is easy to understand the importance of biomass and minihydraulic centrals during the period 1984-1995. Without the contributions of these types of energies in the aforementioned decade it would have been very difficult to manage an important development in the renewables in Navarre during the following decade 1995-2006.

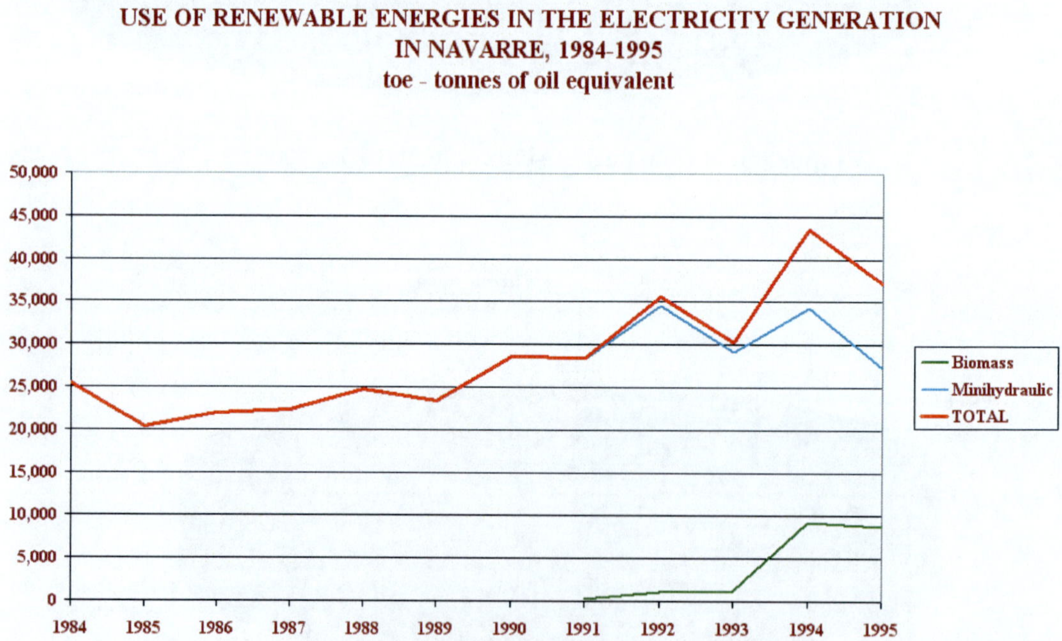

Source: Own elaboration from the Energy Balances of Navarre in 1984-2006 (Government of Navarre, 1984-2006; Government of Navarre, 1994-2006).

Figure 7. Use of renewable energies in the electricity generation in Navarre, 1984-1995.

In 1985, the Government of Navarre, aware of these poor figures, passed a law that embodied the first energy plan of the region (Government of Navarre, 1995). It is sought, among other things, to:

a. Diversify energy sources.
b. Reduce energy dependency, by exploiting the region's abundant renewable energy sources.
c. Develop new infrastructures to enable the use of new energy sources.
d. Enhance the conservation and efficient management of energy.

These objectives had a strong influence in the renewable energy policy, which will be analyzed later in detail. In 1995, the Navarrese authorities approved their First Regional Energy Plan 1995-2000 (FEPN) which are going to discuss in the next section.

3.2. The Renewable Energy Boom in Navarre: 1995-2006

The renewable energy boom in Navarre in the period 1995-2006 cannot be explained without the approval of the Navarre's First Regional Energy Plan (FEPN) in 1995, which was really innovative in the European scenario, because it involved a great interest in the development of renewable energies in Navarre within a sustainable energy policy. Particularly, this Plan emphasized the importance of wind energy in the deployment of renewables in Navarre. Undoubtedly, this milestone was one of the most decisive incentives which explain the success of the renewable energies in Navarre. Thus, Navarre became one of Spain's leading regions in the development of a renewable energy promotion plan, setting itself very ambitious aims for the end of 2010.

Navarre's First Regional Energy Plan (FEPN) was implemented with great success between 1998 and 2002, predominantly in the areas of wind and solar-PV energy, with a moderate increase in other types. The growth pattern is summarized in Figure 8.

The US rating agency Standard & Poors, in a current investigation of standard of living in Europe, ranked Navarre, whose primary source of renewable energy is wind power, uppermost among the 17 autonomous regions of Spain. Navarre, Europe's sixth largest producer of wind power, currently sustains approximately 70 percent of its electricity needs from renewable energy sources, wind farms being used most extensively, and has a 900-megawatt capacity of installed wind power, ranking it ahead of the UK, Sweden, and France (Stewart, 2006).

Navarre's progress in wind energy far surpasses that of other Spanish regions. Thus it ranked slightly ahead of Italy (788 MW), which took fourth place in Europe for wind power production in 2003. Furthermore, Navarre owned 2.5% of the installed power in the world and the 20% of the installed power in Spain (IDAE, 2003). When it comes to solar-PV energy, Navarre owned the largest plant in Spain, with 50% of the installed power. Therefore, it is possible to highlight that the importance of renewables during the period 1995-2006 was mainly based on the wind, biomass and minihydraulic energies, with interesting contributions from the solar-photovoltaic and thermo-solar energies.

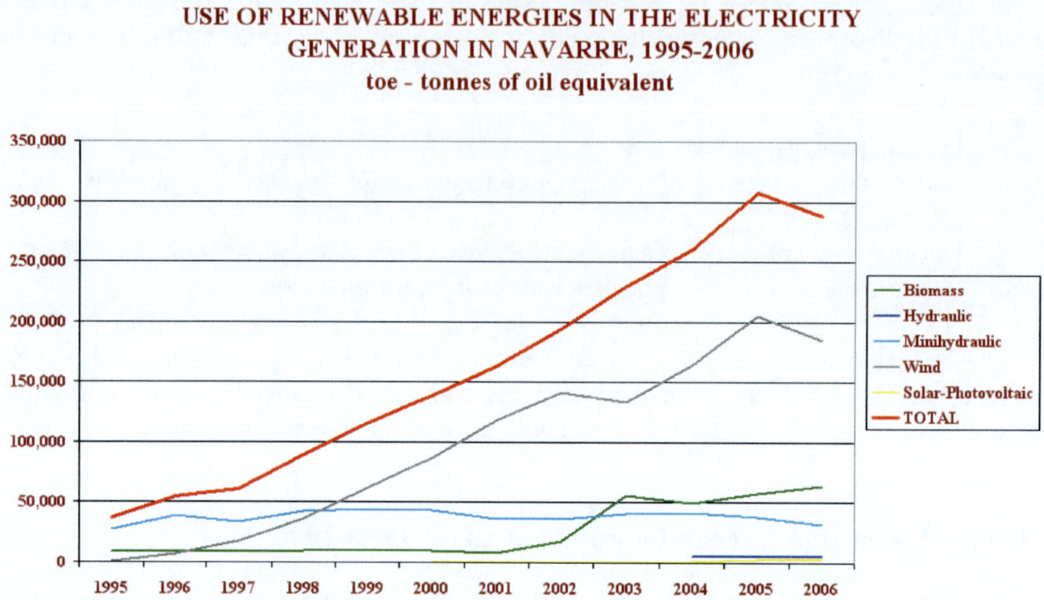

Source: Own elaboration from the Energy Balances of Navarre in 1984-2006 (Government of Navarre, 1984-2006; Government of Navarre, 1994-2006).

Figure 8. Use of renewable energies in the electricity generation in Navarre, 1995-2006.

Similarly, other landmarks have highlighted the importance of Navarre as a European region pioneering in renewables (Fairless, 2007):

a. Navarre lacks thermal, nuclear, coal, oil, gas fields, or hefty hydro-electric power stations, but does possess considerable renewable resources, which the Government of Navarre pursued to drop its foreign energy dependence. Nevertheless, Navarre is a prosperous region in Spain because it has other strengths (Stewart, 2006).
b. Taking into account the aforementioned characteristics of Navarre, the Spanish Government chose Navarre as the location for two important national renewable energy centers: the R&D Centre (CENER - www.cener.com) (CENER, 2006) and the Training Centre (CENIFER – www.cenifer.com) (CENIFER, 2005).
c. Furthermore, the region is steadily emerging as a touchstone for European standards; witness the Navarrese participation in the Wind Energy Potential I project (WEP I) within the European ECOS-Ouverture Initiative.
d. Similarly, Navarre earned the best regional policy in Europe award, at the European Conference for Renewable Energy, held in Berlin in January 2004.
e. Moreover, the success in renewables has implied the creation of a set of companies related to renewable sector in Navarre with an international influence worldwide, for instance, Gamesa, Acciona, Ecotecnia, Derna and MTorres, among others (Faulin *et al*, 2006; Pintor *et al*, 2006).

3.3. Consolidation of the Renewable Energy in Navarre

After the tremendous development of the renewable energies in Navarre in the period 1995-2004 within the First Regional Energy Plan (FEPN) 1995-2000 and its revision for the period 2000-2004, the main tenets of the new Energy Plan 2005-2010 are based on the successful achievements of the previous plan. The Energy Plan 2005-2010 was approved by the Government of Navarre in 2007, what implies that their results were partially known before its implementation. We can mention up to four essential differences which are associated to the new plan:

a. Navarre has reached an important leadership in the promotion of renewable energies, and it is highly desirable to maintain this situation in the future.
b. A significant progress in the appropriate use of electricity production technologies has been achieved, considering also the use of the renewable energies.
c. An important investment in transportation infrastructures in order to distribute electricity and gas has been carried out. This financial effort has involved an energy delivery with quality and competitiveness characteristics.
d. Greater energy intensity has been produced along with new environmental demanding.

Therefore, the Energy Plan 2005-2010 focuses its interests on the next policies:

1. Development of the Navarrese industrial companies associated to the renewable energies technologies, encouraging new entrepreneurship initiatives related to energy efficiency and environmental sustainability (continuous support in R&D).
2. Keeping the leadership of Navarre in the strategic development of renewable energies.
3. Development of active policies of reduction of CO_2 emissions, according to the Kyoto protocol.
4. Assuring the energy supply quality to the rural areas of Navarre in order to facilitate their economic progress.

Hence, if we delve into the energy situation in Navarre in 2003 and in the estimated situation for 2010 we have two really different scenarios according to the results described in Table 2. The savings in CO_2 emissions estimated in Table 2 are related to a lower volume in CO_2 emissions in the probable scenario and the optimist scenario versus an equivalent increase in the energy consumption based of non-renewable energies. Nevertheless, that consumption increase involves a greater failure in the fulfilment of the Kyoto protocol objectives. Assuming that Navarre has to follow the same objectives as Spain of an augment in emissions of 15% in 2010 in relation to the 1990 level, there is an important separation from those objectives according to the prognosis of the Government of Navarre depicted in Table 3. The expected CO_2 emissions are up to six or sevenfold greater than the Kyoto objectives depending on the chosen scenario.

Furthermore, the most important rise in energy production is given by the solar energy, with an expected increase in relation to the most probable scenario in about 1580% in generated power, and by biogas with an increase percentage around 466%. The previous situation involves a greater increase of investment in both energies, mainly in solar, with an implementation increase around 142%, and a 46% increase in public funding.

Table 2. Current situation and predictions of the Energy Plan of Navarre 2005-2010 according to the type of renewable energy

Type of energy	**Wind Power**			**Hydro/Minihydro**			**Solar Energy**		
Scenario	2003	Scenario 1	Scenario 2	2003	Scenario 1	Scenario 2	2003	Scenario 1	Scenario 2
Installed Power (MW)	715.21	1,400	1,530	195.13	305.13	335.13	4.5(1)	40(2)	120(3)
Generated Power (Ktoe)	133.20	261.80	287.30	40.53	57.70	63.40	0.55	9.19	20.60
Saved Emissions (Kt CO_2)	228.50	397.20	435.75	69.49	87.50	95.90	1.04	17.28	35.69
Associated Investment (M€)	n.a.	467.90	583.50	n.a.	56.60	88.40	n.a.	214.71	556.11
Public Funding	n.a.	46.79	58.35	n.a.	5.66	8.84	n.a.	37.15	54.55

Type of energy	**Biomass**			**Biodiesel**			**Biogas**		
Scenario	2003	Scenario 1	Scenario 2	2003	Scenario 1	Scenario 2	2003	Scenario 1	Scenario 2
Installed Power (MW)	25	40	60	30.62	61.25	87.50	1.60	7	9
Generated Power (Ktoe)	14.64	24.08	36.10	n.a.	50.30	70.20	0.85	4.82	6.20
Saved Emissions (Kt CO_2)	25.11	36.52	54.78	n.a.	156	222.80	1.46	7.30	9.39
Associated Investment (M€)	n.a.	22.50	52.50	n.a.	24	48	n.a.	8.64	11.84
Public Funding	n.a.	2.20	5.20	n.a.	2.40	4.8	n.a.	n.a.	n.a.

Note. Scenario 1: Most probable scenario. Scenario 2: Optimist scenario. The most probable scenario is associated to the energy generation according to the 1,400 MW power in the electric network. The optimist scenario assumes a electric network extension to make easier the energy delivery.

Source: Own elaboration from the Horizon 2010 Energy Plan of Navarre (Government of Navarre, 2007a).

(1)Corresponding to the solar-photovoltaic energy, plus the amount of 3,024 m^2 in thermo-solar cells.

(2)Plus 77,045 m^2 in solar cells.

(3)Plus 100,000 m^2 in solar cells.

Table 3. CO_2 emissions increase for 2010 in relation to 1990

	2003	Scenario 1	Scenario 2
Emissions increase (Kt CO2/1990)	1,643.20	2,922.80	2,549.59
Emissions increase (% /1990)	n.a.	102.83	89.70

Source: Own elaboration from de Horizon 2010 Energy Plan of Navarre (Government of Navarre, 2007a).

Moreover, the wind energy, which is the most important renewable energy in Navarre, it is expected that it increases in a 96% the generated power, while biomass, hydro power, and biofuels will only augment globally around 50%. Taking into account the information provided by Figure 9, it is easy to realize that the wind power is going to be the most popular renewable energy in Navarre, for the next years. In fact, the installed wind energy capacity in Navarre in 2006 (Figure 9) was higher than in other countries such as Australia, France, Greece, Sweden, Ireland, Belgium, Norway, Poland or Finland (Government of Navarre, 2007b). Nevertheless, it is thought that this growth of installed power in wind energy will slow down in the following years. Thus, the electricity production by wind power will decrease from the 74% in 2006 to 71% in 2010. That is to say that once the renewable energies development has been consolidated in Navarre, a most homogeneous expansion according to the source type is needed, searching a better growth of all the renewable energies having less sway from an economic point of view. Thus, since 2005 it is forbidden in Navarre the construction of new windmills, and therefore, the only way to renew the current situation according to the present legislation involves the substitution of the old aerogenerators for new ones with a bigger power to produce electricity.

Thus, Table 4 presents the investments done by the two energy Plans from 1995 to 2006 and the importance of public funding in those plans. Considering that information, the public investment in renewable energies has been increasing little by little, being a usual support to reach the renewable boom in Navarre. For instance, the First Energy Plan assumed a 10% of public investment from the total investment in renewables. Nevertheless, it cannot be thought that the public help has been the key factor to explain the importance of the investment done in renewables in the Navarrese region.

On one hand, if we analyze the different investments in renewables, two important aspects have been revealed related to the energy policy: energy saving and efficient use of energy. Hence, these two aspects involved a tremendous reduction of the energy wasting in the years 2005-2008. On the other hand, the consolidation of the renewable energies development has implied a reduction in the total investment in the aforementioned period, going down from 934 M€ to 801 M€, although the public support goes up in percentage because some kinds of renewable energies have received more funding now than in the past according to their smaller development. Furthermore, taking into account the infrastructures consolidation in order to improve the electric networks needed to transport the energy produced by renewables, the Energy Plan 2005-2010 assumes an investment of 259 M€ in comparison to the 68 M€ of the first Plan (which is an increase of 280%). Likewise, it is observed a parallelism between the expected investments in renewables and the objectives of the Energy Plan 2005-2010.

Table 4. Developed investments according to the Energy Plans of Navarre 1995-2004 and 2005-2010, in M€

Investment	Total Investment 1995-2004	Public Investment 1995-2004	Total Investment 2005-2010	Public Investment 2005-2010
Renewable Energies Enhancement	933.82	80.3	801	97.85
Usual Production of Electricity	336.70	33.67	353.16	38.01
Efficient Use and Saving of Energy	39.39	5.78	411.06	91.8
Electric Networks and Infrastructures	68.01	5.1	258.6	13.2
Public Organizations (CENER and others)	23.67	11.75	0	0
TOTAL	**1,401.59**	**136.5**	**1,823.82**	**240.86**

Source: Own elaboration from the Horizon 2010 Energy Plan of Navarre (Government of Navarre, 2007a).

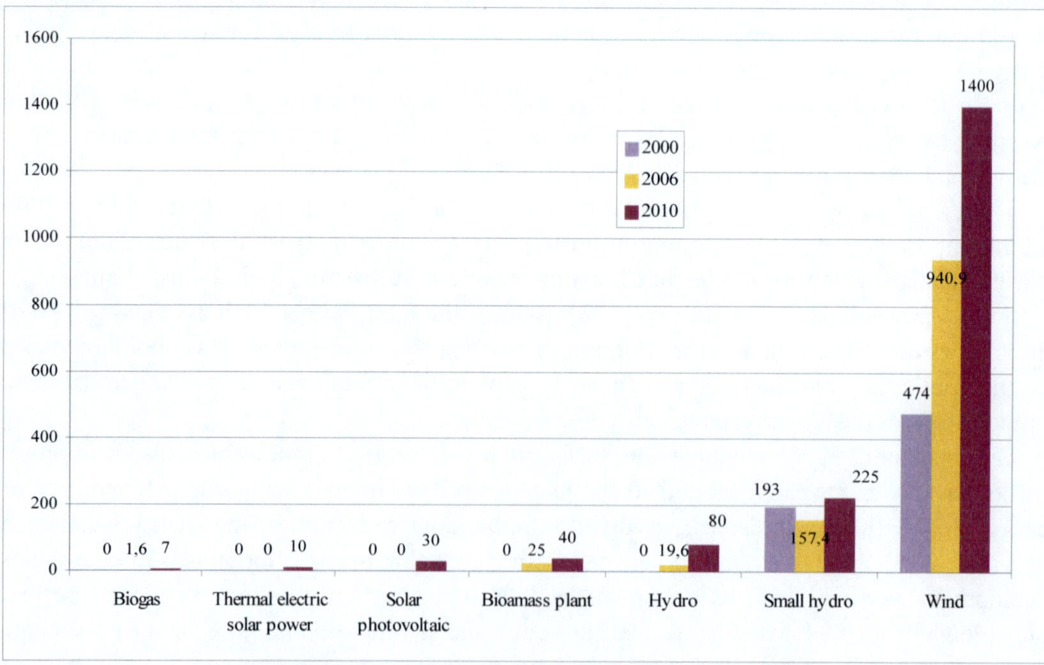

Source: Horizon 2010 Energy Plan of Navarre (Government of Navarre, 2007a).

Figure 9. Installed renewable energy capacity in Navarre. Data in MW.

3.4. Driving Factors behind the Renewable Energy Boom in Navarre

We will now attempt to explain the success of Navarre's 1995-2000 first Regional Energy Plan 1995-2000 and its sequel the Navarre's Renewable Energy Plan 2005-2010, and describe how the specific characteristics of this region contributed towards it. Success was, perhaps, mainly due to a combination of circumstances involving the collaboration of social agents and the local and regional authorities in Navarre (Martinez Erauskin, 2008; Fairless, 2007). This co-operation was essential, as evidenced by the difficulties experienced by Holland (Agterbosch *et al.*, 2004) or the United Kingdom in exploiting the biomass (Upreti, 2004). Other authors have also highlighted the importance of the confluence of four key variables to reach the success in the development of renewables (Toke *et al.*, 2008; Wolsink, 2007) particularly in wind energy. These variables are the next ones: i) the planning regime, ii) the financial support system, iii) the value attached to landscape preservation, and iv) the degree of local ownership of wind power schemes. Undoubtedly, one of the key factors which has been present in the Navarrese success with renewables was the financial support system, but it is not possible to explain that success considering only this factor. The particular idiosyncrasy of the political structure of the Government of Navarre along with the type of entrepreneurship present in that region, the social perception of the renewables, and the local acceptance of projects in renewables can be considered as other important factors in the current success of this type of energies.

The support of renewables in Spain has been implemented using feed-in tariffs, which made the generation of renewable energy profitable. There are two tariff options for green electricity. One is an annual guaranteed fixed tariff which remains the same throughout the year (for wind energy from 6.62 €cents/kWh in 1999 to 6.28 €cents/kWh in 2002 or 6.48 €cents/kWh in 2004 and later). The other is an average pool price plus a production incentive or premium of 50% of the average pool price, approved under the Spanish Royal Decree RD436/2004 (for wind energy from 6.67 € cents/kWh in 1999 to 7.46 in 2002 or 6.67 in October 2004 and later). Tariffs are adjusted annually in accordance with the variation in average electricity prices, but always within a price corridor of between 80% and 90% of the pretax consumer price for electricity, as stipulated for the electricity sector under a 1977 law. In addition, Spanish Royal Decree RD2818/1998 requires premiums to be reviewed every four years, thus generating additional uncertainty. However, this policy of funding renewable energies which has lasted more than 10 years, seems as if it is going to change in the medium term in the funding intensity for some of the renewable energies: the Spanish Government plans to enact a new act reducing the public support to photovoltaic energy production from 0.45€ to 0.29€ per kilowatt (Diario de Navarra, 2008).

Nonetheless, this tariff policy is common to all Spanish regions, and cannot, therefore, be used to explain the Navarrese success nor the current situation in some European countries (Reiche and Bechberger, 2004; Meyer, 2003). Other explanatory factors are the pooled use of strategies based on supply support (supply-push) and demand (demand-pull) policies, along with social policies. These policies explain the growth of renewables in Navarre, especially in the case of wind energy. Moreover, the successful development of wind energy was the catalyst that provided the stimulus for exploiting other renewable sources. Thus, we can summarize the determining factors underlying the development of this sector in Navarre in the following terms:

a. Specific characteristics of Navarre:
   i) Navarre presented a major deficit in energy generation capacity.
   ii) Major wind and solar potentials were discovered in Navarre.
   iii) Navarre is a small region, and the energy policies of the regional government are highly efficient, thanks to their precise knowledge of local conditions.
b. Appropriate regional policy:
   i) The Government of Navarre was quicker to back renewable energy sources than other Spanish regions, and set itself more ambitious goals. This policy was the outcome of the awareness of lack of natural resources for energy generation.
   ii) A long-standing policy to back renewables. The regional authorities followed their initial policy with a solid decision to rely on renewables to solve Navarre's energy problem. Yue *et al.* (2001) reported that the main hindrance to the growth of wind energy was lack of government aid.
   iii) The Government of Navarre helped companies by granting direct subsidies in the periods 1994-1999 (20% of gross investment) and 2000-2007 (30% of gross investment). This financial aid helped all kinds of industrial companies in Navarre, including those belonging to the renewable sector.
   iv) A specific regional policy to support the construction of installations for renewable energies (wind parks, solar parks or biomass factories). This policy involved € 400 million worth of investment, on the part of the Government of Navarre for the promoters of such installations. Measures were also implemented to support the communities affected. In many cases, they were to take the form of investment in areas surrounding wind parks, and included payment for the use of the land on which the parks are built. These support measures involved payments from the promoters of wind parks to local communities in the period 1995-2004. Such investment was crucial in breaking down reluctance at that level (Reiche and Bechberger, 2004). Likewise, the local authorities were aware of the potential benefits of renewable energy projects implementation and their impact on the living standard of the local communities (Del Rio and Burguillo, 2008).
   v) Furthermore, the regional energy policy in Navarre with the support of the Spanish Government managed to set up in Pamplona (Navarre) the R&D Centre - CENER (CENER, 2006) in 2002 and the Training Centre –CENIFER in 2003. On one hand, CENER is a reference international center for renewables whose main research areas are the next ones: i) wind energy ii) solar-photovoltaic energy iii) solar-thermal energy iv) biomass energy v) bioclimatic architecture vi) network integration of renewable energies. Moreover, CENER has also worked in many far-reaching projects related to renewables: *BioSouth* project (forest management for energetic uses), *E-TOOL* project (software development of economic and energy analysis to renovate buildings), *WindHyGen* project (systems of electricity production based on wind energy and hydrogen technologies), among others. On the other hand, CENIFER is the reference training center for renewables in Spain, having three important entities: a) *Centro Integrado de F.P. Superior en Energías Renovables* (Integrated Center of Professional Training in Renewable Energies) b) *Centro Nacional de Formación Profesional Ocupacional en Energías Renovables* (National Center of

Professional Training for Workers in Renewable Energies) c) *Fundación para la Formación en Energías Renovables* (Foundation for Training in Renewable Energies). These entities are devoted to train people respectively in renewables in three different arenas: undergraduate students, workers in companies in the renewable sector or related to it and other interested people. The main goal of CENIFER is training people belonging to other fields to qualify a good number of professionals in the renewable sector (Martínez Erauskin, 2008). The specificity of the CEFINER in relation to the types of renewable energies in which they have to train, their characteristics and the potential demand of training by companies in the renewable sector was studied by the RN2002 report (Pintor *et al.*, 2003), managing to predict some of the current structural characteristics of CENIFER. That report was an interesting contribution to the suitable design of CENIFER.

c. Entrepreneurship abilities:
    i) The presence of a group of Navarrese investors with a strong entrepreneurial spirit who, foreseeing the success of the renewable energies, were prepared to invest in them. This group of investors included Esteban Morrás, awarded with the Poul La Cour prize in November 2004, and one of the key figures in the development of wind energy in Europe. He was also one of the driving forces behind the creation of the EHN company, the first Navarrese firm to venture into renewables.
    ii) The birth of EHN, founded in 1989 as a mixed firm with both public and private capital, has proven historical. EHN's main activity from 1993 was the production of aerogenerators and the installation of wind parks. The support of the Government of Navarre, as EHN's principal stockholder, along with the management of its CEO, Esteban Morrás, ensured the success of this policy.
    iii) After the success of EHN, other companies, such as Gamesa, Ecotecnia, or Mtorres, and more recently, Acciona raised Navarrese renewables to international status. These companies succeeded in developing their own technology, producing excellent aerogenerators, and, eventually, ranking among the top ten companies in the international market.
    iv) The support of Iberdrola, the electricity company that maintained a monopoly over the Navarre energy market in the early nineties. In other European countries, such as Denmark or Germany, the opposition of electricity companies delayed the development of renewables (Farstad and Ward, 1984).

d. Social Conscience:
    i) For a proper understanding of the success of the Navarre's energy policy, due recognition must be given to the support shown by the Navarrese public. This came about, chiefly, as a result of their awareness of the scarcity of energy resources in Navarre, about which they were well informed, thanks to a specific campaign run by the Government of Navarre.
    ii) A specially designed survey was held in Navarre to poll public opinion on the subject of Navarre's energy problems and the issue of renewables. The outcome of the survey was favorable to wind energy and other renewables. It also revealed the importance of building wind parks on clearly visible sites, to increase public support for that kind of energy. This set of policies, implemented

in the period 1994-2000, met with success. The approval of people residing in vicinity of wind parks was crucial for the success of the Navarre Government policy (Wolsink, 2000; Ek, 2005; Toke *et al.*, 2008).

Once we have analyzed the regional energy policy to promote the renewables we are going to describe their environmental and economic benefits in Navarre. The empirical literature has put much more emphasis on the environmental benefits while economic analysis has not received a comparable attention. With the exception of the diversification and security of energy, these benefits have usually been mentioned, but their analysis has been too general and a focus on the regional and, even more so, the local level, has been lacking.

4. Impact of Renewable Energy and its Deployment on Sustainability

Renewable energy sources are essential to tackle climate change and sustainability problems. These sources are uniquely well suited to respond to the limitations of current energy model. Furthermore, renewable sources of energy can contribute to an overall strategy of sustainable development. On one hand, this action will be effective, if finally the renewables should replace the traditional non-renewable energies which are emitter from gases causing the greenhouse effect (GGE). On the other hand, the renewables help to reduce the dependence from the imported energies, assuring the sustainable safety of the energy supply. Accordingly, a detailed analysis of the energy sustainability in the past can help to understand better the future role of renewables (Daly, 1999).

During the last decades, there is a great concern in mankind about the risks which the human life is facing when the fragile equilibria of the terrestrial ecosystems are menacing by a tremendous development of the human action worldwide. Thus, the exponential increase of human population during the last centuries and the corresponding production increase at a greater pace have worried the world politicians since the fifties of the last century. Many authors think that this situation can cause a global crisis because of the use of more physical space by human beings to the detriment of other species (Meadows, 2004). This huge industrial development produced by the mankind in the last centuries has been caused by the use of exosomatic energies which have fossil origins. These fossil energies, therefore non-renewable, present limits in their use because of their exhausting nature. Moreover, the extensive use of fossil energies has generated big stocks of pollution remains which are, finally, non-recyclable. Therefore, the human action on the environmental resources should take into account the physical limits of Earth to manage a sustainable style of life in our world.

According to the previous comments, we are going to analyze in this section the sustainability of the energy system of Navarre. We will begin making some theoretical considerations about sustainability, and will follow with the calculation and study of the primary energies contained inside the final energies used in Navarre and their progression. Similarly, we will pay special attention to the development in use of renewables versus non-renewables and to the evolution of CO_2 emissions. Later, we will make factor decomposition according to the factors which control the aforementioned evolutions using a specific statistical procedure, in order to know the positive and negative effects in sustainability terms. The previous analysis we will help to distinguish the renewability and CO_2 emissions evolutions in relation to the energy generation, the final energy sources substitution and the global increase of energy use.

4.1. The Sustainability Problem

4.1.1. The Concept of Sustainability

First of all, it is necessary to highlight that the world in which the mankind lives is physically finite and limited in our possibilities of change. We have constraints in the amount of resources we use and in the number of sinks we need to absorb the waste material we produce. Therefore, a sustainable world implies a reasonable use of the limited resources and the constraints we have in order to keep the environment in a good condition for the next human generation (Jacobs, 1993). Moreover, it is possible to classify *energies* and *materials* as renewable and non-renewable. *Non-renewable materials* are those which are in a limited amount in the earth such as the rocks or the minerals. *Renewable materials* are those which are involved in the life cycle using solar energy from the plants photosynthesis, from the atmospheric CO_2 and from other elements (nitrogen, calcium, phosphor,...) from soil. The limits of use of renewable materials are given by their disposal and in their utilization rhythm. When the use rhythm is overcome by the natural environment rhythm, they finally exhaust. Therefore, a sustainable world only makes use of renewable materials at a lower rhythm than their renovation pace. Similarly, *non-renewable* sources of energy are those are present on earth in a limited amount, such as oil, coal, uranium or gas; while *renewable energies* are those which come directly from the sun, such as the photovoltaic energy or the solar heat, o indirectly, such as the wind or the tides. Thus, the limits to the energy uses are given by the available energy amounts and the solar energy which can be capture from the outer space, without compromising their use by other living beings.

Furthermore, the earth is a sink of waste materials generated by living beings. It transforms those waste materials into other products which can be reused to enter another time in the food chain. This natural recycling process needs specific amounts of time to finish their cycles and it implies the existence of limits in waste production procedures. Thus, *a sustainable world in materials would be that in which the heritage of materials to use was not diminished, that is to say, a sustainable world would be specifically based on renewable materials at a lower pace than that their natural replacement*. Similarly, *a sustainable world in energy would be that it is directly or indirectly based on solar energy to its correct operation*.

Moreover, the waste production rate should be less than the nature recycling capacity rhythm in order to assure the sustainability of that process. Otherwise, the waste remains would increase with time making more difficult the renewability of the process. The appearance of non-biodegradable products in the chemical industry, noxious products and nuclear products can be considered as threats to reach a sustainable balance. Thus, *a sustainable world in sinks would be that which eliminates the noxious, nuclear and non-biodegradable wastes, and that produces other wastes in an amount which can be recycled by nature*. A similar definition could be done to *a sustainable world in space terms, which would be a world in which the human population size and their economic activities would not imply the disappearance of the remaining living beings.*

According to the statistical use of the information provided in this section, we will center on the sustainability study about the waste renewability and release, without paying attention to other aspects of the non-renewability problems. The final conclusions about the sustainability of the Navarrese energy systems will be associated to the increase or decrease of the use of non-renewable energies and their CO_2 emissions.

4.1.2. The Energy Balances Use

The classical analysis of final energies shows neither the total needs of energy nor the emission they release. That is the reason why to answer to those questions an analysis about primary energies is needed. Any employed energy has involved the energy consumption to make it at disposal from its natural origin. Those disposal processes imply the loss of energy as heat. The knowledge of primary energies behind the final energies will help to understand better the energy needs. Therefore, the use of information from the energy balances helps to recognize the role of the primary energies. Knowing that Navarre is a region with deficit in energy terms (see section 2), we have used not only of the Navarre's energy balances but also the Spanish energy balances to estimate the primary energy values to understand better the final energy consumption (Government of Navarre, 1984-2006 & 1994-2005; International Energy Agency, 2007).

Furthermore, the energy balances explain the links and interrelationships between the energy sectors showing the final energy consumption per economic sector. The information given by the balances helps to improve the primary energies approximation provided by the final energies taking into account different limitations. The first limitation is based on the ignorance of the energy expenditure developed in the production of the industrial inputs of the energy sector. For instance, the calculation of the primary energies in the transportation sector should also include the energies used in the construction and maintenance of roads, harbors or airports, or in the cars production. The second limitation is the ignorance of the consumed energies in the production and transformation of energy worldwide, because they are used in an aggregate way.

Trying to solve the aforementioned problems and knowing that the balance energy method has its own limitations, we have studied the EROEI (Energy Returned on Energy Invested) method which has been revealed as a good procedure to calculate the total energies involved in any energetic process. This method applies the Life Cycle Analysis method into the study of the production processes considering inputs and their transformation in those processes. Nevertheless, the main disadvantage of this procedure is that the EROEI method is specific for each production infrastructure and should be concretely developed for each production process. This drawback involves an important limitation to the application of EROEI method in real cases. In fact, we are not going to use this method in this paper.

4.2. The Use of Primary Energies and their Effects

4.2.1. The Primary Energies Calculation Procedure

One of the main concerns in our study is the calculation of the final energies matrix $EF\ (m \times n)$ where m is the number of economic sector and n is the number of final energy sources. Moreover, it is possible to calculate from the energy balances the matrix $E\ (i \times i)$ where i is the number of primary and final sources. Each column of the matrix E represents the direct use of primary energy units per source unit of final energy. Applying the usual procedure in input-output tables (Leontief, 1986), we obtain the transformation matrix $T\ (n \times p)$ in which each row represents the direct and indirect use of units for each primary energy p contained in each source of final energy n, having previously defined $\overline{T} = (I - E)^{-1}$ with I the matrix unit and E is the aforementioned matrix. Thus, T is built from $\overline{T}$ having ignored the rows corresponding to exclusively final sources such as electricity, and the columns corresponding to exclusively primary sources, such as the nuclear energy, the hydroelectric production, and others. Finally, the product of T and EF yields the matrix $EP(m \times n)$ which provides the use of each type of primary energy per economic sector: $EP = T * EF$. We will use the previous calculations in the following paragraphs.

4.2.2. The Evolution in the Use of Primary Energies

After a discussion of the methodology to use in the energy consumption estimation in Navarre, we are going to present the details of the energetic behavior of Navarre as a region inside Spain in the period 1984-2006. Thus, the use of primary energies is not rising to the same pace than the consumption of final energies in Navarre in the mentioned period, being multiplying by 1.9 at the end of that period. The temporal evolution of final and primary energies is parallel, but with a greater use of primary energies than it would be initially expected. The main cause of this difference in the evolution of those types of energies is the electricity generation by new methods which are less intensive in energy consumption to produce other types of energies (Figure 10).

Considering the results depicted in Figures 11 and 12 in the period 1984-2006, the next comments can be done: a) the use of coal has slowly decreased with a light rise in the last three years b) the hydraulic energy has decreased slowly in the whole period c) the biomass energy maintains its level during that period d) the gas energy rises use slowly with a big increase since 2002 although in 2004 is dramatically stopped e) the oil consumption is important following the gas consumption. Moreover, the most important energy in Navarre in 2006 is undoubtedly oil, following by gas. Wind energy increases tremendously its influence in the whole period but with less absolute weight than oil, while coal and hydro-power decrease in absolute terms. Simultaneously, nuclear energy and the local-waste-generated energy disappear during that period. Nevertheless, some new (then) renewable energies appear in Navarre: thermo-solar, photo-voltaic solar, biogas and biodiesel.

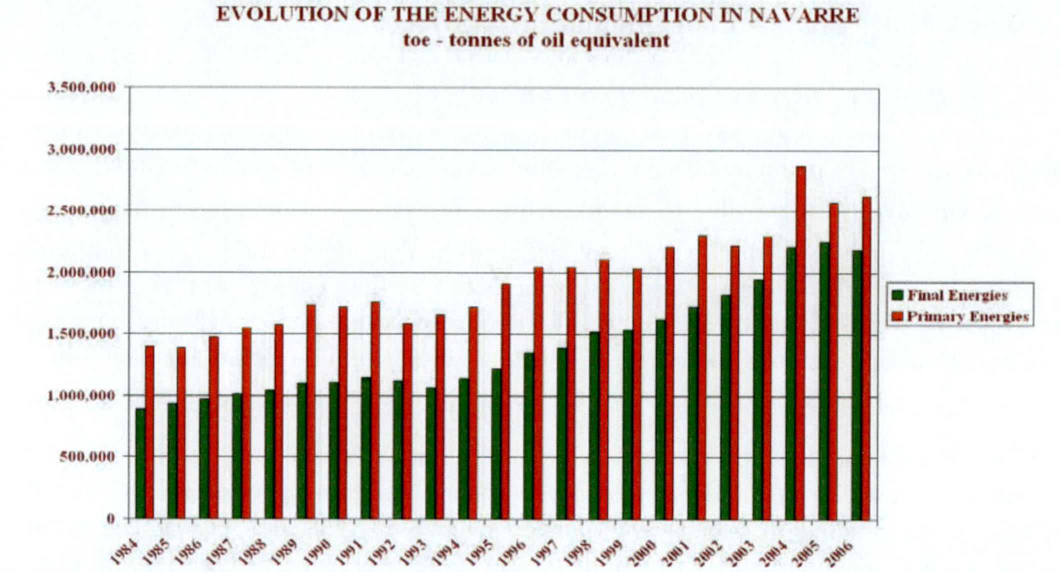

Source: Own elaboration from the final energy balances in Navarre, 1984-2006; the Energy Balances in Spain, 1984-2005; and the Extended Energy Balances in Spain, 1984-2005 (Government of Navarre, 1984-2006).

Figure 10. Evolution of the energy consumption in Navarre in the period 1984-2006.

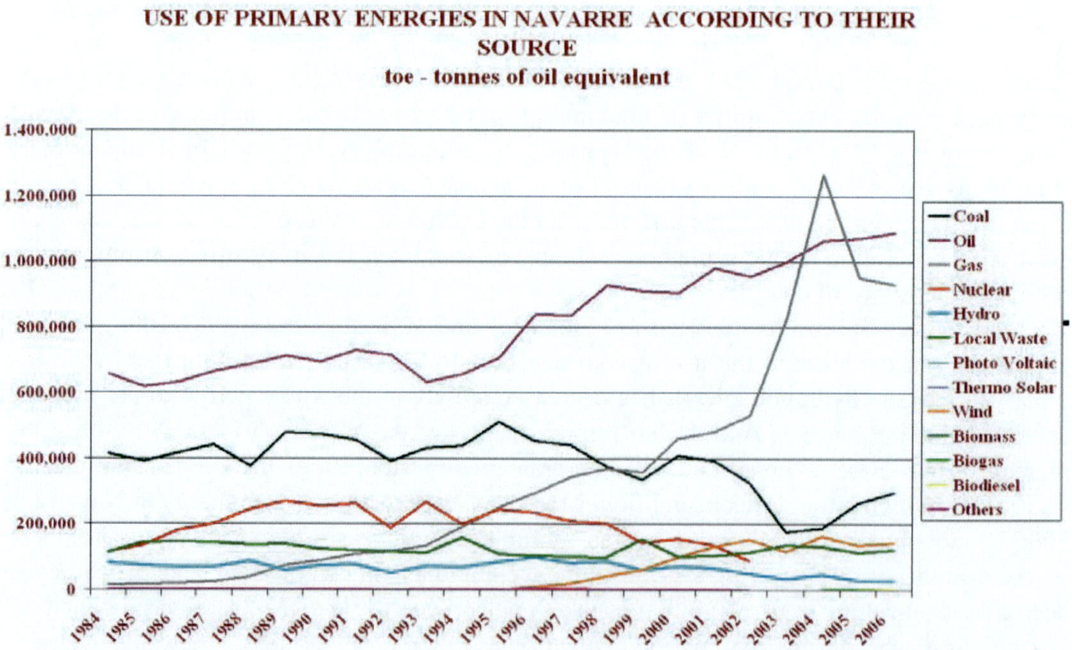

Source: Own elaboration from the final energy balances in Navarre, 1984-2006; the Energy Balances in Spain, 1984-2005; and the Extended Energy Balances in Spain, 1984-2005. (Government of Navarre, 1984-2006).

Figure 11. Use of primary energies in Navarre according to their source in the period 1984-2006.

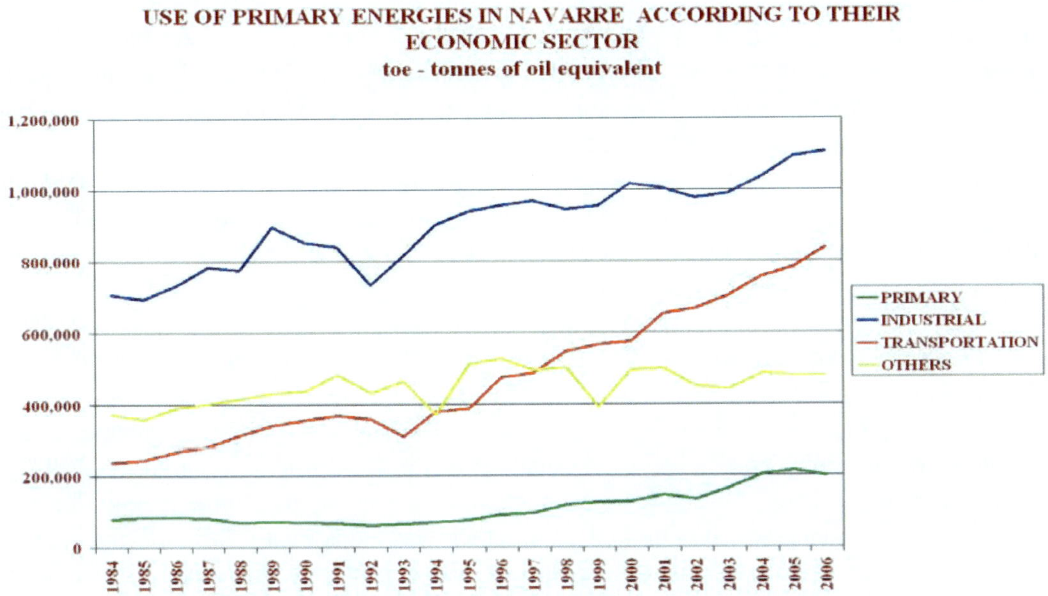

Source: Own elaboration from the final energy balances in Navarre, 1984-2006; the Energy Balances in Spain, 1984-2005; and the Extended Energy Balances in Spain, 1984-2005. (Government of Navarre, 1984-2006).

Figure 12. Use of primary energies in Navarre according to their economic sector in the period 1984-2006.

Thus, according to the information provided by Final Energy Balances in Navarre (Government of Navarre, 1984-2005) the evolution of the primary energies consumption according to the economic sector describes a similar profile than the history of the final energies (Figure 12). Nevertheless, this similarity is not complete and we are going to comment some differences. The industrial and the transportation sectors present a greater difference in the primary energy consumption than in the final energy consumption, because the industrial sector needs a more intensive use of primary energies as oil. Similar explanations can be done in the comparison of the transportation sector with other sectors.

4.2.3. The Use of Primary Energies according to their Renewability

Studying the primary energies evolution according to their renewability, we observe a small number of changes in the whole period. Observing the data in Figure 13, there are some changes about the energy policy in the period 1984-2006, reaching a small diminution for renewables in the last studied years. Thus, the renewables decreased their use in the period 1984-1995, going down from the 14% to 9%. Later, they increased their use in the period 1995-2002, with a small reduction since 2005. Consequently, the popularity in the renewable energy use has changed in the whole period while non-renewable energies increased their use with different rhythms in the same period. It is clear, then, that the importance advance of renewable energy has not involved a substitution of the non-renewables but a conjoint use. This assertion is perhaps, one of the most important conclusions of the complete evolution of renewables in Navarre.

Concerning the energies study according to their economic sector (Figures 14), we can highlight that the *transportation sector* has been hardly swayed by the renewables, because this sector makes use almost exclusively of non-renewable energies (gasoline and diesel). Nevertheless, the *primary sector* shows a similar profile to renewable and non-renewable energies with greater influence of renewables than for the transportation sector. Moreover, the *industrial sector* shows a good improvement in renewables because this sector uses them to produce electricity in a more sustainable way. Finally, the last sector, called *others*, shows a balanced profile because of the introduction of some of the renewable energies in energy production of the activities of the sector entities.

4.2.4. The CO_2 Emissions Associated to the Energies Consumption and the Energy Self-sufficiency

After having discussed the energy scenarios of renewable versus non-renewable energies in Navarre, we are going to comment the behavior of the CO_2 production associated to the energy generation (Figure 15). The increase of the CO_2 emissions due to the energy use was increased in the period 1990-2006 in an 81.9%, which was equivalent to a yearly increase of 2.8%. This augment was lower than the energy global consumption increase and the fossil energy consumption. It is interesting to pay attention in the tremendous increase of CO_2 emissions in 2004 due, mainly, to the start of energy production with natural gas.

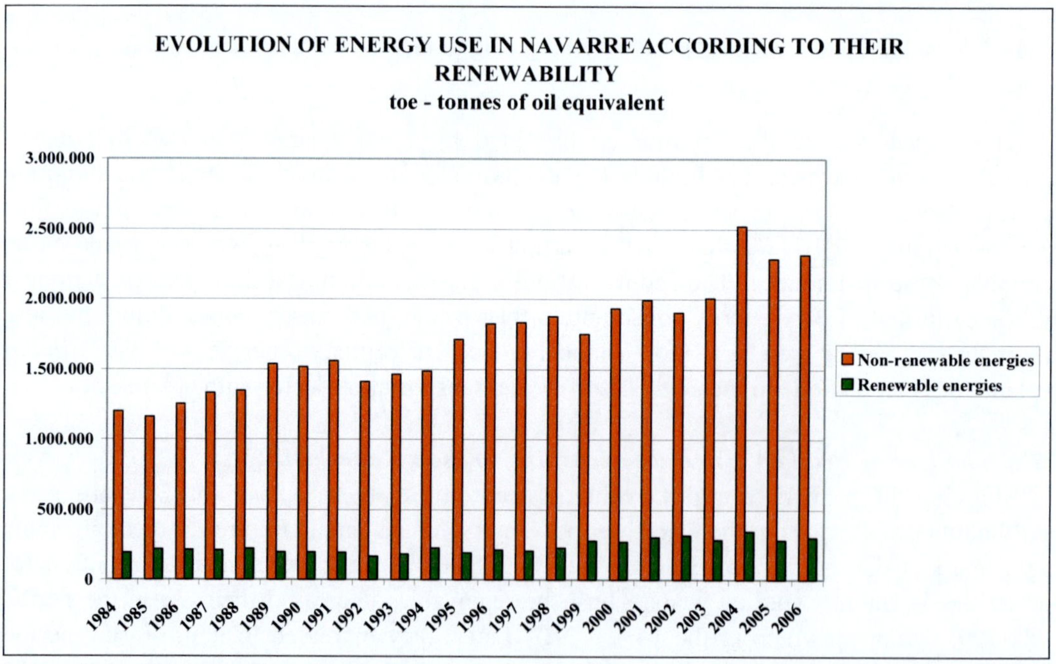

Source: Own elaboration from the final energy balances in Navarre, 1984-2006; the Energy Balances in Spain, 1984-2005; and the Extended Energy Balances in Spain, 1984-2005. (Government of Navarre, 1984-2006).

Figure 13. Evolution of energy use in Navarre according to their renewability in the period 1984-2006.

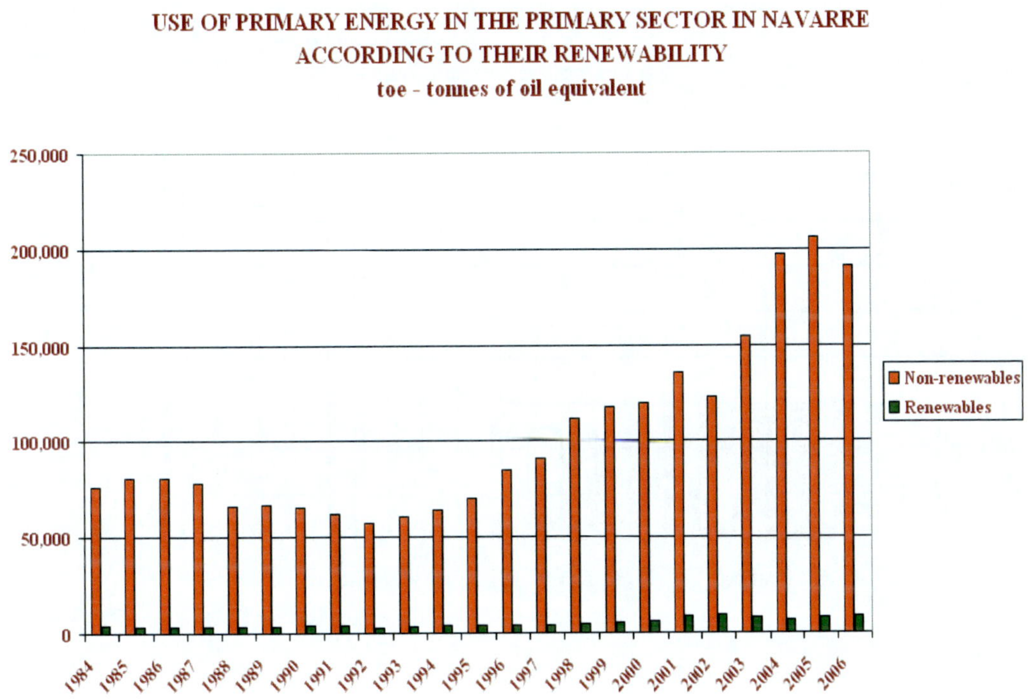

Figure 14.a)

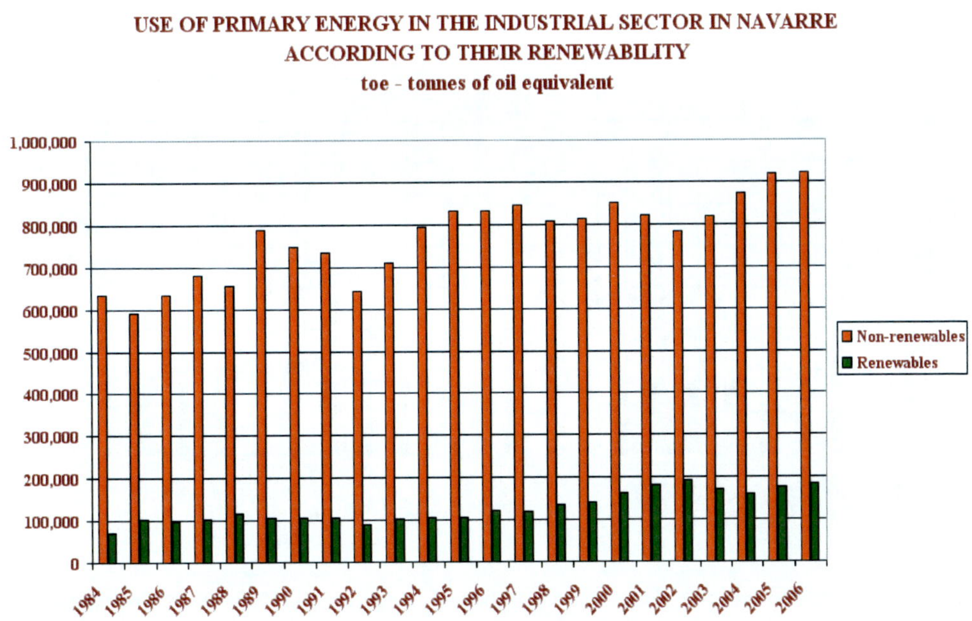

Figure 14.b)

Figures 14. (Continued)

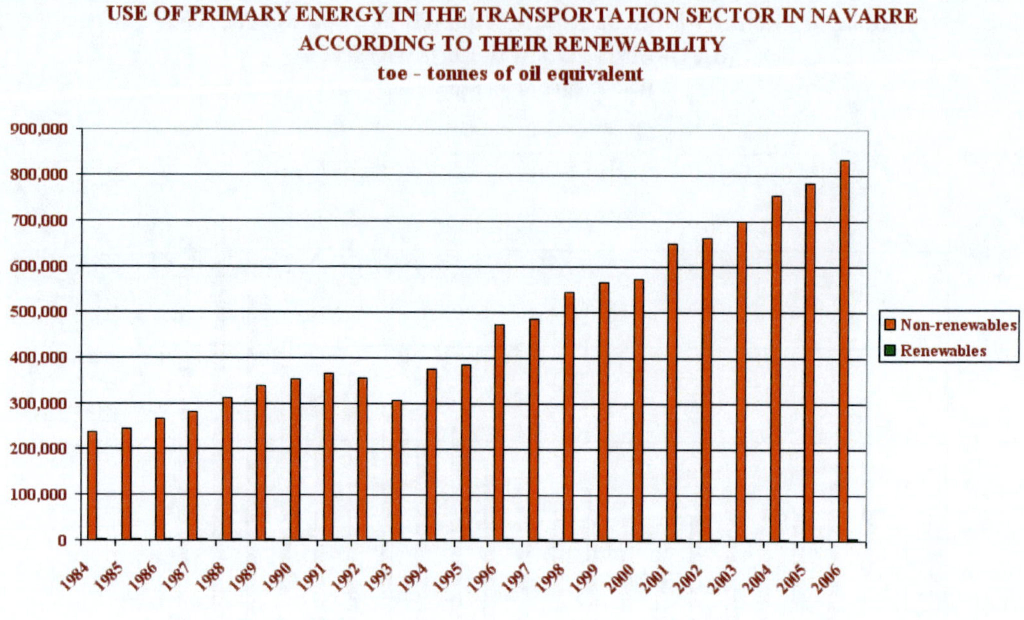

Figure 14.c)

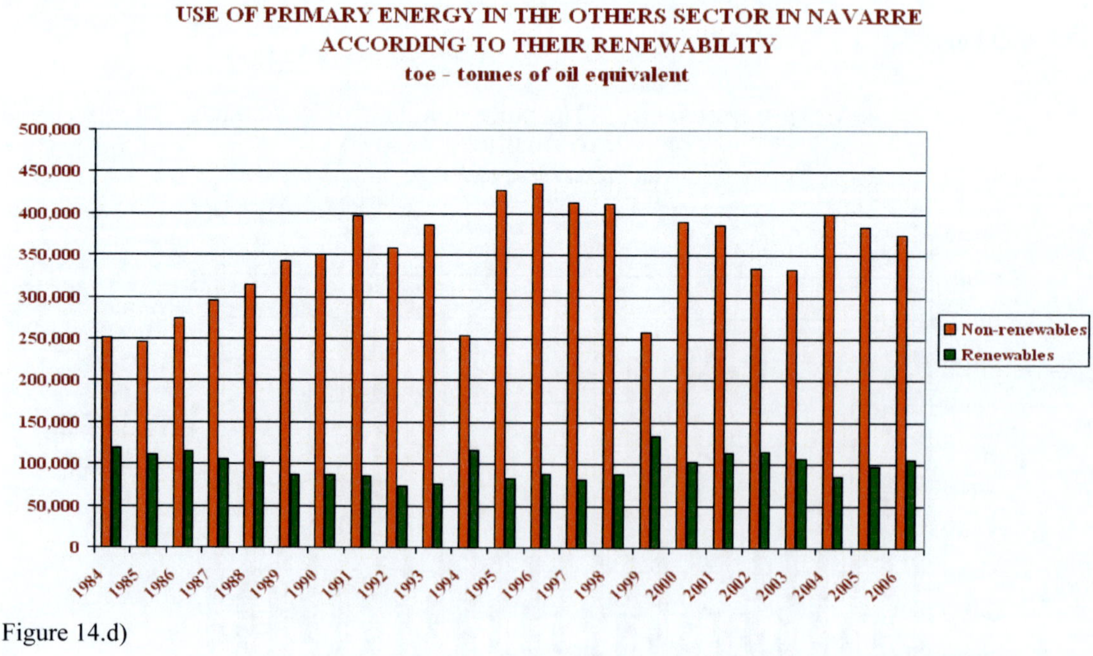

Figure 14.d)

Source: Own elaboration from the final energy balances in Navarre, 1984-2006; the Energy Balances in Spain, 1984-2005; and the Extended Energy Balances in Spain, 1984-2005. (Government of Navarre, 1984-2006).

Figures 14. Evolution of energy use by economic sectors in Navarre according to their renewability in the period 1984-2006.

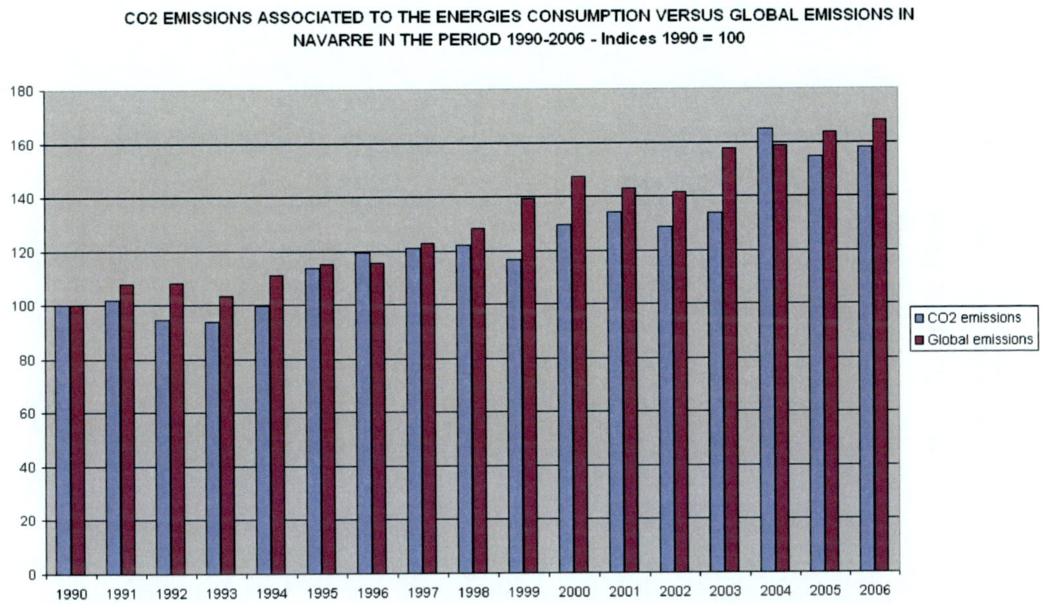

Source: Own elaboration from the final energy balances in Navarre, 1984-2006; the Energy Balances in Spain, 1984-2005; and the Extended Energy Balances in Spain, 1984-2005. (Government of Navarre, 1984-2006).

Figure 15. CO_2 emissions associated to the energies consumption versus global emissions in Navarre in the period 1990-2006.

Another important problem to tackle in relation to energy policy in Navarre is its energy self-sufficiency (Sanz, 2009). The self-sufficiency percentage of primary energies in Navarre is up to the 13% out of the total energies per year. Nevertheless, the Navarre primary energies profile is U-shaped (Figure 16) with peaks in the years 1985 and 2005. Considering the whole period 1984-2006, the self-sufficiency percentage ranges from 5% in 1995 to the 12.7% in 2005 having 10% in 1984 and 12.2% in 2006, which represents an important variation in the energy supply in Navarre. The improvement of the self-sufficiency in Navarre during the years 1996-2006 is closely connected to the success of the renewables.

Notwithstanding the global behavior of energies, we realize that the situation of the main final energy, electricity, is different. Since 1995, the electricity production in Navarre has tremendously increased because of the success of wind energy. Thanks to the development of the centrals of combined cycle with natural gas, the electricity production reached a peak in 2003 managing an energy surplus for Navarre. Nevertheless, this appearance of the energy self-sufficiency in Navarre is due to the importation of natural gas, but it speaks volumes about the importance of the use of the renewables in Navarre as an economic and political fact (Figure 17).

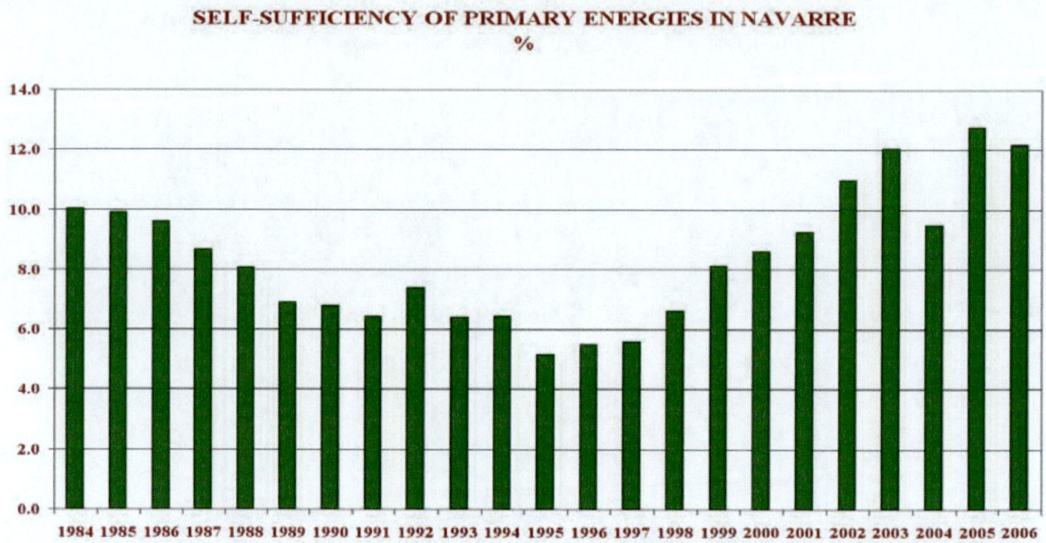

Source: Own elaboration from the final energy balances in Navarre, 1984-2006; the Energy Balances in Spain, 1984-2005; and the Extended Energy Balances in Spain, 1984-2005. (Government of Navarre, 1984-2006).

Figure 16. Self-sufficiency of primary energies in Navarre in percentage in the period 1984-2006.

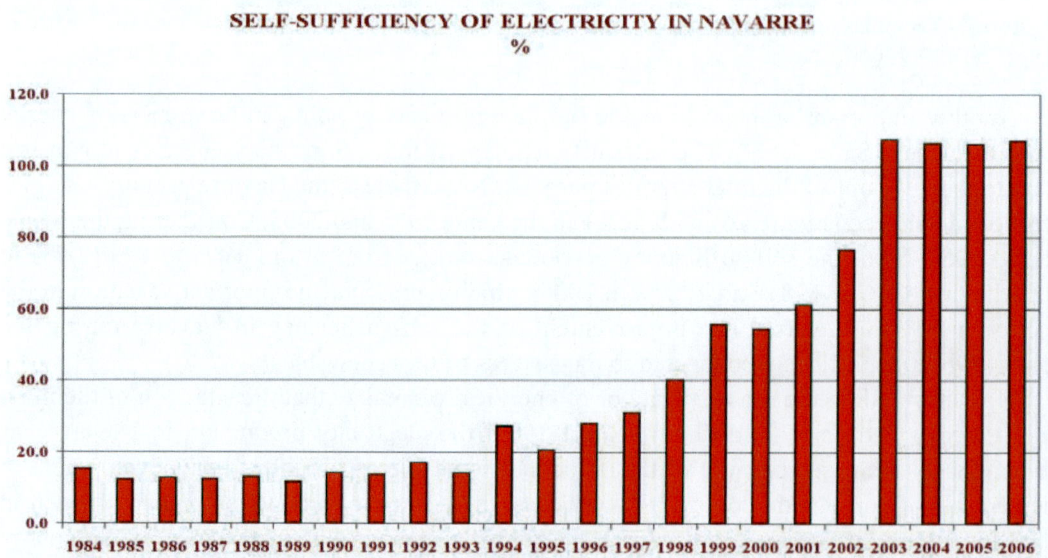

Source: Own elaboration from the final energy balances in Navarre, 1984-2006; the Energy Balances in Spain, 1984-2005; and the Extended Energy Balances in Spain, 1984-2005. (Government of Navarre, 1984-2006).

Figure 17. Self-sufficiency of electricity in Navarre in percentage in the period 1984-2006.

4.3. The Method of Calculation in Factor Decomposition

According to the comments made in the previous sections, it is important to find a concrete and specific way of calculation of the level of final energies, in a similar way we have obtained the amount of primary energies. In that case, we used an appropriate method of factors decomposition. Now, we are going to try to generalize the method for final energies using a simple decomposition method with the Laspeyres' methodology (Alcántara, 1995; Diewert, 1993). This method is trying to isolate energy factor to understand better the temporal evolution of energies.

Thus, we have already discussed in section 4.2.1 the final formula for the primary energies: $EP = T * EF$ and the final energies con be break down into two factors: i) matrix $A\left(m \times n\right)$ which in each column depicts the final energy type proportion per economic sector in relation to the total energy used per economic sector ii) matrix $B\left(n \times n\right)$ which is a diagonal matrix with the global use of the final energies per economic sector. Therefore, $EF = A * B$ and, then, $EP = T * A * B$.

It is important to highlight that matrix T represents the energy efficiency in the energy sector, matrix A represents the changes in the energy use due to final energies substitution, and matrix B represents the changes per sectors in the used final energy volume. Supposing that two of those three elements are unchanged per period, we can calculate the variable corresponding to each factor in the primary energy use. Thus, we would obtain the weight of the *transformation effect* (ET), the weight of the *substitution effect* (ES), and the weight of the *consumption effect* (EC), between the years *1* and *t*. Therefore, we can write:

$$
\begin{aligned}
&EP = ET + ES + EC \quad where \\
&\quad ET = (T_t - T_1) * A_1 * B_1 \\
&\quad ES = T_1 * (A_t - A_1) * B_1 \\
&\quad EC = T_1 * A_1 * (B_t - B_1)
\end{aligned}
$$

Furthermore, the decomposition calculations of this method produce a residual effect ER due to the interaction of the three considered factors, and the previous expression EP can be written in the next way:

$$EP = ET + ES + EC + ER$$

We are going to follow the Sun's (1998) method to delete formally the residual effect ER, according to the philosophy of distributing the residual effect among the other three factors. Thus, the Sun's method produces the next results in the decomposition factors formulae:

$$
\begin{aligned}
ET &= (T * A_1 * B_1) + ½ (\Delta T * \Delta A * B_1) + ½ (\Delta T * A_1 * \Delta B) + 1/3 (\Delta T * \Delta A * \Delta B) \\
ES &= (T_1 * \Delta A * B_1) + ½ (T_1 * \Delta A * B_1) + ½ (T_1 * \Delta A * \Delta B) + 1/3 (\Delta T * \Delta A * \Delta B) \\
EC &= (T_1 * A_1 * \Delta B) + ½ (T * A_t * \Delta B) + ½ (T_t * \Delta A * \Delta B) + 1/3 (\Delta T * \Delta A * \Delta B)
\end{aligned}
$$

Table 5. CO_2 emission factors in Europe in the period 1990-2004

CO_2/EP			
Mt/Mtoe	**1990**	**2004**	**1990-2004**
Coal	3.881	3.885	3.883
Oil	2.648	2.553	2.600
Gas	2.225	2.298	2.262
Total	**2.974**	**2.776**	**2.875**

Source: Own design taking information from the International Energy Agency, 2006.

where

$$\Delta T = T_t - T_1;\ \Delta A = A_t - A_1;\ \Delta B = B_t - B_1$$

We have defined three time periods to know in detail the new changes over time. These periods are the next ones:

a. *1984-1994,* period characterized by a poor production in electricity with the support of traditional energies (hydro and biomass).
b. *1995-2002,* period of great expansion of electricity generated by wind power.
c. *2003-2006,* period known by the intensive use of wind energy, along with other types of energy production: combined cycle procedures, solar PV, biogas and biodiesel.

Trying to mitigate the influence of annual changes, we have calculated the CO_2 emission factors in Europe in the period 1990-2004, calculating the average of the first three years and the last years for each period, with the exception of the last one because of its small size. We have aggregated the source by renewability to explain better its role in the energy use in Navarre. The improvement of sustainability will imply a reduction of the non-renewable energies use. Therefore, applying these procedures we will know the weight for each factor in the CO_2 emission changes due to the energy system (Table 5). Those factors are obtained from the average of the emission proportions per Mtoe of primary energies used in the European countries belonging to OECD (Organization for Economic Cooperation and Development) in the years 1990-2004 (International Energy Agency-IEA, 2006, 2007).

It is important to highlight that the emissions of each type of fossil energies are different. Therefore, a substitution among those sources modifies the CO_2 emission volume according to the fossil energy used. It implies that the sustainability level can be changed according to the origin of the primary energies. Now, we are going to apply this method to the energy use in Navarre.

4.4. Change Factors in the Energy Use in Navarre, 1984-2006

We have observed in the previous paragraphs that the use of primary energies had increased in 89.6% in the period 1984-2006 (3.3% in a yearly basis). Thus, we can study the factor decomposition due to this increase in Table 6. Considering the whole period, we can analyze a decrease in the primary energies use due to the *transformation effect* (ET), that is to say, due to the transformation procedures of the primary energies into final energies, and the substitution of the energy sources in that process. Nevertheless, that positive effect were compensated by a greater *substitution effect* (ES), and mainly by a greater the *consumption effect* (EC), due to an increase of the final energies consumption. Table 7 shows this process detailed by energy sources.

In order to study specifically the primary energy use, we are going to split up the whole period 1984-2006 into three subperiods: 1984-1993, 1994-2002 and 2003-2006 according to the sensitivity of those years to the results of the local energy policy.

Taking into consideration the different types of energies, there have been important decreases in the use of coal, nuclear energy, hydro energy, biomass and other energies. In all cases, except in biomass, those reductions are due to the *transformation effect* (ET). Similarly, it is important to mention that Navarre stopped buying electric energy from Spain in 2003, and therefore, she stopped using primary energies which are associated to the production of that kind of electricity. Nevertheless, in the biomass case, there have been an important decrease in use because it were partially replaced by other energies (*substitution effect* (ES)), but compensated in the associated demand because of the greater consumption of electric energy (*consumption effect* (EC) and *transformation effect* (ET)).

In the oil case, it experienced a demand decrease associated to the *transformation effect* (ET) and to the substitution of its derivatives, mainly the gas (*substitution effect* (ES)). Nevertheless, this decrease was compensated by the consumption increase of gasoline and diesel (*substitution effect* (ES)). But, this behavior has been different depending on the considered subperiod. For instance, in the second subperiod the electricity production has increased with the success of wind energy, while, the third subperiod the natural gas was the cause of the electricity productions increase. The subperiods 1994-2002 and 2003-2006 can be seen in the Tables 8 and 9, respectively.

The subperiod 1994-2002 presents the same general characteristics than the global period and it is essential to understand the general outlook in the Navarre energy policy: changes in the electricity generation and oil substitution in the final use. The subperiod 2003-2006 shows very small effects in the reduction of primary energies with the success of renewables. There is an important *substitution effect* (ES) which depletes the oil consumption because of the expansion of natural gas with the greater demand of gasoline and diesel. Moreover, there has been a great reduction, in relative terms, in the coal demand, the hydro energy and the biomass caused by the important increase of the gas natural use.

Table 6. Increase factors in the primary energy use per period in Navarre

toe	ET	ES	EC	Total
1984-2006	**-447,535**	**130,974**	**1,449,527**	**1,132,966**
1984-1993	-51,576	39,510	248,093	236,026
1994-2002	-429,520	48,639	755,113	374,231
2003-2006	137,311	18,091	311,965	467,368

Source: Own elaboration from the Energy Balances of Navarre in 1984-2006 (Government of Navarre, 1984-2006; Government of Navarre, 1994-2006; IEA, 2007).

Table 7. Increase factors in the primary energy use per energy source in the period 1984-2006 in Navarre

toe	ET	ES	EC	Total
Coal	-421,823	18,750	237,555	-165,518
Oil	-109,272	-257,067	805,646	439,308
Gas	304,635	418,404	211,405	934,445
Nuclear	-234,996	28,752	63,105	-45,897
Hydro	-102,709	17,904	38,908	-143,138
Urban waste	0	0	0	0
Photovoltaic	983	136	279	1,398
Thermo solar	0	1.120	211	1,331
Wind	90,518	12,507	25,739	128,764
Biomass	27,916	-110,973	64.783	-18,274
Biogas	1,356	187	386	1,929
Biodiesel	0	747	397	1,144
Others	-4,144	507	1,113	-2,524

Source: Own elaboration from the Energy Balances of Navarre in 1984-2006 (Government of Navarre, 1984-2006; Government of Navarre, 1994-2006; IEA, 2007).

Another important fact to consider is the disappearance of the nuclear energy, of urban waste and other energies in Spain, and there are not going to be use longer in Navarre. Finally, the two great transformations in energy supply in Navarre, which are *wind energy* and *natural gas energy* in combined cycle plants, have finished their strength as energy engines of Navarrese economy (Figure 18). This characteristic shows the birth of a new stage in the energy policy in Navarre.

Table 8. Increase factors in the primary energy use per energy source in the period 1994-2002

toe	ET	ES	EC	Total
Coal	-324,226	27,466	119,048	-177,711
Oil	-43,076	-114,910	403,390	245,404
Gas	111,154	143,259	114,009	368,422
Nuclear	-222,713	20,911	49,133	-152,669
Hydro	-62,882	8,887	20,818	-33,178
Urban waste	-939	113	264	-561
Photovoltaic	145	11	26	182
Thermo solar	0,11	0,01	0,02	0,13
Wind	107,578	7,160	16,464	131,202
Biomass	7,800	-44,569	31,232	-5,537
Biogas	0	56	131	188
Biodiesel	0	0	0	0
Others	-2,362	254	597	-1.511

Source: Source: Own elaboration from the Energy Balances of Navarre in 1984-2006 (Government of Navarre, 1984-2006; Government of Navarre, 1994-2006; IEA, 2007).

Table 9. Increase factors in the primary energy use per energy source in the period 2003-2006

toe	ET	ES	EC	Total
Coal	115,408	13,502	-7,870	121,040
Oil	126	-77,547	168,466	91,045
Gas	18,632	62,864	98,001	179,496
Nuclear	0	0	0	0
Hydro	-12,645	1,921	6,943	-3,781
Urban waste	0	0	0	0
Photovoltaic	2,379	71	278	2,727
Thermo solar	0	1,579	41	1,620
Wind	16,202	7,523	27,697	51,422
Biomass	-5,321	-11,746	10,253	-6,814
Biogas	2,531	53	213	2,797
Biodiesel	0	19,872	7,943	27,815
Others	0	0	0	0

Source: Source: Own elaboration from the Energy Balances of Navarre in 1984-2006 (Government of Navarre, 1984-2006; Government of Navarre, 1994-2006; IEA, 2007).

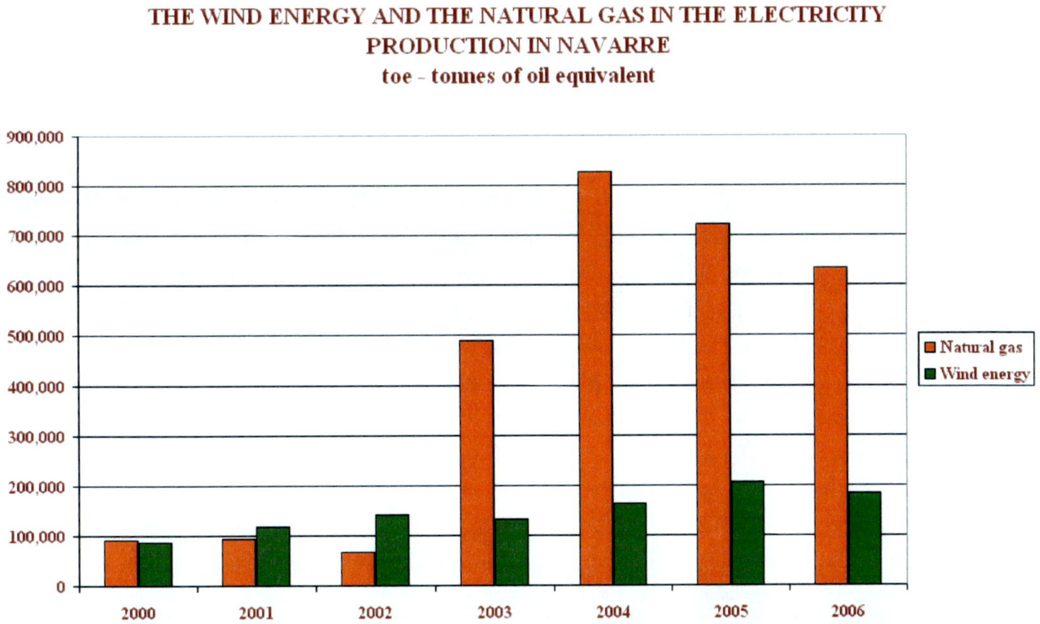

Source: Source: Own elaboration from the Energy Balances of Navarre in 1984-2006 (Government of Navarre, 1984-2006; Government of Navarre, 1994-2006; IEA, 2007).

Figure 18. The wind energy and the natural gas in the electricity production in Navarre in the period 2000-2006.

Table 10. Increase factors in the renewable energies use per period

toe	ET	ES	EC	Total
1984-2006	**18,065**	**-78,373**	**130,702**	**70,394**
1984-1993	-22,407	-35,401	29,397	**-28,411**
1994-2002	51,703	-28,342	68,935	**92,296**
2003-2006	3,146	19,272	53,369	**75,786**

Source: Source: Own elaboration from the Energy Balances of Navarre in 1984-2006 (Government of Navarre, 1984-2006; Government of Navarre, 1994-2006; IEA, 2007).

4.5. Change Factors in the Renewability of the Primary Energy and in the CO_2 Emissions in Navarre

Now, we are going to analyze the energies renewability and the CO_2 emissions in Navarre using the same time periods than in the previous sections. We have aggregated the energy sources to have a more accurate outlook of the energetic changes in Navarre in the period 1984-2006 in order to understand better the sustainability of the energy supply in this region. If we study the whole period, we realize of the existence of an increase in the renewable energies use in Navarre (Table 10). The *transformation effect* (ET) is positive in that period, although the *substitution effect* (ES) of the final energies is negative as a result of the fall in the biomass use. Nevertheless, the whole effect is an increasing use of energies due to the rise of the final energies utilization (*consumption effect* (EC)).

Concerning the subperiod 1984-1993, the increase of the final consumption implies a rising in the use of renewable energies. That increase is balanced with a fall in the relative use of hydroelectricity and biomass. Nevertheless, the last two subperiods present an increase in absolute terms with two characteristics: the *transformation effect* (ET) has disappeared in the subperiod 2003-2006, and the *substitution effect* (ES) has become positive because of the lack of energy acquisition outside Navarre. Finally, according to the renewability outlook, it is meaningful to study the evolution of non-renewable energies use. Thus, according to the information provided by Table 11, an increase in the use of non-renewable energies can be observed. This situation can be partially explained by the depletion of the *transformation effect* (ET) in the period 1994-2002 and the disappearance of the *substitution effect* (ES) in the period 2003-2006.

Table 11. Increase factors in the non-renewable energies use per period

toe	ET	ES	EC	Total
1984-2006	**-465,599**	**209,347**	**1,318,824**	1,062,571
1984-1993	-29,169	74,910	218,697	264,438
1994-2002	-481,223	76,980	686,178	281,935
2003-2006	134,166	-1,181	258,597	391,581

Source: Source: Own elaboration from the Energy Balances of Navarre in 1984-2006 (Government of Navarre, 1984-2006; Government of Navarre, 1994-2006; IEA, 2007).

As a conclusion, it can be said that an important increase in the renewable energies use has been associated to a greater increase of the use of non-renewable energies, making Navarre a region in a difficult situation according to the sustainability standards. The positive effects of the development of renewables have been the structural changes performed to substitute final energies sources. Nonetheless, those changes have been compensated by the increase of final energy consumption and during the subperiod 2003-2006 those positive transformation effects have disappeared.

In a second stage, we are going to analyze how the energies use in Navarre has impacted in the the CO_2 emissions in Navarre. Thus, the *transformation effect* (ET) has played a very positive role in the reduction of the CO_2 emissions in the whole period 1984-2006. Nevertheless, the *substitution effect* (ES) and the *consumption effect* (EC) have compensate the good behavior previously observed in all the subperiods. Similarly, it is important to highlight the disappearance of the reduction role of the *transformation effect* (ET) during the period 2003-2006. The substitution of coal and oil by natural gas and new renewable energies in the electricity production explain those depletion effects. Those positive effects can be also extended to the CO_2 emissions, and in a similar way, those positive effects have been depleted by the *substitution effect* (ES) and the global use of energies (*consumption effect* (EC)) (Table 12).

To sum up, the increase of the use of renewable energies in Navarre can be considered as a positive element in the decade 1995-2006. Nevertheless, the previous assertion is only partially true because in that period the renewability of the used energies and the CO_2 emissions have worsened their situation in relation to 1984 (Figure 19). In some aspects, Navarre can be seen less sustainable than two decades ago. Moreover, some positive aspects of renewables in the period 1994-2002 have recently disappeared. Perhaps, the main cause of this situation has been the intensive use of final energies, which has compensated the positive aspects of the renewables. That is to say, that the intensive use of renewables has not replaced the use of fossil energies but, conversely, it has been like catalyst which has involved a greater use of renewable and non-renewable energies.

Table 12. Increase factors in the CO_2 emissions due to energy use per period

toe	ET	ES	EC	Total
1984-2006	**-1,232,960**	**350,862**	**3,495,304**	**2,613,206**
1984-1993	-249,686	113,748	573,752	437,814
1994-2002	-1,119,536	131,937	1,768,969	781,370
2003-2006	481,860	42,304	411,697	935,861

Source: Own elaboration from the Energy Balances of Navarre in 1984-2006 (Government of Navarre, 1984-2006; Government of Navarre, 1994-2006; IEA, 2007).

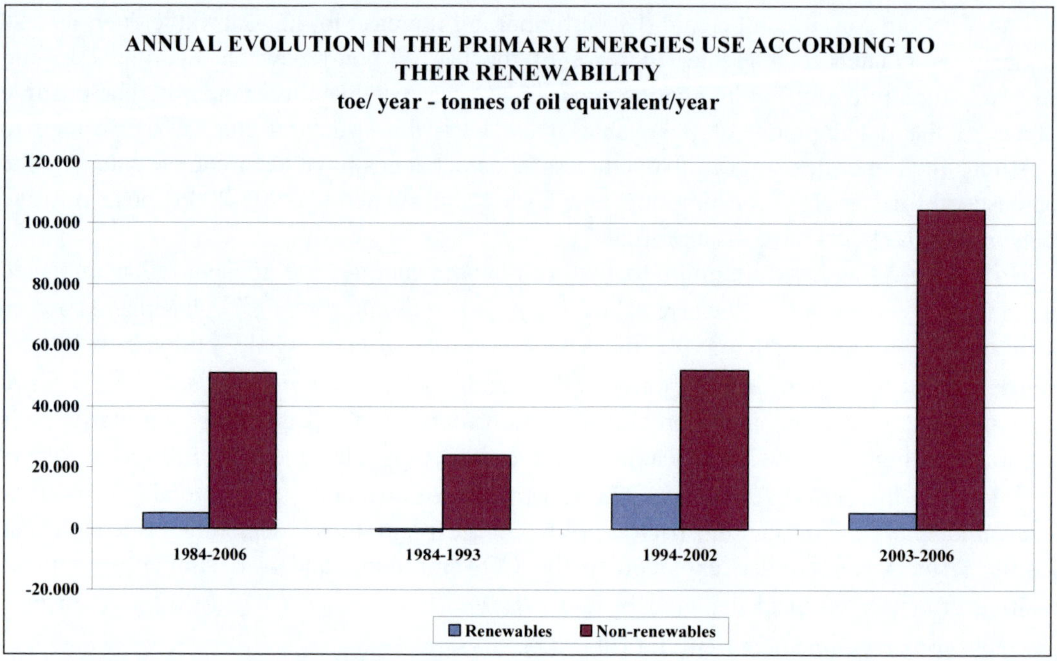

Source: Source: Own elaboration from the Energy Balances of Navarre in 1984-2006 (Government of Navarre, 1984-2006; Government of Navarre, 1994-2006; IEA, 2007).

Figure 19. Annual evolution in the primary energies use according to their renewability.

5. Impact of Renewable Energy Deployment on Regional Economic Development

5.1. The Contribution of Renewable Energy to Local and Regional Development: A Brief Literature Review

The development of renewable energy presents not only environmental benefits but also economic advantages at regional level. Traditionally, the existing literature has made emphasis on the environmental benefits, analyzed in the section 4, while economic impacts have not received a similar attention. Nevertheless, after a period where the main focus was on the environmental analysis, the economic benefits are gaining increasing attention (Del Rio and Burguillo, 2008), particularly at local and regional levels. These effects could be very significant to many regions with a relatively large share of rural, dispersed and declining population and with scarcity of regional development with the exception of a waning agricultural sector.

Given the narrow link between poverty and energy, some literature reviews have focused their interest on the impact of renewables into the economic development of underdeveloped and developing countries (see, for example, the literature review made by Del Rio and Burguillo, 2008). Nevertheless, it has also been relevant to consider the regional impact of renewables in developed countries such as Spain (Faulin *et al.*, 2006), Ireland (Komor and Bazilian, 2005) or the UK (ADAS, 2003). In general, the empirical evidence shows that implementation and development of renewables favours a productive diversification of economic activities and the creation of new firms in manufacturing and service activities, and contributes to the reduction of regional income disparities. Logically, these impacts differ according to the type of renewable energy technology considered.

Traditionally, the economic impact of renewable development has been primarily analysed in terms of employment. The literature on the employment impacts of renewable energy adoption generally highlights a positive effect (Thothathri, 1999; Ministerio Medio Ambiente-MMA, 2000; Menéndez, 2001; European Commission, 1996, 2000, 2003, 2006; Kamen *et al.*, 2004, Faulin *et al.*, 2006). The studies generally show that renewable energies are more labor intensive than conventional energy production for the same amount of produced energy (Del Rio and Burguillo, 2008). Nevertheless, some authors such as Hillebrand *et al.* (2006) observe that after an expansion effect on employment, it could appear a contraction effect due to the increase in electricity prices as a result of the greater costs of renewables and consequently a reduction of production and employment in the electricity sector. The contraction effect could be higher than the expansion effect, leading to a net negative effect on employment in energy sector (Hillebrand *et al.*, 2006).

Anyway, not all renewable energies contribute in the same manner to the employment creation. In the short-term, wind energy is likely to have the greatest impact. However, this impact may not only be short-term but also long-term on places far from the project location if the manufacturing takes place outside the local community. Then, it is very important to be able to develop at regional level a competitive enterprise renewable sector, including not only the construction and O&M stages but also the manufacturing activities. In contrast, biomass is likely to have a more permanent economic impact during the whole life cycle of the project due to the fact that transport and processing of the biomass fuel accounts for the greatest share of the total cost, which are spent at regional level (Del Rio and Burguillo, 2008). Furthermore, the ability to provide a decentralised employment source is a great advantage of renewables, particularly at regional level, positively affecting geographical areas with scarce employment opportunities.

Nonetheless, employment impact is not the only economic effect that could be considered in the renewable development at regional level. Other significant economic impacts are: income generation effects due to the payment to local farmer and communities for hiring their land, demographic impact, educational impacts, a higher productive diversification of the area, income distribution and social cohesion, industry creation, and local R&D (Del Rio and Burguillo, 2008). Consequently, the greater the benefits for the local communities and social agents, the greater the possibilities for the social acceptance of the renewable projects (Wolsink, 2000, Agterbosch *et al.*, 2004, 2007; Ek, 2005; Faulin *et al.*, 2006).

5.2. The Effect of Renewable Energy on Regional Development in Navarre

After a literature revision about the effect of the renewables at regional level in terms of economic development, it is possible to consider the main effects of renewables on the regional development in Navarre.

5.2.1. The Upsurge of a New Enterprise Sector: Phases and Development

A new enterprise sector has been created in Navarre, under the aegis of renewable energy, as a consequence of official support and the great development of renewables in Navarre. In fact, in 2002 the Navarre renewable sector comprised 54 companies, engaged either in industrial activities or services, in relation to the roughly 400 or 500 renewables companies in Spain as a whole, according to figures supplied by the Spanish Ministry for the Environment (Ministerio de Medio Ambiente-MMA, 2000). In 2006, the renewable sector in Navarre comprises 88 companies, in relation to the 1,027 renewable companies in Spain as a whole (ISTAS, 2008). Nevertheless, the development of this new enterprise sector has not been a continuous process and consequently it is possible to establish the following phases:

a. The first couple of years 1994-1995 were characterised by inactivity due to the non-existence of plans to promote new energy sources. Only 22 companies were created with a total of 200 jobs.
b. During the second stage, 1995-2000, Navarre made major strides forward with the creation of new renewable companies, thanks to the FEPN (First Energy Plan of Navarre). The number of firms is nearly tripled, with a total of near 900 jobs.
c. The third stage, 2001-2002, was marked by a decrease in the creation rate of new companies. The employment rate, however, rose steadily during this period until 1,446 jobs. Thus, in 2002, the 54 firms engaged in activities relating to renewables provided around 0.6% of the working population of Navarre, and 2.3% of the active population in industrial activities (INE, 2008). The degree of specialisation of the Navarre economy in renewables becomes patent when these employment figures are viewed in relation to the 3,522 jobs (0.03 % of Spain's active population) in renewables at national level (MMA, 2000).
d. The fourth stage, 2003-2006, was marked by a new increase in the creation rate of new companies. In fact, in four years the number of renewable firms increases 63%, from 54 to 88 firms and the employment in 183%, from 1,446 to 4,099 jobs. Figure 20 shows the distribution of firms according to the type of renewable energy.

Thus, in 2006, the 88 firms related to activities of the renewables provided around 1.4% out of the active population of Navarre and 5.6% out of the active population in industrial activities (INE, 2008). In Spain, at the same time, there were 89,000 jobs, what means the 0.4% of the working population in the country and the 2.7% of the active population in industrial activities (INE, 2008). Then, in relative terms, the employment importance is double in Navarre than in Spain.

5.2.2. Characteristics of the Renewable Energy Enterprise Sector in Navarre

This sector has some specific characteristics closely related to the drastic development of the renewable sector. The most significant characteristics are the following:

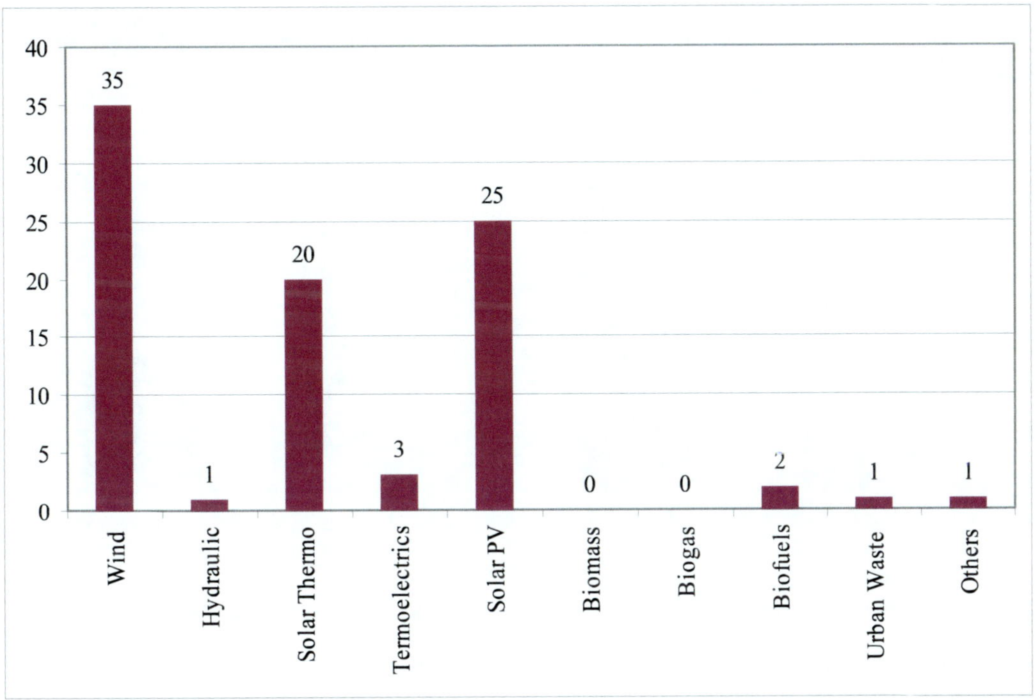

Source: Government of Navarre (2007c).

Figure 20. Distribution of firms in Navarre according to the type of renewable energy.

5.2.2.1. A Significant Sector in the Regional Economy and Very Intensive in Employment

The renewable sector has become, in a few years, a significant sector in the regional economy, traditionally focused on the agricultural and car industrial activities. In terms of employment, the total employment by Navarrese firms are estimated in 6,278, while the total employment in the renewable sector in the region is estimated as 4,099 workers, what means that one out of three Navarrese firms workers are working out of the regional borders (Government of Navarre, 2007c). This positive relationship between renewable energy and employment has been empirically analysed in the literature review (Thothathri, 1999; MMA, 2000; European Commission, 1996, 2000, 2003; Pintor *et al.*, 2006; Faulin *et al.*, 2006; Del Rio and Burguillo, 2008) as it has been shown in previous pages.

Furthermore, it is important to highlight that many of the renewable firms has been established in rural areas where the economic perspectives were not very positive. Some of the renewable activities need water, wind, timber or wood remains, and consequently they establish in rural areas. Then, this new renewable sector has drastically contributed to a more harmonised economic development in the whole region. For example, in Northestern Navarre, one of the less industrialised areas in Navarre, two factories, related to renewables, will be established in the near future (Acciona and Gamesa).

Finally, in terms of turnover, the total turnover of the firms in 2006 was 3.585 M€ while in 2002 was 0.780M€. To sum up, the contribution of the renewable sector to the Navarrese economy can be calculated as the 5% of the GDP, more important that other traditional sectors such as agriculture or textile sector.

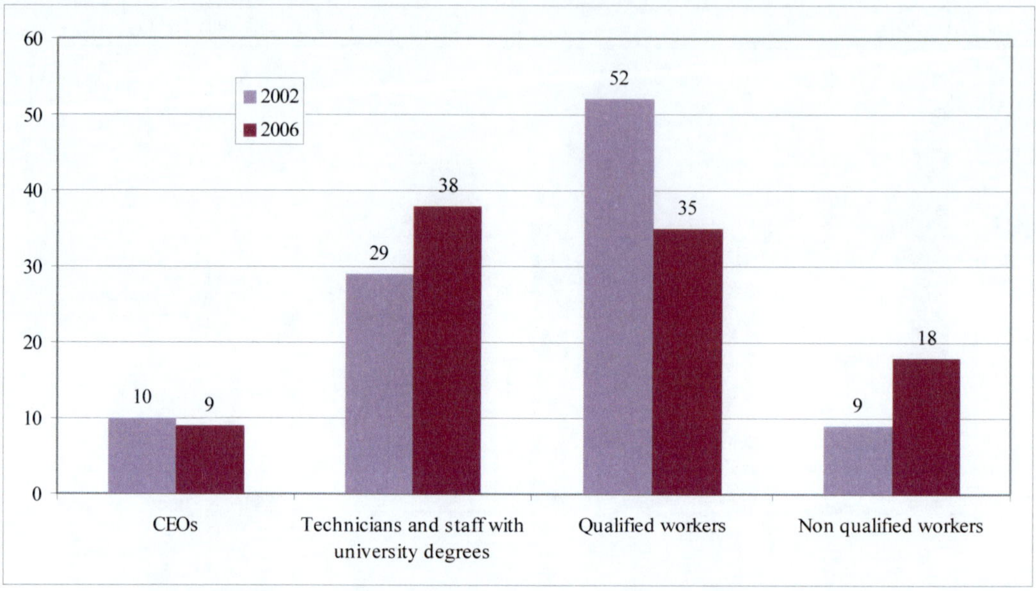

Source: Own elaboration from Faulin *et al.* (2006) and Government of Navarre (2007c).

Figure 21. Distribution of workers by skills level in the Navarre renewable sector.

5.2.2.2. A Young Workforce

As the previous section has described, enterprise development in renewable energy is a recent phenomenon. Also, some jobs require certain physical conditions. Consequently, the employees and workers hired by the companies in this sector are comparatively young: 39% of them being under the age of 30, and 88% under 40. Only 3% has more than 50 years old.

5.2.2.3. A High Qualified Workforce

One of the most outstanding features of the renewable sector is their level of training. Unskilled workers total barely 18% of the workforce in 2006 (Government of Navarre, 2007c). In 2002, the percentage was even lower: 9% of the workforce (Faulin *et al.*, 2006). Thus, in 2002, 42% out of the companies in the renewable sector were unable to hire more staff because of the lack of skilled workers, especially for jobs requiring technical skills. Figure 21 shows the qualification of the workforce in 2002 and 2006.

As you can see, the number of unskilled workers has increased in a context of a dramatically rise in the jobs number. Consequently, the requirement of a qualified workforce could be a constraint in the development of the renewable sector. To overcome this constraint, in 2003 it was created and established in Navarre the CENIFER (Integrated National Centre for Training in Renewables), described in a section 3.4. In fact, some the firms who have decided to establish in the region have considered the workers skills as the most relevant factor to take account in.

5.2.2.4. A Significant Imbalance

The most outstanding feature of the renewable sector is the significant imbalance in terms of energy type. There are two main types of renewable energy in Navarre: wind and solar-PV. Together they were being developed by 94% of the companies dedicated to renewable in

2002. If we are to focus on the role of wind energy in the sector as a whole, we will find that it accounts for 81% of the employment and 95% of turnover in 2002 (Faulin *et al.*, 2006). This distribution can be observed in Figures 22 and 23.

Nevertheless, it has been a greater increase during the last years corresponding to other energy sources, different from wind. This situation has involved the reduction of the importance of wind energy in relation to the renewable energies as a whole. Thus, the wind energy has multiplied by 2.7 its number of workers in the period 2002-2006, while the solar photovoltaic energy has multiplied that number by 7.5 in the same period. This situation has been depicted by Figures 24, 25 and 27 during the year 2006, where that tendency can be observed.

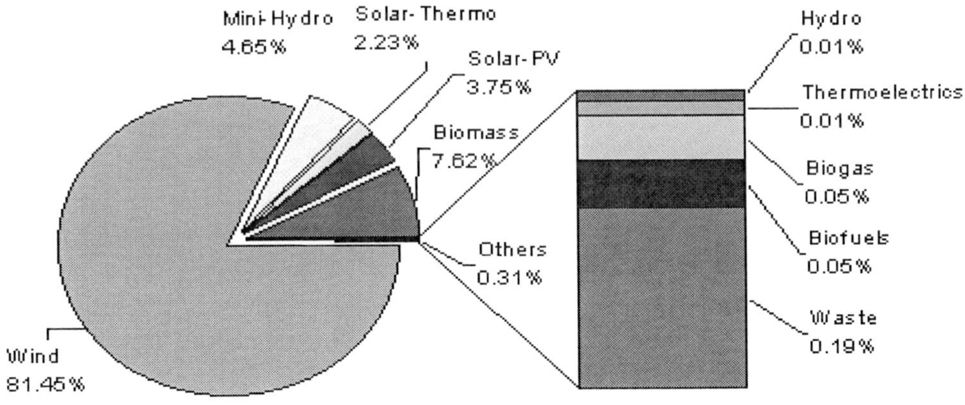

Source: Faulin *et al.* (2006).

Figure 22. Workers (%) in renewable energy in Navarre in 2002.

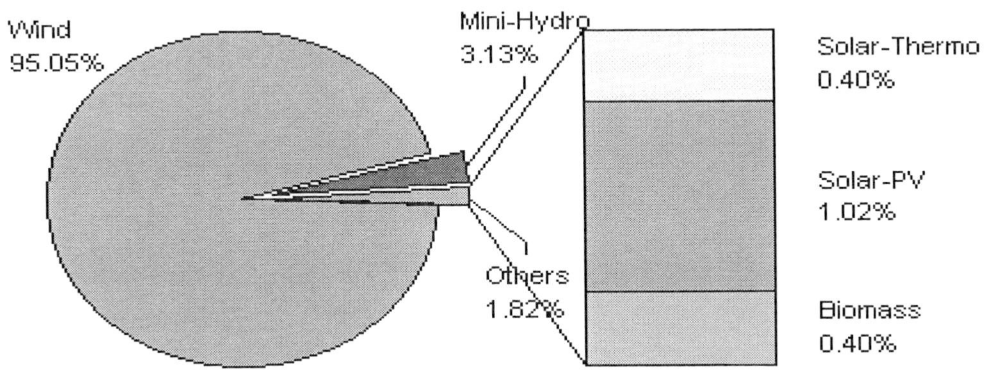

Source: Faulin *et al.* (2006).

Figure 23. Turnover (%) per renewable energy type in Navarre in 2002.

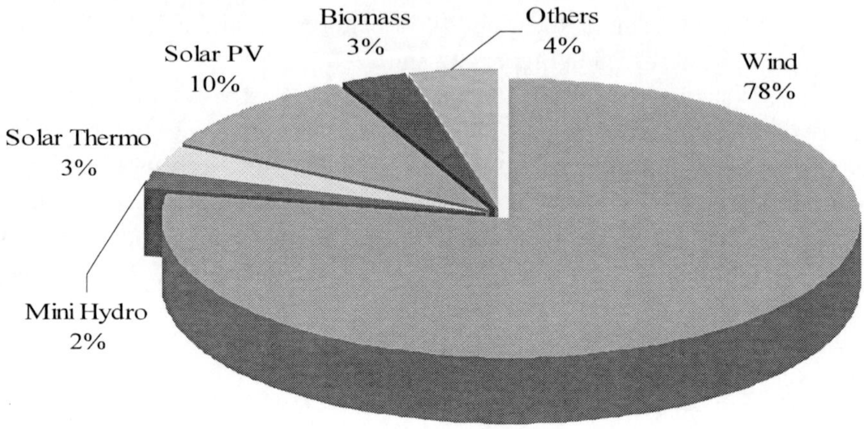

Source: Government of Navarre (2007c).

Figure 24. Workers (%) in renewable energy in Navarre in 2006 according to the Government of Navarre's (2007c) study.

Conversely, if we study the same situation in Spain in the period 2002-2006, it is observed a lower concentration of the sector in one specific energy (Figure 26). Hence, the wind energy in Spain presents a 37% of the employment in the renewable sector, with the solar photovoltaic energy in second place. In contrast, the wind energy in Navarre reaches the 78% while the solar photovoltaic hardly reaches a 10%

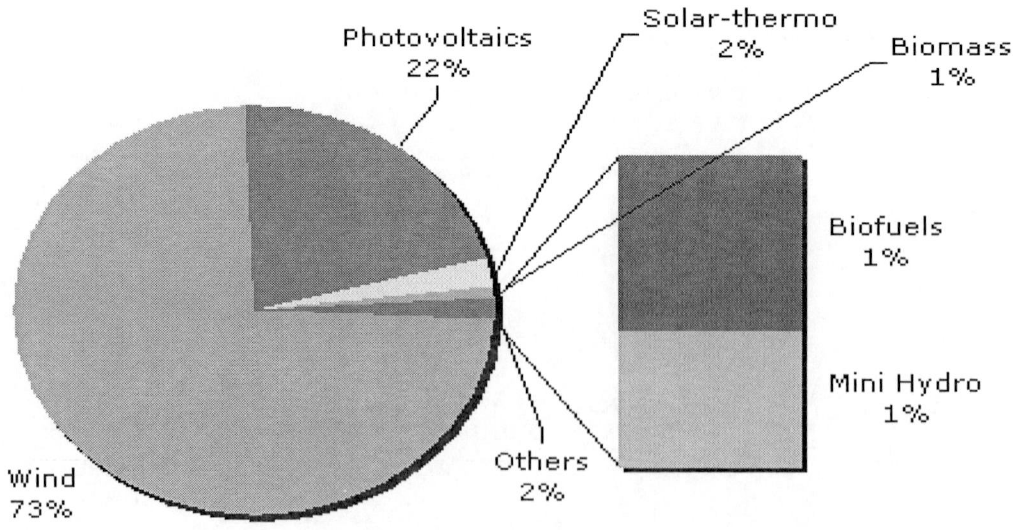

Source: Government of Navarre (2007c).

Figure 25. Turnover (%) per renewable energy type in Navarre in 2006.

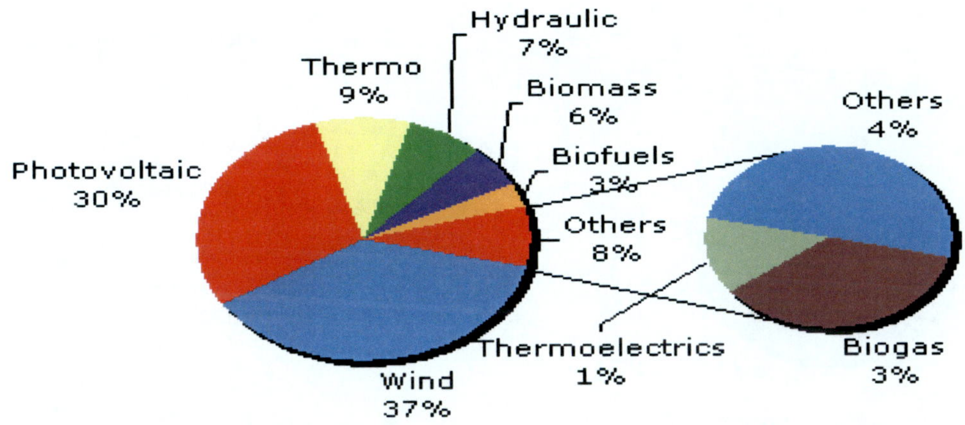

Source: ISTAS (2008).

Figure 26. Workers (%) in renewable energy in Spain in 2006.

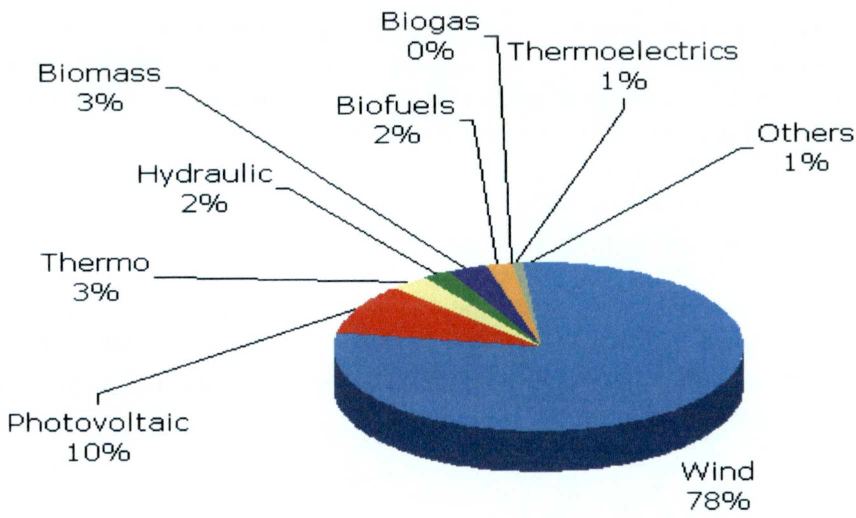

Source: ISTAS (2008).

Figure 27. Workers (%) in renewable energy in Navarre in 2006 according to the ISTAS' (2008) study.

5.2.2.5. Productive Specialization of Renewable Firms

Production of capital goods, energy production and equipment installation services are among the most productive activities associated with renewables in Navarre. Companies producing this kind of goods tend to be large and have an excellent level of technology. Currently, a number of the companies sited in Navarre are leading the wind park development sector (Acciona Energía – Acciona Group, Eólica Navarra- Enhol Group, Gamesa and Iberdrola). A similar situation can be described with manufacturing companies of wind turbines, components and spare parts (Acciona Windpower, Ecotecnia, Derna, Gamesa, Ingeteam and M. Torres). Figure 28 shows the productive specialisation of the Navarrese renewable firms in 2006.

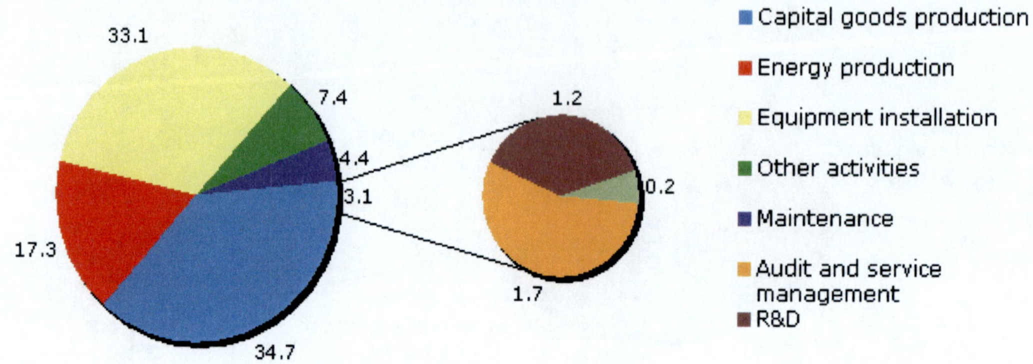

Source: Government of Navarre (2007c).

Figure 28. Productive specialization of the renewable firms. 2006.

5.2.2.6. Internationalization of the Renewable Firms

Finally, an important and recent characteristic of the renewable sector is the internationalization commitment of its companies, mainly the manufacturing companies of capital goods in this sector. Some data can show the importance of this internationalization. For instance, considering the total turnover of the Navarrese companies in renewables in 2006 (3,585 M€), its 31% (1,106 M€) was invoiced in Navarre, being the remaining 69% activities of those Navarrese companies in other places outside Navarre. Furthermore, in 2006, the Navarrese companies in renewables made the 20% of their investment in the region itself, while it was reaped the 52% of the total profit of those companies. Concerning exportations, there is a 12.5% of companies in the renewable sector which export, mainly, those devoted to the capital goods production, and in general, the biggest ones. Thus, the 49% of the turnover corresponds to the exportations worldwide. Figure 29 describes the exportations distribution according to geographical destinations.

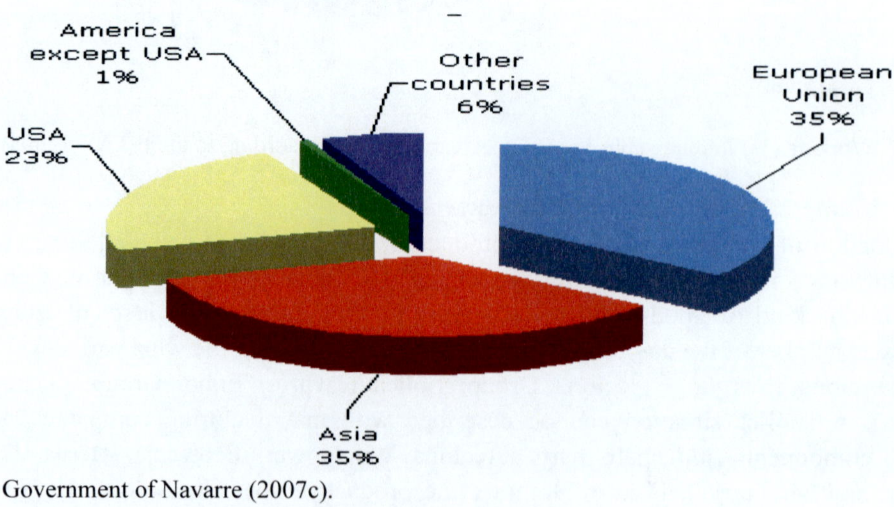

Source: Government of Navarre (2007c).

Figure 29. Exportations distribution in the renewable sector, 2006.

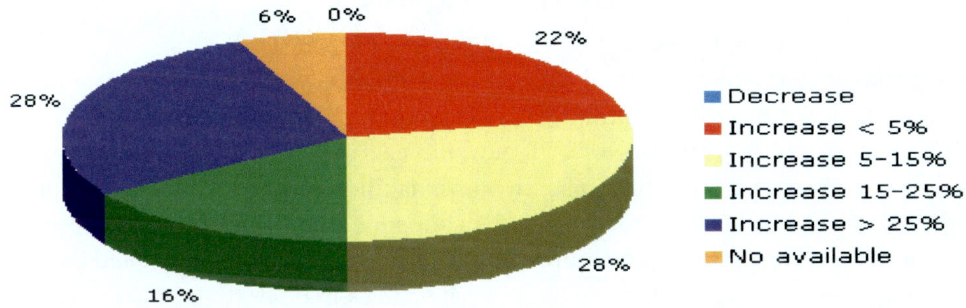

Source: Government of Navarre (2007c).

Figure 30. Prediction of the turnover increase for renewables in Navarre for the period 2007-2010.

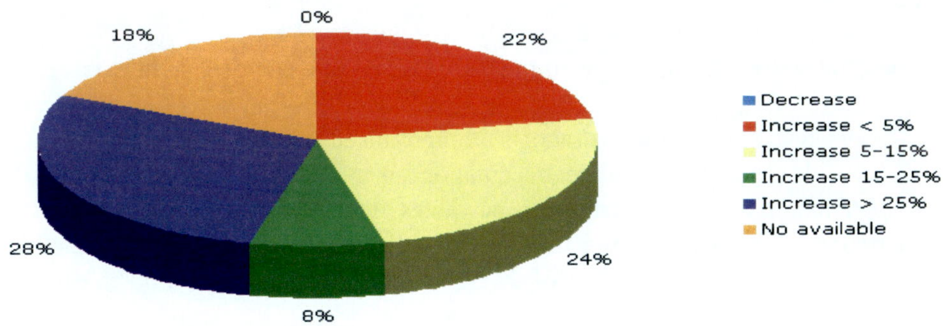

Source: Government of Navarre (2007c).

Figure 31. Prediction of the employment increase for renewables in Navarre for the period 2007-2010.

Consequently, we can assert that the renewable sector is a mature and competitive sector with a clear international presence, trying to amplify its influence worldwide because of the limitations of the Spanish market. This situation implies the development of big investments in mature and highly competitive markets, such as the European or the American ones; or in developing markets such as the Asian ones. This internationalization process has been possible because of the companies' profitability in their primigenial region: Navarre.

This maturity scenario does not imply that the sector increase has been stopped. So, following the methodology developed by Pintor *et al.* (2003), a Government of Navarre's (2007c) report describes the increase expectations of the renewable sector in Navarre for the period 2007-2010. Figures 30 and 31 present those increase expectations in relation to turnover and employment for the years 2007-2010.

According to Figures 30 and 31, more than the 25% of companies in the renewable sector consider that the turnover and employment are going to rise more than a 25% in the period 2007-2010. This positive analysis in the short term is focused on the good expectations of renewables in Spain in the following years. Therefore, it is thought that for 2010 the number of created jobs in Spain in the renewable sector would be between 44,970 (TERES II estimation, European Commission, 1996) and 94,925 (Ministerio de Industria, Turismo y Comercio, 2005). Similarly, the predictions for 2020 ranges from 228,435 and 270,788 jobs in the renewable sector in Spain as a whole. At European level, the expectations are also very optimistic (EREC, 2004).

6. Conclusions

Taking into account the performed analysis in this paper, we are going to establish some conclusions which we have divided into the next epigraphs. First of all, we will discuss the strengths and weaknesses of the sector of renewable energies in Navarre. Secondly, we will analyze the environmental implications, in sustainability terms, of the development of renewables. Thirdly, our analysis focuses on the economic implications and the development of renewables in Navarre. These conclusions can help to establish some main tenets to discuss the energy policy to the success development of renewables in other regions in the world.

6.1. Strengths and Weaknesses of the Sector of Renewable Energies in Navarre

On one hand, among the strengths of the use of renewable energies are the next ones:

i) The existence of institutional agents, compromised with development of renewables, in a wide context of public acceptation of renewables. Thus, favorable tariffs legislation has made easier the renewables development. Furthermore, the initial participation of public administration supporting the private sector by the creation of companies and the acquisition of foreign technology has been another key factor of success.
ii) Similarly, the commitment of municipalities and local agents has been essential in the acceptation of projects related to renewable energies. This situation has implied the management of important economic resources to rural municipalities, along with a great social support to this type of energies.
iii) Another important trait of renewables in Navarre is the achievement of a high index of energy self-sufficiency. Thus, Navarre has been supplied by renewables in a 13.48% of the total of primary energies, which is twice in percentage (6.9%) than in Spain. Furthermore, the 70% of the electric production in Navarre has been generated by renewables. Let us recall that the European directive 2001/77/CE demands the 29.4% for Spain in 2010. This improvement in self-sufficiency, which was tremendous during the years 1995-2003, has slowed down in the years 2004-2006 both in the primary energies generation and in electricity production.
iv) The development of a consolidated and competitive industrial sector, being successful in the international arena has been another important triumph of renewables. This has implied the concentration in Navarre of the most important economic activities in renewables, reaching a significant level of experience to compete in the international markets. In order to do that, Navarre has performed a great impulse in R&D activities in the last years, above average in Spain. This, along with the previous experience in the development of the car and machinery industries, the existence of a good communications network and the positive experiences by the public administration of supporting newborn companies and developing industrial policies, has involved the tremendous impulse of renewables of the nineties.

v) The existence of one specific type of renewable energy, the wind energy, to push the development of the remainder of renewable energies, is another characteristic of the Navarrese case. Due to the geographical characteristics of Navarre and the existence of industry related to aerogenerators production, the wind energy has been the spearhead of the development of renewables in the region. This managerial strategy is known as concentrated or selective strategy, focused on one type of renewable energy and its favorable weather conditions.

vi) The existence of research centers and institutions related to renewable energies, such as National Center of Renewable Energies (CENER-CIEMAT), and the Integrated Center of Professional Training in Renewable Energies (CENIFER)

On the other hand, we can mention the next weaknesses of renewable energies sector in Navarre:

vii) The renewable energies are excessively dependant on the public funding, because they are not profitable yet, in comparison to the conventional energies, chiefly if we do not consider the externalities associated to the latter in relation to the former. This situation makes necessary the existence of a stable legal regulation to guarantee the profitability of renewables.

viii) The renewables have not been integrated in the usual consumption of energy, but they take part of a different energy system economically supported by public institutions.

ix) The development of renewables is still scarcely diversified. According to the selective strategy mentioned in point v) the 74% of the electrical energy generated by renewables in Navarre in 2006 was produced by wind. This scenario would imply the design of a different energy policy of enhancement of renewable energies different from wind in order to even the renewables supply in Navarre.

Similarly, in the context of strengths and weaknesses of renewable energies to analyze their future expansion, it is necessary to consider the next constraints:

a. *Physical, geographical and environmental limitations*. The renewables need a set of physical and geographical characteristics associated to weather conditions to be really feasible.
b. Likewise, there are other important limitations which are specific from renewables, such as *the intensive use of materials, the space occupancy, the countryside impact and the invasion of the animals' ecosystem* of the area selected to set up renewables facilities.
c. *Technical and technological limitations* in the development of more efficient machines with greater generating power.
d. *Constraints related to the transport capacity of the electric network*. The capacity of transportation of the electric network in an international or national perspective could be a limitation in the short term of the renewables impulse in some regions or countries.

6.2. Development Implications in Sustainability of the Renewable Energies in Navarre

The improvement of energy system in Navarre implies the continuous reduction of the non-renewable energies use and the CO_2 emissions. A partial achievement of these goals is the reduction of the current CO_2 emissions until the threshold calculated in the following way: the CO_2 emissions in 1990 plus a 15% extra. Furthermore, it implies to pay attention to the space use and the control of the habitats corresponding to other living beings. Thus, we can set up the next conclusions:

i) The important breakthrough of the renewables in Navarre does not involve the improvement of sustainability in energy terms, because the non-renewable energies have increased their consumption, and also, the CO_2 emissions.
ii) The changes in the transformation procedures of primary energies have helped to limit the use increase of non-renewable energies between 1995 and 2002. Nevertheless, this effect has been widely compensated by the substitution effect of sources, and mainly, by the global increase of the energy consumption. This increase has been much greater to the renewable energies use along with the non-renewables.
iii) After the depletion of the positive effects of changes in the transformation procedure in 2002; the period 2003-2006 implied many changes in the energy transformations and in the sources substitution..
iv) This change during the period 2003-2006 is based on the decrease in the electricity generation using wind energy, and simultaneously, in the increment of central of natural gas combined cycle.
v) Similarly, the energy factors have acted on the CO_2 emissions: the outcomes partially positive in the energy transformation until 2002, have been compensated and vanished by the sources substitution and the global increase of non-renewable energies consumption, increasing the CO2 emissions among 2003-2006.
vi) The Government of Navarre estimations for 2010 show an augment in fossil energies use and the existence of a breach in the fulfillment of the Kyoto protocols, even in the most optimist scenario.

6.3. Economic Implications of Development of the Renewable Energies in Navarre

Concerning the economic point of view, the development of renewables has allowed the production and consolidation of a very competitive economic sector, having the 5% of GDP in Navarre. The apparition of this sector implies in economic terms the following positive elements:

i) A big number of jobs has been created because the sector is intensive in work. This increase has been spread in geographical terms. Moreover, many companies in the sector need to be set up close to the resource they use: water, wind, soil, wood remains ...). Many companies in the sector have been organized to cause a cohesion development of the region.

ii) Jobs having different levels in training and degrees, with a bigger weight of qualified posts, knowing the importance of technology and R&D. All the previous discussion has involved an increased need of regulated and continuous training, which the local authorities have developed in conjoint with other local research teams. This training has been provided by the University centers in Navarre and CENIFER.

iii) Many jobs have been created to young workers, making that more than the 85% of total workers in the renewable sector have less than 40 years old. The quick expansion of the sector explains the youth of the workforce.

iv) The expectations of the renewable sector until 2010 show an important increase of workforce, investment and turnover, not only because of favorable political and public support but also because of the well-known competitiveness of the companies in Navarre at international level.

v) Nonetheless, it is important to highlight that the aforementioned increase appears in an unsustainable scenario, environmentally speaking. These effects in a more sustainable situation, with a real substitution of non-renewable by renewable sources, would be mitigated or even would disappear.

Notwithstanding, the main challenge that the sector is facing in economic terms is its capacity of maintaining in Navarre the most strategic activities which have more added value, along with an important innovative R&D activity. In order to do that, it could be interesting the beginning of potential collaborations between companies in the renewable sector, for instance, in the wind sector among companies which are producing aerogenerators with similar technologies. Other potential collaborations would be in the construction and development of new wind parks worldwide. Thus, the regional public support could be very helpful to enhance the R&D effort in order to make easier collaboration agreements among companies with the purpose of create synergies in the sector of renewables.

6.4. Lessons Learned of the Renewable Energies in Navarre

Concerning the lessons learned in the process here explained, the current development of renewables is seen as a need everywhere in order to reach the sustainability goals. We have tried in this paper to explicate the Navarre's experience, a Spanish region without fossil energy resources, which decided during the nineties to invest to the renewable energies. This decision has become historical and a model worldwide. We consider the Navarrese experience could show interesting meditations to the successful advance in other regions in the world. Particularly, we believe that the next points are meaningful to understand our points of view:

a. An appropriate energy policy at the regional level which combines the next fundamental traits:

   i) A public support of the investment projects by means of the bought of electric energy generated by renewables in order to compensate the producer the greater cost of this type of energy. In this situation, it is important the design of a suitable legal structure to develop a more stable sector.

ii) Another way to present the public support to renewable energy companies would be the creation of other companies (public, private or mixed) to facilitate the attraction of multinational enterprises in the sector.

iii) The collaboration of the regional authorities in the municipalities in which the renewable energy companies are going to install the renewable projects sharing the profits of those projects.

iv) A support of the renewable sector consolidation of renewable energies in the following two aspects. First of all, supporting and enhancing the R&D effort in the sector. Secondly, developing training policies to readapt workers from other economic sectors facing the growing training demands of the renewable sector. Both aspects are essential to avoid the companies move to other regions in the consolidation phase of the renewable sector in Navarre.

b. Generally speaking, the Navarrese experience shows the suitability of using, in an initial stage, a specific renewable energy source as a catalyze to implement the set of renewables as a whole. This primigenial energy is going to enhance the development of a new economic sector. In a second stage, it would be necessary to balance the renewable energy development in the region with the help of appropriate energy policies. This balance would result essential to the sector consolidation. Otherwise, the future of the sector would be uncertain. In the Navarrese case, the catalyze has been the wind energy.

c. It is important the establishment of a regional plan of renewable development inside a regional energy plan having a global outlook which includes universal goals of sustainability improvements. This plan should include some of the next main tenets:

i) Trying to perform substitutions activities between non-renewable and renewable facilities keeping in mind the expansion of renewables implies the progressive disappearance of non-renewable sources.

ii) R&D development which involves the continuous improvement in the energy efficiency, not only in the productive activities but also in the consuming units.

iii) The design of demand policies which control the increase in the final energy use in the next aspects: 1) Systematic information to final user about the existent options to save energy performing some specific activities. 2) Transport regulation in energy terms, enhancing the use of public transportation and making territory regulations which diminish the transport needs. 3) Tax policies which discourage the unnecessary energy uses.

d. Finally, the analysis of the Navarrese experience shows the importance of creating a competitive economic sector, in order to create wealth and regional cohesion using the renewable energies. But this positive experience could be shadowed by the lack of sustainability of the energy model designed in this way. Creating a more sustainable world should be one of the key principles of each energy policy rule in the future.

ACKNOWLEDGMENTS

It is not easy to mention here the long list of people who helped us in the development of this paper. Nevertheless, we are going to cite those who helped us in a more intensive way making easier the data obtaining and the understanding of the regional energy policies of the Government of Navarre during the last decades. Thus, we can mention the following people:

i) Gaspar Domench and Fernando Señas from the Energy Section of the Industrial Promotion Service of the Department of Innovation, Enterprise and Employment belonging to the Government of Navarre; and Pedro Melero from the Energy General Secretary of the Ministry of Industry, Commerce and Tourism of the Spanish Government because of their help in the elaboration procedures of the Energy Balances of Navarre and Spain respectively.
ii) Cruz Martínez Erauskin helped us to understand the importance of the CENER and CENIFER centers as catalysts in order to make easier the development of renewables and training people to solve real problems in renewables. Begoña Urien and Cristina Urdanoz help us to find the right people in the right place in the Department of Innovation, Enterprise and Employment.
iii) Jordi Roca Jusmet because of his comments to the first version of the factor decomposition analysis in the energy use.
iv) Last but not least, we would like to thank Justo García Ortega from the Public University of Navarre because of his scientific support helping us to understand the importance of renewables in the current world. The obtaining of some data and their subsequent processing would have been impossible without his help.

REFERENCES

ADAS (2003): *Renewable energy and its impact on rural development and sustainability in the UK*. K/BD/00291/REP URN 03/886. Wolverhampton, UK: ADAS Consulting Ltd. and University of Newcastle.

Agterbosch, S., Glasbergen, P. and Vermeulen, W. (2007): "Social barriers in wind power implementation in the Netherlands: Perceptions of wind power entrepreneurs and local civil servants of institucional and social conditions in realizing wind power projects", *Renewable and Sustainable Energy Reviews*, 11: 1025-1055.

Agterbosch, S., Vermeulen, W. and Glasbergen, P. (2004): "Implementation of wind energy in the Netherlands: the importance of social-institutional setting", *Energy Policy*, 32: 2049-2066.

Alcántara, V. and Roca, J. (1995): "Energy and CO_2 emissions in Spain. Methodology of analysis and some results for 1980-1990", *Energy Economics*, 17: 221-230.

Bancaja (2003): *Capital humano y actividad económica*, Bancaja e Instituto Valenciano de Investigaciones Económicas, Valencia. Spain (in Spanish).

CENER (National Center of Renewable Energies) (2006): *Memoria de Actividades CENER 2006*. Downloaded from www.cener.com (last access 12 June 2008) (in Spanish).

CENIFER (Integrated National Centre for Training in Renewables) (2005): *Las energías renovables en Navarra y su impacto en el empleo*. CENIFER, Government of Navarre, Pamplona (in Spanish).

Club Español de la Energía (2007): *La eficiencia energética de la Península Ibérica 2007.* Everis Report. Downloaded from www.everis.com (last access 25 May 2008) (in Spanish).

Daly, H. E. (1999): *Ecological economics and the ecology of economics: essays in criticism*, Cheltenham (UK)-Northampton (Mass.), Edward Elgar.

Del Rio, P. and Burguillo, M. (2008): "Assessing the impact of renewable energy deployment on local sustainability: Towards a theoretical framework", *Renewable and Sustainable Energy Reviews*, 12: 1325-1344.

Diario de Navarra, (2008): "La rebaja de la subvención estatal a la energía fotovoltaica amenaza el empleo en el sector", *Diario de Navarra*, July 24, 2008, p.16 (in Spanish).

Diewert, W.E. (1993): "Chapter 5: Index Numbers" in Diewert, W.E. and Nakamura A.O.: *Essays in Index Number Theory*. Vol 1. Elsevier Science Publishers.

Ek, K. (2005): "Public and private attitudes towards "green" electricity: the case of Swedish wind power", *Energy Policy*, 33: 1677-1689.

EREC- European Renewable Energy Council (2004): *Renewable Energy Target for Europe: 20% by 2020*. Internal Report. Renewable Energy House. Downloaded from www.erec-renewables.org (last access 25 May 2008)

European Commission (1996): *TERES II, The European Renewable Energy Study*, Advisory on Research and Development, Luxembourg.

European Commission (2000): *The Impact of Renewables on Employment and Economic Growth* (IREEG), D.G. TREN. Brussels.

European Commission (2003): ALTENER Programme. *The impact of renewables on employment and economic growth*. Brussels.

European Commission (2006): *Study on renewable energies in Spain. Recommendations for strategic guidelines after 2006*. CE160AT024. Final Report. March 2006.

European Union (2008): *EU Funds for Energy and Environment. Opportunities in 2008 and beyond*. Downloaded from http://ec.europa.eu/regional_policy/index_en.htm (last access 26 June 2008)

Eurostat (2006): *Regional Gross Domestic Product ESA 95*. Downloaded from http://epp.eurostat.ec.europa.eu (last access 26 June 2008)

Fairless, D. (2007): "Energy-Go-Round", *Nature*, 27 June 2007: 1046-1048.

Farstad, H. and Ward, J. (1984): *Wind Energy in Denmark* in Baumgartner, T. and Burns, T.R.: *Transitions to Alternative Energy Systems. Entrepreneurs, New Technologies and Social Change*. London, Westview Press.

Faulin, J., García, J., Lera, F. and Pintor, J.M. (2003): "Expansión de las energías renovables a nivel regional", *Boletín Económico del ICE*, 2787: 9-21 (in Spanish).

Faulin, J., Lera, F., Pintor, J.M., Garcia, J. (2006): "The outlook for renewable energy in Navarre: An economic profile", *Energy Policy*, 34: 2201-2216.

Government of Navarre (1984-2006): *Balances de energías finales.*. Balances corresponding to years 2004 and 2005 can be downloaded in (last access 25 May 2008): *http://www.cfnavarra.es/INDUSTRIA/areas/energia/Balances%20energeticos%20de%20 Navarra%202004-2005.pdf* (in Spanish).

Government of Navarre (1994-2006): *Balances de producción de energía eléctrica.* Balances corresponding to years 2004 and 2005 can be downloaded in (last access 25 May 2008): *http://www.cfnavarra.es/INDUSTRIA/areas/energia/Balances%20energeticos%20de%20 Navarra%202004-2005.pdf* (in Spanish).

Government of Navarre (1995): *Primer Plan Energético de Navarra - First Energy Plan of Navarre (FPEN).* Approved by the Government of Navarre in January 1996 (In Spanish).

Government of Navarre (2007a): *Navarre. Energías Renovables. Horizonte 2010.* Government of Navarre (In Spanish).

Government of Navarre (2007b): *Plan Energético de Navarra, 2005-2010.* Government of Navarre (In Spanish).

Government of Navarre (2007c): *Situación de las energías renovables en Navarra en 2006.* Government of Navarre, Pamplona (In Spanish).

Hillebrand, B., Buttermann, H., Behring, J., and Bleuel, M. (2006): "The expansion of renewable energies and employment effects in Germany", *Energy Policy*, 34: 3484-3494.

Hogan, W.L. (2002): "Energy Modeling for Policy Studies", *Operations Research*, 50: 89-95.

IDAE (2003): *Energías Renovables en España*, V Document, Spanish Ministry of Industry and Energy, Institute for Energy Saving and Diversification (IDAE) (in Spanish).

International Energy Agency- IEA (2006): *World Energy Outlook, 2006.* Downloaded from http://www.worldenergyoutlook.org/2006.asp (last access 9 June 2008)

International Energy Agency- IEA (2007): *Energy Balances of OECD Countries - Energy Balances Vol 2007.* Downloaded from http://www.worldenergyoutlook.org (last access 9 June 2008).

ISTAS (2008): *Energías renovables y generación de empleo en España: presente y futuro.* Instituto Sindical de Trabajo, Ambiente y Salud - ISTAS. Madrid (in Spanish).

Jacobs, M. (1993): *Green Economy: Environment, Sustainable Development and the Politics of the Future*, University of British Columbia Press, Georgetown, ON, Canada.

Kamen, D., Kapadia, K., and Fripp, M. (2004): *Putting renewables to work: how many jobs can the clean energy industry generate?* RAEL report. Berkeley: University of California.

Komor, P. and Bazilian, M. (2005): "Renewable energy policy goals, programs and technologies", *Energy Policy*, 33(14): 1873-1881.

Leontief, W. W. (1986): *Input-Output Economics.* Oxford University Press, 2nd ed., New York.

Martínez Erauskin, C. (2008): *Personal interview with the authors about the success of renewable energies in Navarre in the period 1995-2008.* Interview held on June 25, 2008.

Meadows, D. H.; Randers, J.; and Meadows, D.L. (2004): *Limits to Growth: The 30-Year Update*, Chelsea Green.

MEC, CENER, SOCINTEC, AND F^2I^2 (2005): *RENOVALIA- Situación de las energías renovables en España.* Final Report. Ministerio de Educación y Ciencia (in Spanish).

Menéndez, A. (2001): *Renewable Energies, Sustainability and Employment Creation: An Economy based on the Sun.* Asociación Los Libros de la Catarata. Madrid (in Spanish).

Meyer, N. (2003): "European schemes for promoting renewables in liberalised markets" *Energy Policy*, 31: 665-676.

Ministerio de Industria, Turismo y Comercio (2005): *Plan de Energías Renovables en España 2005-2010.* Instituto para la Diversificación y Ahorro de la Energía. Internal Report. (in

Spanish) Downloaded from http://www.mityc.es/Desarrollo/Seccion/Energia/Renovable/Plan/ (last access 9 June 2008)

Ministerio de Medio Ambiente - MMA (2000): *Estimación del empleo medioambiental en España*, Draft Report, Madrid, Spain (in Spanish).

Moreno, B. and López, A.J. (2008): "The effect of renewable energy on employment. The case of Asturias (Spain)", *Renewable and Sustainable Energy Reviews*, 12: 732-751.

Navarrese Institute of Statistics (2008): *Cuentas económicas. Población y demografía.* Downloaded from http://www.cfnavarra.es/estadistica/ (last access 26 June 2008) (in Spanish).

Pintor, J.M., Faulin, J., García, J. and Lera, F. (2003): *Report Renewables Navarre 2002: Prospección de futuro del subsector empresarial de las energías renovables en Navarra: potencial económico y potencial humano*. Internal Report for the Employment Service of the Government of Navarre. Pamplona, Spain (in Spanish) (Acronym: RN2002).

Pintor, J.M., Lera, F., Garcia, J., Faulin, J. (2006): "Energía eólica y empleo: el caso de Navarra como paradigma", *Revista del ICE*, 829: 253-271 (in Spanish).

Reiche, D. (Ed.) (2002) *Handbook of Renewable Energies in the European Union.* Case Studies of all Member States. Frankfurt/Main, Peter Lang.

Reiche, D. and Bechberger, M. (2004): "Policy differences in the promotion of renewable energies in the EU member states", *Energy Policy*, 32: 843-849.

REN 21 - Renewable Energy Policy Network for the 21st Century (2007): *Renewables 2007. Global Status Report.* Downloaded from http://www.ren21.net/pdf/RE2007_Global_Status_Report.pdf (last access 26 June 2008)

Sánchez-López, B. (2006): *Empleo en PyME del sector de las energías renovables e industrias auxiliares en España.* Instituto Sindical de Trabajo, Ambiente y Salud - ISTAS. Internal Report (in Spanish).

Sanz, M. (2009): *Speech of His Excellency Miguel Sanz Sesma, President of the Government of Navarre, on Renewable Energies.* Harvard University, January 23rd, 2009. Downloaded from http://www.navarraconfidencial.com/portal/modules.php?name=Noticias&file=article&sid=2953 (Last access January 27th, 2009)

Spanish National Statistics Institute - INE (2008): *INEBASE Database.* Downloaded from http://www.ine.es (last access 26 June 2008) (in Spanish).

Stewart, J. (2006): "Windmills of the Green Mind." *Geographical.* 78: 56-58.

Sun, J. W. (1998): "Changes in energy consumption and energy intensity: A complete decomposition model", *Energy Economics* 20: 85-100.

The Economist (2008): "The power and the glory. A special report on energy", *The Economist*, 21st June 2008.

Thothathri, R. (1999): "The wind brought jobs and prosperity", *New Energy*, 4: 28-30.

Toke, D., Breukers, S., and Wolsink, M. (2008): "Wind power deployment outcomes: How can we account for the differences?", *Renewable and Sustainable Energy Reviews*, 12: 1129-1147

Upreti, B.R. (2004): "Conflict over biomass energy development in the United Kingdom: some observation and lessons from England and Wales", *Energy Policy*, 32: 785-800.

Wolsink, M. (2000): "Wind power and the NIMBY-myth: institutional capacity and the limited significance of public support", *Renewable Energy*, 21: 49-64.

Wolsink, M. (2007): "Wind power implementation: The nature of public attitudes: Equity and fairness instead of 'backyard motives' ", *Renewable and Sustainable Energy Reviews*, 11: 1188-1207.

Yue, C., Liu, C. and Liou, E.M.L. (2001): "A Transition toward a Sustainable Energy Future: Feasibility Assessment and Development Strategies of Wind Power in Taiwan", *Energy Policy*, 29: 951-963.

In: Energy Policy: Economic Effects, Security Aspects... ISBN 978-1-60692-894-3
Editor: Noah B. Jacobs

Chapter 2

SAUDI ELECTRICITY COMPANY: RESTRUCTURING AND DEREGULATING

Isa S. Qamber
Electrical and Electronics Engineering Department, University of Bahrain
Tareq A. Al-Muhawesh
Saudi Electricity Company, Department of CA, Dammam, Saudi Arabia

ABSTRACT

Since the start of utilizing the electric energy in its basic shape, until today, the way in delivering energy to customers have gone through various stages according to the needs, capability, knowledge, technology, and rules. Starting without regulators until it is found necessary to have regulators govern the small utilities before more than a century ago. Then, vertically integrated utilities became the most popular practice all over the world with little differences in detailed practices. Despite those differences, typically, such utilities own three major business lines namely generation, transmission and distribution covering a specific geographic area. Such utility is self dependent in all aspects from technical support to auxiliary services and billings. However, some of utilities are getting financial support (subsidies) from their governments to avoid high bills on customers [1].

INTRODUCTION

The power sector is modeled at the level of the wholesale markets. Each country has its own wholesale market which is interconnected to other wholesale markets by a transmission network. The electricity industry throughout the world is currently undergoing a significant transition towards restructuring and deregulation. It is well known that governments all over the world, in the recent past, are seriously adapting and implementing privatization through restructuring or deregulating the electrical industry. The reasons behind deregulation differ from one country to another. However, the paramount reason is to avoid uncertainty of the future. This uncertainty is due to expected tremendous load demand growth, depletion of natural resources, and negative environment impact. Other reasons could be seeking competition, inefficiency of the existing electrical utilities, avoiding subsidies from government's budget, possibility of getting power cheaper elsewhere.

Due to various reasons and needs, as explained in the previous paragraph, restructuring of electric industry had started in different countries all over the world since 18th century until now and it passes through so many stages which differ according to the utility set up. But the most common stages can be listed as follows:-

- Rule to be established by the government, regulator, or owners.
- Reform the utility or utilities.
- Privatization.
- Unbundling or separating the three main components (Generation, Transmission, and Distribution).
- Allowing Independent Power Producer's (IPP) to participate in generation.
- Allowing bulk customers to choose their suppliers.
- Establishing Independent System Operator (ISO) or Regional Transmission Organization (RTO) or Power Exchanger (PX) or a combination of two or all of them.

It should be noted that each country has its own timetable in going through some or all of the above listed stages in addition to other detailed steps suiting the uniqueness of the industry.

Deregulation Policy [2]

In the present study, the deregulating process calendar presented through the years 1978 till 2008. This calendar will help us in reflecting the reasons behind the question raised now a days (Why deregulations?). Therefore, in order to help any utility to decide either to deregulate or not, it should be aware of the following facts:-

- It is wrong to conclude that privatization is needed in all cases.
- Each utility has its own characteristics which affect the decision.
- The country culture, rules, and size of industry where the utility exists has major effect in the decision.
- The system parameters and the location have another effect.
- Small market cannot be put in competition.
- The following list gives some reasons to support the deregulation decision which could be applicable to a good number of utilities in the world:
  i) Avoiding subsidies from government's budget.
  ii) Shifting from monopoly market will lead to more fair prices for each service.
  iii) Efficient power plants are expected to be built by IPP's to compete with the market which will reduce the cost.
  iv) Innovations and renewable energy will find better atmosphere.

v) Inefficiency of the existing electrical utilities.
vi) Uncertainty of the future due to expected tremendous load demand growth, depletion of resources, negative environment impact.
vii) Privatization will force the companies' management to be more toward increasing shareholders investments.
viii) International energy company concept – a company whose focus is becoming more global and more multi-purpose.
ix) Cost reduction due to competition.
x) Fuel supply stability.
xi) Improve customer services.

Despite the above positive points, still some difficulties may face the deregulating process which could be solved or avoided or reduced by extra efforts and practices:-

- Electricity markets transactions complexity and data availability due to different parties involved which need sophisticated information technology with open access to everyone all the time [3].
- Due to the fact that energy cannot be stored, up to nowadays technology, balancing supply with demand on real time between parties is a must which is a real challenge to the system operator. Any shortcomings could lead to partial or total blackout compared to a busy tone in communication service or using the stored gas when we have gas supply shortages [4].
- Chance for miss handling of network by the Transmission Provider (TP) to increase profits being a natural monopoly company. In other words, we can say that the Transmission Company (TRANSCO) has a natural monopoly characteristics. So, it is advised to overcome the misusage problem by establishing ISO to be independent from transmission network owner.
- Market power could be forced by some of the service providers and control the prices rather than to be controlled by the market prices. So, a very well established market is needed with corrective tools and supported by a well stated rules by regulatory to eliminate the market power negative effect.

Literature Review

To provide a good review of different opinions and to know the limitations of previous studies in the electric industry deregulating process; this section will summarize each reference according to its order in the bibliography.

Jeffrey Hein in his master thesis in reference [5] proposed a new model for the entire United States' electrical industry to resolve most of the problems associated with the restructuring efforts at the time of his thesis. In the present chapter of our study has the same direction of Mr. Hein but with different framework suitable for Saudi Arabia electrical industry uniqueness and requirements.

A good comparison between AC and DC power flow models is carried out by the authors of reference [6] which concluded that DC models are good approximation for AC models with less complications and it is advisable to replace them in the market type studies but not in the operational studies. This conclusion encourage us in this study to propose the MW LP-based OPF to be used for market studies such as finding the future generation potential areas.

As per reference [7] the transmission is a very complicated issue in the deregulation process due to the need of reliable interconnected system either with one path or with parallel paths for short distances or long distances. The question will be raised who pays for what portion from the overall cost? The book's author analyzed different pricing methods like: Flat Fee Method, Postage Stamp, MW–Mile, and Contract Path Method. The advantages vs. the disadvantages of those methods are discussed. Also, the author addressed the problem of transmission congestion due to the intendancy of buying from the cheapest generated power even if it is very far. This can be controlled by means of: Planning, Price incentives, and Congestion Management. The book's author believes that this is the time of deregulation. Though he has no solid results from practices all over the world, he is speculating that the deregulation will promote competition and cause reduction of costs, increase reliability, and increase availability of more energy services.

Reference [8] looks to the problem from different angle. Instead of giving attention to the cost techniques only, he suggests technical solutions like discussing the static compensator to solve the congestion problems. Other example is the idea of converting the existing generator to synchronous condenser to avoid high cost of adding generation capacity. The book's author discusses the transmission transfer capacity and encourages the utilization of Capacity Benefit Margin (CBM) technique to ensure access to generation from interconnected systems for meeting the required reliability and reducing the generated capacity. Dynamic Line Rating (DLR) is explained as an important and an effective in increasing transmission capacity with preference to the technique of Tension Monitoring because the way it rates the line based on actual clearance and allows the operation of the line near to its design parameters. The book also, calls for modern teaching methods to encounter the deregulation matters to be included in the graduate education level which encourage utilities to hire employees with master degrees.

Reference [9] believes that advancement in technologies allow small producers to compete with others in a win-win atmosphere. The author considers three types of markets: Completely Market Driven atmosphere, Transitional Market, and Embryonic Free Market. International case studies with problems and solutions are addressed in this book. The author Lio says "The market should dictate when new generation is needed and where it is located." The author also, presented analytical tools for the modeling of the competitive power market and its analysis, presented issues related to transmission open access, and presented the application of Flexible AC Transmission System (FACTS) to utilize the capability of the system. Lio said "The explosion in the use of information technology has seen the introduction of computer-base work management systems, assets management systems, and control systems to manage system operation."

Reference [10] discusses most of the topics of the previous book but with more emphasis on the transmission level and power quality without case studies which make it more toward generalization. So, it is not advised to use the book as troubleshooter, on the other hand it can be taught in the universities due to the generality and being dealt with basic techniques.

Robertson reported that "market forces aren't doing enough to encourage the construction of extra generation to guard against shortages." [11]

In the same article, he points out a very logical conclusion "*small population make it hard to open the market, because there aren't enough customers to attract power companies*."

In Plano, people knowing that they are paying the old prices for the product which becomes cheaper. "*The January deregulation was accompanied by a state mandate for a 6-percent reduction in electricity rates, anticipated to save an average of $4 a month for customers. But the Legisture also authorized fuel factor adjustments to be made twice a year by the companies to respond to fluctuations in natural gas prices*." [12]

In Columbia, Wendy the writer of article [13] is referring to an official report that "*recommended splitting B.C. Hydro into separate generating, transmitting and distribution divisions and moving to "market" pricing that it said could lead to increase of 30 percent for residential customers and up to 60 percent for industrial customers.*"

[14-21] are selected articles on the subject where most of them are strongly against deregulation due to any drawback in the implementation experienced in their countries. Others are realizing this point relating the drawback to specific reasons and proposing some solutions.

"*Instead of abandoning deregulation, California should find a way to inject real competition into the market*" [14]

"*If all the troubled companies shut up shop tomorrow, the UK could end up some 10,000 MW short of what it needs during winter weeks. A California-style brown-out is not out of question*" [15]

"*Partial deregulation doesn't work, "he said "and deregulation without some sort of government structure doesn't work, either*." [16]

"*it should be no surprise that Dynegy is seeking a return to its past as an energy generating and transporting company as the owner of a utility*" [17]

"*Texas-New Mexico Power and Light Co. have donated the electricity required to decorate Bogotá's downtown area for the holidays. "he said "But energy deregulation changed all that. The City Council was told that the power company could not give away electricity*." [18]

"*There was nothing to be gained by deregulation here, because South Dakota already had some of the lowest electric rates in the country." [19]*

"*The chairman of the state Assembly Energy Company released a report Friedy charging that the state's energy deregulation policy has put New Yorkers at risk." [20]*

"*Last year being the first time for cities to deal with deregulation, the city of Midland entered into a one-year contract with TXU Electric to serve the city's electric needs. The one-year contract was made because of an uncertain market." [21]*

Even though, those articles are seeking tangible proves and or scientific analysis, still can expose opinions and events which can lead to the correct direction and effective decisions. So, it worth full to analyze and study the concerns to end up with the best implementation of deregulation process.

The electric power consumption in Saudi Arabia is growing at annual rate of 07% compared to the world rate of more than 03% as in reference [22].

Dr. Lee & Dr. Heydt in reference [23] provide a procedure to give the right signals to suppliers & consumers to balance the power quality with its related costs.

Saudi Arabia is implementing a new hierarchy in the Saudi Electric Company (SEC) by 2003. This is reported as the first step of joining all Saudi Arabia under one company, the second step will be releasing generation business to private sector. [24]

Another factor which is indirectly related to deregulation through the optimization of load transfer for better utilization of transmission capability is the future plan of interconnecting gulf network expected to take place on year 2006 as per [25]. Until the time of releasing this study, the interconnection is under construction but not yet operated which means the schedule is expected to be delayed for some time.

Article [26] listed eleven lessons learned from California crises. The author concluded that deregulation process is not out of dangers in terms of generation shortage or grid congestion. He called also for greater regulatory resources.

Reference [27] proposed a Refined Immune Algorithm (RIA) as a decision tool to perform the optimal generation expansion planning in a deregulated industry. Despite the traditional generation planning methodologies (linear & non-linear programming, etc...), a new algorithm based on artificial intelligence has been developed combined with Tabu search list to avoid invalid searches.

Security Constrained Unit Commitment (SCUC) is used by reference [28] to calculate the optimal unit commitment with flexible operation conditions. The proposed algorithm covers both the generating units and the network flow constraints. The article's authors emphasize the importance of fuel diversity to minimize the cost satisfying other constraints like system parameter and environmental effects.

Even though ref. [29] had been reported for Japan's electrical restructuring purposes, it can be a good reference for other countries after tailoring it to its unique characteristics. The analysis and the comparison of the restructuring processes in several countries give the reader the main guidelines in forming electricity market structure. This reference was a good base for this study in proposing the Saudi electrical supply industry framework.

It is worthwhile to conduct comparative analyses of electricity reform in a group of countries with some common features. This has been done by reference [30] which compared OECD Pacific countries in terms of reform policies, legislative process, market structure, and market design. Those countries namely: Australia, Japan, NewZeland, and Korea are all electrically isolated from other countries. Such common feature which is also valid for Saudi Arabia makes the report a good source of experiences to learn from.

Reference [31] analyzes California crises and links it directly with the price capping in curtail prices and deregulating the wholesale price which make it high enough not to encourage new comers into the market.

Advice from the author of article [32] to Japan is to avoid a wait-and-see approach to deregulation due to the fact that electricity is so fundamental service in our daily life, however, he expects Japan to be very slow in deregulation process affected by California crisis.

Article [33] proposed a model based on game theory in new partially regulated environment to clarify the mechanism of power trading in determining the electricity price and transaction volume, where multiple utility companies and two IPP's negotiate their sales prices with a given set of large customers in the deregulated environment.

Game theory is a useful tool for analysis of effective power market as concluded in article [34] where two test systems were used to demonstrate the approach.

Dr. Richard Schuler with his Economics background conclude on reference [35] that the market clearing prices are inversely related to the number of generating firms bidding to enter a single market. Also, the author briefly summarizes the effect of transmission constraints on the clearing price as it is considered to be additional charges to generation costs.

Reference [36] recommended a securitization tool to be used to recover the stranded costs which are generated in the transition period from regulated to a competitive market. The article reported that securitization has a unique advantage of reducing the level of stranded costs while recovering it.

Changes in the value relevance of book value and earnings in response to deregulation are tested in reference [37] by examining changes in coefficient magnitudes and incremental explanatory power for 9 years. Book value founded to be decreasing while earnings increasing.

Transmission Congestion Management's (TCM) role in power operation was emphasized by reference [38] especially in a deregulated industry. Lower prices of larger generators under competitive market could direct the power flow to cause transmission congestion. The paper analyzed TCM types including NCM and ZCM. It concluded that different types of TCM to be adapted depending on the uniqueness of the industry and the stage of the restructuring process.

Acceptable limits of non participation to the load frequency control problem are called for in reference [39] to allow any unit to enter to the system in any deregulated industry.

As per reference [40], simple marginal prices are not covering startup and other generation costs needed to satisfy market participants. The same reference proposed Lagrange Relaxation (LR) based method to cover generators' costs at all times. After running sample system, the article concluded that demand elasticity has impact on prices and profits at peak hours.

The author of reference [41] calls for immediate researches to design an efficient industry, edible, and resistant to market power. Also calls for enhancement of customer services and enhancement of operational techniques.

Reference [1] discussed most of the key issues in deregulation and its models with more emphasis on pricing, congestion management, tagging, energy trading, hedging, and stranded costs. The book also, gives full illustration about OASIS Functions. The author gave the following link between RTO & ISO "RTO could act as an ISO (non-profit institution), stand-alone transmission companies (for-profit TRANSCO), hybrids or any other form that could accomplish the aims of Order No. 2000. The flexibility in determining the future structure of RTOs is one of the appealing advantages of the final rule."

The information used from reference [42] is purely from what is published only in the annual reports of SEC to avoid any conflict of interest since this study is reported by an engineer in SEC which I am proud of.

The last three references [43, 44, and 45] are the latest presentations related directly to the study subject presented to SEC management and outside seminars. These presentations are proposing the future framework of Saudi Arabia Electrical Industry (SESI).

SAUDI ARABIA AND DEREGULATION

Since deregulation process has already started in Saudi Arabia, management and staff of Saudi Electricity Company (SEC) are recommended to work together to form an efficient industry framework. The efficient design is required to keep prices down and satisfy the future expected high demands without the need of financial support while keeping in mind environment protection and reliable and safe power supply to customers.

The study of putting a framework for SEC encourage power system expansion in the right direction. However, three main questions arise during the course of this study:-

- Where will be SEC future generation expansion?
- What is the future shape of Saudi electricity supply industry?
- Which is cheaper to transmit electricity or fuel from one province to another province?

Answering any of the above questions needs gathering of information and forming a complete optimization model and testing it, is by itself a complete study.

This study started about Saudi Arabia electricity industry. It gives SEC data at year 2005 including some of the current practices and parameters. The ongoing deregulation activities by SEC are also explained.

The most popular market models are reviewed and more attention is given to the participants as the critical links in any deregulated system. As a result, SEC framework is proposed, including the pricing methods and trading techniques.

The complete linear model of the Mega Watt (MW) Linear Programming (LP)-Based Optimal Power Flow (OPF) [46]which is used to know the future generation potential areas in any utility is proposed. The study also explains the available power flow optimizing tools. As well, Economic Load Dispatch (ELD) is considered through the study as the basic optimization tool to dispatch the power system.

The model established in through the study is applied to the eastern province of Saudi Arabia. Case studies are investigated to know the potential generation areas and finally the results are analyzed. Two cases have been considered in reference [46], where a third case is considered in the present chapter.

Finally, and out of the study Saudi Arabia is undergoing deregulation of the electrical industry to introduce and increase competition among the power providers at the wholesale generation level and they may be encouraged to go further to the customers' level by giving them the right to choose their suppliers. So, the timing of this work gave an advantage to the researcher to:-

- Monitor the real steps taken by SEC in the deregulating process.
- Analyzing the pioneers' practices in the world in market structure of the electrical industry.
- Start from where others end in selecting the participants and the interface points.
- Avoid wrong practices in the decisions taken for separating system operator from transmission owners or not.
- Be clearer about what problems to be solved.

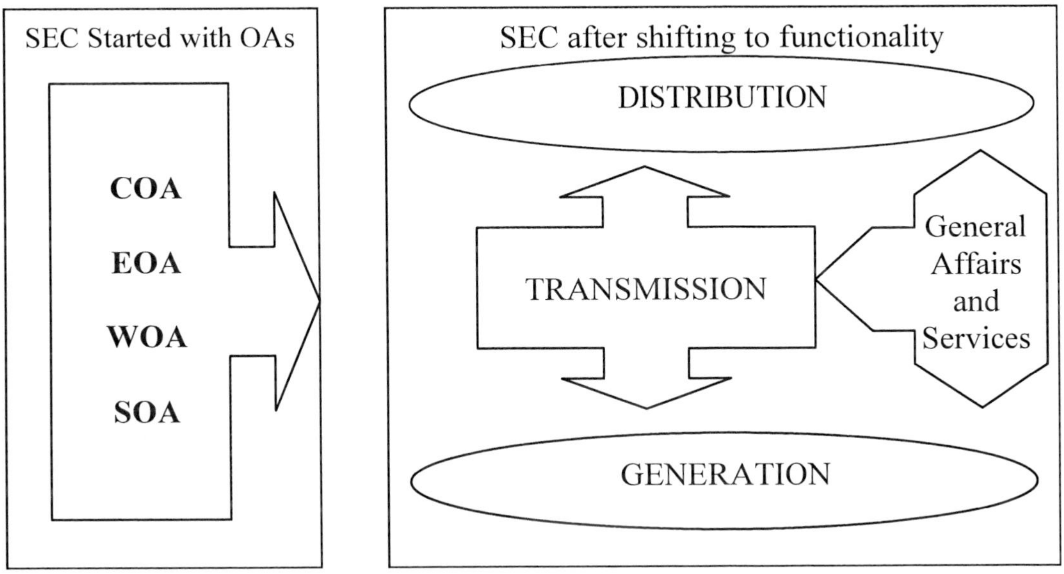

Figure 1. SEC Shifting Stage from OA's to Functional Organization.

FORMING SEC [42]

Forming SEC by joining all existing electrical companies in year 2000 was the first step toward deregulating the electricity industry in Saudi Arabia. The second step was shifting from isolated four operating areas doing individually all activities to new organization operating as separated business lines according to activities of generation, transmission, and distribution as the main core functions in addition to other corporate affairs and general services as illustrated in Figure 1.

As the first stage in deregulating process; SEC is functioning in years 2003 and 2004 with the new organization shown in Figure 2.

During this stage; the following activities are carried out:-

- Each of the three main business lines (generation, transmission, and distribution) has interface and a planning role with other BL's, as well as the general service's organization and corporate services.
- Setting procedures to support the main BL's to act independently.
- Distributing the fixed assets among BL's.
- Fixing the interface points between BL's.
- Studying the sell and buy energy prices at the interface points.
- The three main BL's practicing as separated companies owned by SEC to be ready for the actual unbundling in the future.

- Allowing outsiders to participate in generation capacity and putting for sale with encouraging prices some of the generation units if any investor interested to enter to the market.
- SLA is already approved; agreeing on different cost for different quality level of service for all kind of services between BL's.
- Introducing Enterprise Resource Planning (ERP) to have the most updated comprehensive program to join all the needs of the existing programs. The main purpose is to have an automated business with one database to support the existing procedures and produce all required reports.
- All procedures, polices, and practices in distribution B.L. are unified among the four rejoins in Saudi Arabia keeping in mind the investment sense.
- Customer service culture is emphasized.
- Starting the Customer Call Centre (CCC).
- New co-generation company was established which is Water & Electricity Co. (WEC) where the ownership is shared between SEC and Saline Water Conversion Corporation (SWCC). WEC relations with SEC and others are shown in Figure 3.

As a second stage toward deregulation; legal separation for TRANSCO was done in the first quarter of 2005 to encourage outside investors in the generation. TRANSCO ia required to give equal opportunities and information to inside and outside competitors. During the second stage; the following activities are carried out:-

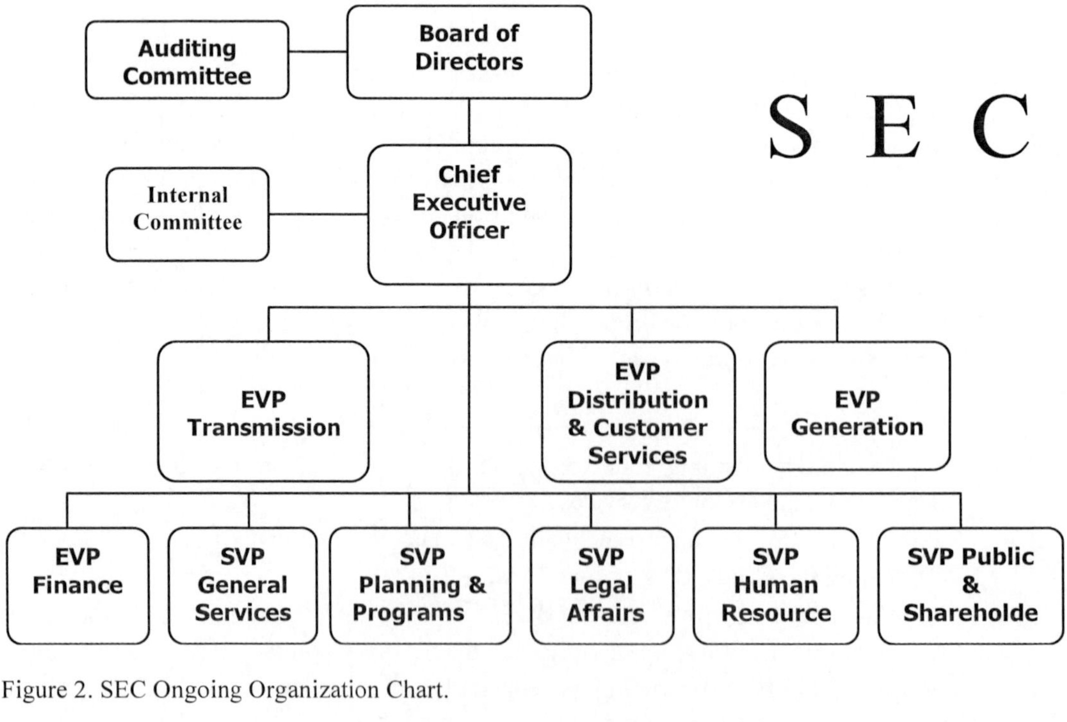

Figure 2. SEC Ongoing Organization Chart.

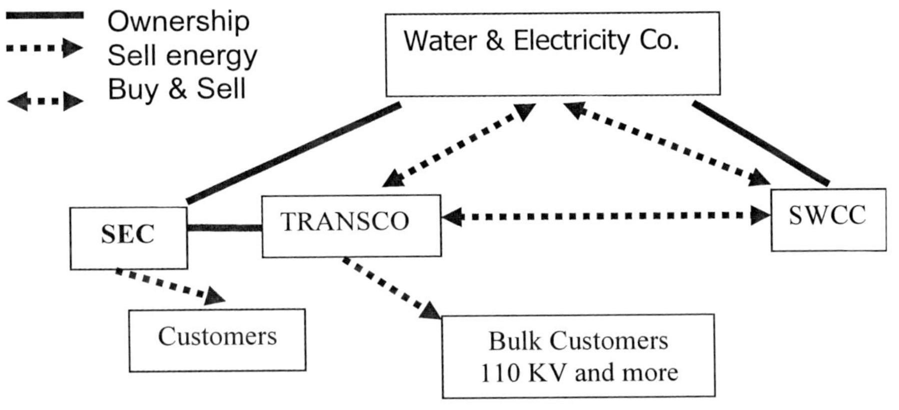

Figure 3. SEC Relationship with SWCC.

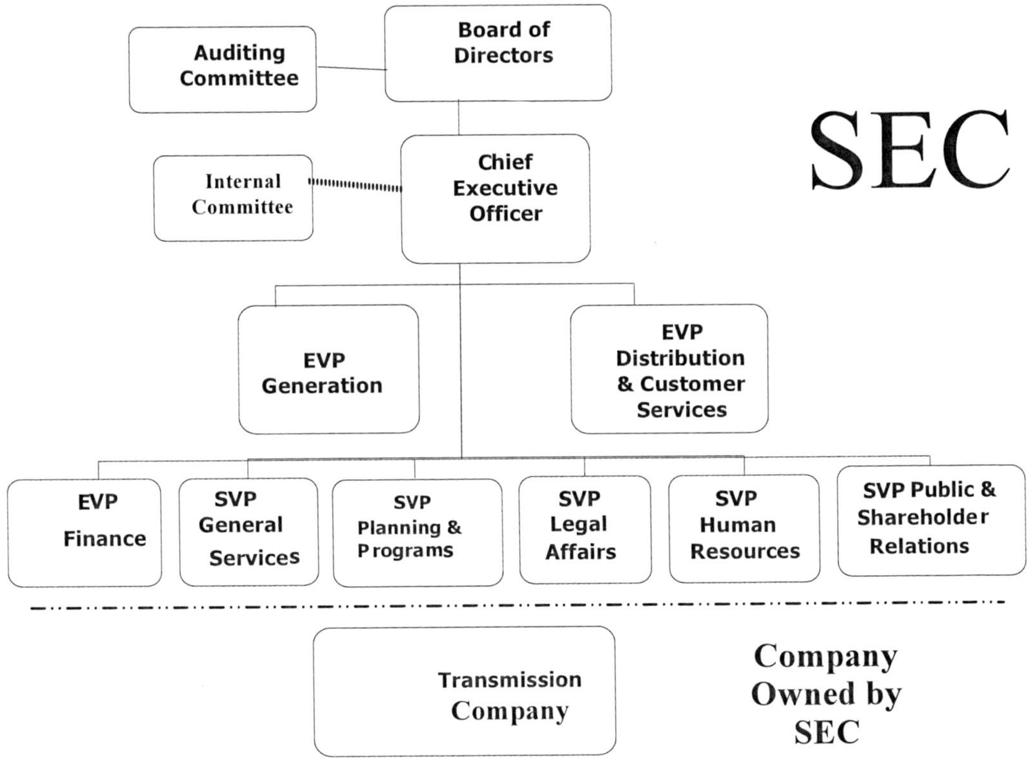

Figure 4.Transmission Separated from SEC.

- TRANSCO is officially a separate company owned by SEC as per Figure 4.
- SLA is practiced as a trial and the actual implementation will be at the start of year 2007.

- To implement ERP; the program selected is SAP and expected to start in end of 2008.
- Starting new contracts between TRANSCO and bulk customers outside the government KWH price.

The above two stages are the ongoing stages and the remaining will be in the SEC plans including the following main highlights:-

- Actual separation of TRANSCO to be carried out.
- All financial matters to be finalized among BL's by January 2007.
- IPP's to enter to the market and contracts to be signed and implemented.
- Generation to be divided into sub generation companies.
- Insure availability of enough spinning reserve.
- Insure competitive market through large number of competitors.
- The last stage will be to divide the distribution B.L. to sub distribution companies.

The future plan as seen by SEC planners is illustrated in Figure 5.

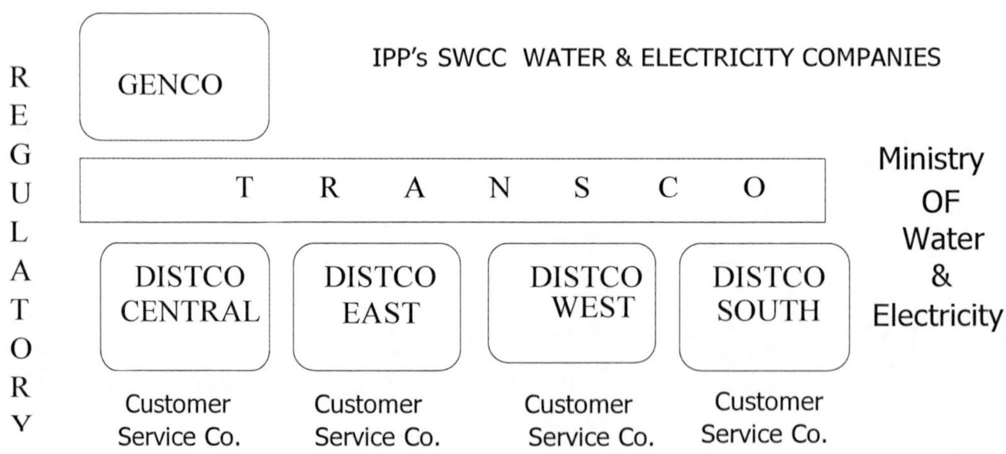

Figure 5. Future SEC Vision.

To conclude this chapter, it can be pointed out that the present regulatory process in Saudi Arabia which involves yearly reviews along with fix income for a specific coming year's period creates a risk. Increased reinforcement expenditure could affect other capital programs which could end for the time being with customer dissatisfaction. Customer dissatisfaction is the expected short term outcome even with the additional efforts and emphasis on customer services as mentioned earlier.

In the coming sections summarizes and analyzes some of the market models with special emphasis on the system operator will be shown.

FRAMEWORK [47]

Electricity is coming in the second level of needs after air, water, and food to support us doing other life activities. It is guaranteed to the degree of being forgotten unless it is disconnected or a high consumption bill is found to be paid at the end of the month. In Saudi Arabia, outages are rare events due to the highly existing reliable system. So, high bill is the part that worries customers in Saudi Arabia. Even though, the Saudi Riyal per kWH (SR2/ kWH) is relatively low compared to others world prices, the bill becomes relatively high due to the fact that Saudis are high consumers and also due to the current Tariff structure.

For the above and for the fact that demand of electricity in Saudi Arabia is rapidly expanded, Saudi Electricity Supply Industry (SESI) is to be restructured in a way to meet the future demand without burdening the government budget, along with reducing electricity bill and at the same time maintaining a safe and reliable system.

Historical Background

SESI was explained in earlier of this study along with the ongoing deregulation activities by SEC. However, it needs to be emphasized in this section that what has been approved or written about the future vision of SESI structure inside and outside SEC. The aim of that is to start from where others end and recognize others' effort.

The three stage plan in SESI restructuring process was presented in Bahrain in 2005 by the executive vice president of transmission B.L. in SEC, at that time, Eng. Tariq Al-Betairi [43]. He presented the plan as follows:-

"Stage 1 By 2002-2006

- Single Buyer Model
- SEC functions more as a holding company
- System of transfer payments creates a 'virtual' industry

Stage 2 By 2006-2010

- Transition to (Wholesale Competition) Market Model
- Market function is separated from technical function in Transmission (TSM); OR
- Either Generation or TSM is divested from SEC

Stage 3 By 2010-2014

- Transition to Full Competition Model
- Private sector participation increases and competition intensifies." [43]

---

[2] 1 US $ = 3.761 SR

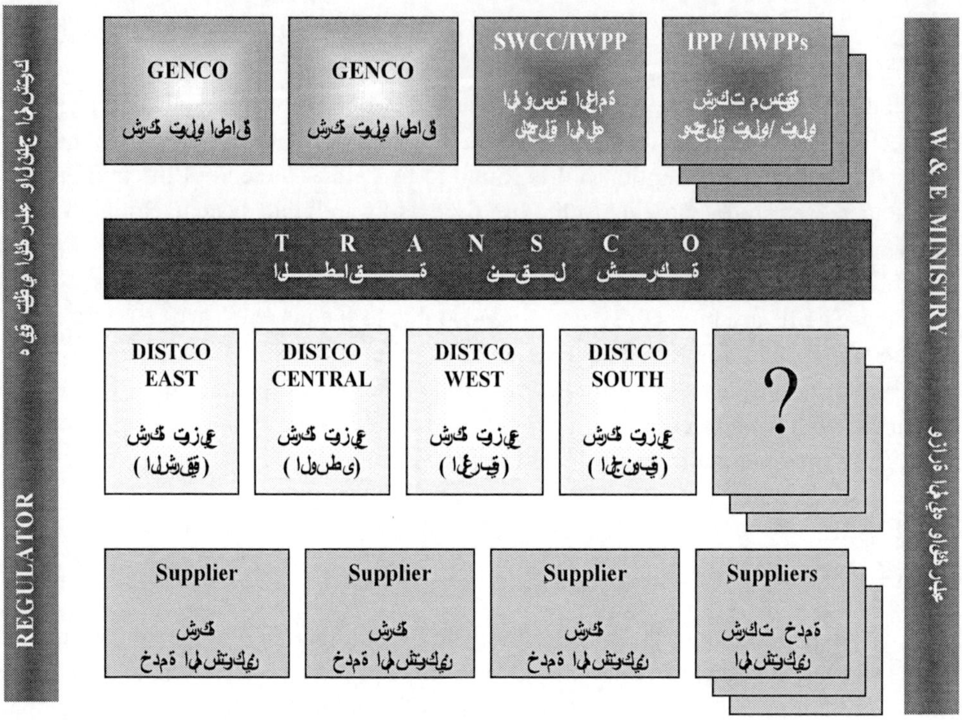

Figure 6. [44] Ultimate Vision of SEC Management for SESI Structure.

The above vision as per reference [44] is also adopted by SEC management as presented by Dr. Amer Al-Swaha, senior vice president of HR in SEC at that time, to top SEC management where the long term vision was illustrated in the reference as per Figure 6.

It is worth to report here that the official reports by SEC Management are always considering TRANSCO as a network owner and operator as shown in Figure 6. Furthermore, the official report never went through any detailed proposal for separating customer services from distribution companies.

Professor Ibrahim Al-Amin (expert in the subject from King Fahad University of Petroleum and Minerals) has set a criterion for grouping power plants of SEC as follows:-

- "EC generating plants to be divided into 3-5 companies irrespective of geographical location.
- MW capacity of each company should be approximately the same.
- Production cost of each company should be as close to the weighted full system average cost.
- East based plants to be fairly divided among the companies.
- Each company should have a fair mix of generating unit type (ST, CC, GT, and DE).
- Generating plants sharing common facilities will belong to the same company (e.g. PP4 & PP4X)." [45]

Proposal's Principals

- It is not recommended to deregulate the whole industry aiming for full competition without analyzing the nature of each participant in the industry. Wires including transmission and distribution lines and equipments are natural monopoly businesses due to the tendency of duplicating the existing lines by new investment of constructing new lines [7]. However, generation and retail businesses are not a natural monopoly and the doors are open for the competition to encourage new technologies and innovations in generation part. This competitive atmosphere is also needed to provide better services in the retail part; including value – added services like price hedging to reduce risk and automating remote services.
- The proposal is expected to support a competitive environment wherever possible and not to be guided by the existing positions. Despite that, after finalizing a suitable structure a well established implementation plan is required to consider the existing resources and apply the most economical possible way of exchanging the duties and responsibilities.
- The ownership of any participant is to be set by the structure in a way to avoid any market power or conflict of interest and to provide OASIS and provide same rights to connect to the grid. So, private participation should be in a competitive manner and transparent with no privileges given to any party among others [1].

For the above, transmission business is to be independent from generation business to allow direct purchases between customers and producers. Also, all functions of system operator, market operator and schedule coordinator are to be carried out by an independent non-profit entity from the whole industry. To support this principle, private sector is to be encouraged to participate in generation and retail sales, but transmission and distribution (wires) are to be non-profit governmental entities.

- The last principle is that Saudi Arabia like others has a unique situation and restrictions to be taking care of like:-
  - The type of fuel used in generation and the fact that ARAMCO is the single source of gas and its availability in Eastern and Central regions only.
  - Gulf interconnection project carried by the Gulf Cooperation Council Interconnection Authority. The first stage will connect Saudi Arabia, Qatar, Bahrain, and Kuwait by 2006/2007. Then UAE and Oman to be connected as the second stage by 2011.
  - The council of Ministers Resolution # 169 which represents the direction of Saudi government to have similar deals among household customers all over the country regardless of the region.
  - High annual percentage growth in power consumption in Saudi Arabia.

The Proposed SESI Structure

Power Supply Providers (PSP's) in generation business are represented by GENCO's, IPP's & IWPP's which are to compete to sell in the wholesale market controlled by Saudi Electricity System Operation & Marketing entity (SESOM). This entity is to be responsible for the duties of system operator, market operator and schedule coordinator. There will be other competition among Customer Service Providers (CSP's) represented by retailers as new entries or the once to be separated from DISTCO's which are called Customer Services Companies (CSC's) to leave DISTCO's with wires only as owners and operators. This will enhance retail services in terms of prices and quality. While PSP's are encouraged to be private generation entities, Saudi Electricity National Grid (SENG) is to be a non-profit government entity owning but not operating the high voltage transmission grid system with open access to all participants in a fair manner. Also, DISTCO's are to be regulated as Saudi Electricity Distribution entity (SED) which is to be a non-profit entity owned by the government. CSC's and other retailers are to be unregulated private entities competing to provide electrical supply and other related services. The already established Saudi Electricity and Co-generation Regulatory Authority (ECRA) is to have an important role in promoting real market competition among generation and retail businesses and make sure to regulate wire businesses (transmission & distribution).

Figures from 7 to 10 show the SESI proposed structure with respect to ownership, physical electricity flow, financial flow, and the full proposed framework.

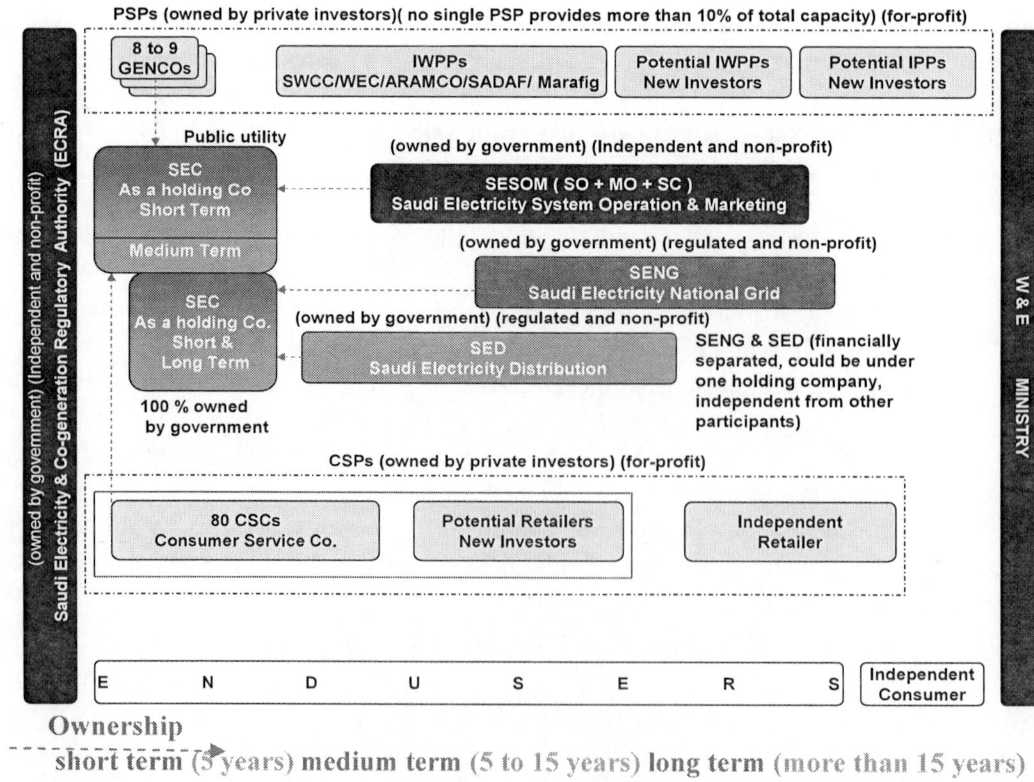

Figure 7. The proposed SESI Framework (Ownership).

From Figure 7, it can be seen that SEC in its current status can be the holding company for SESOM and GENCO's in the short term only. And CSC's to be financially separated from SED to be ready for complete separation from SEC in the medium term. Furthermore, SEC at its long term form to be the holding company for SENG and SED only. By that time, SEC expected to be 100 % governmental and non-profit utility.

Even though, the customer and the retailer can get power directly from PSP's with and without the usage of the regulated wires, the actual operation is under full control by SESOM. As shown in Figure 8, SESOM is to be responsible for releasing dispatch orders to PSP's and conducting the actual operation of SENG network. Beside that, SEASOM gives load shedding orders to SED if needed according to contingency plans.

OASIS is to be a very sophisticated information system to provide all the information required by market participants in the same time [1]. Figure 8 shows all the possible flow of information with directions. Any specific category of participants has a specific authority to access OASIS.

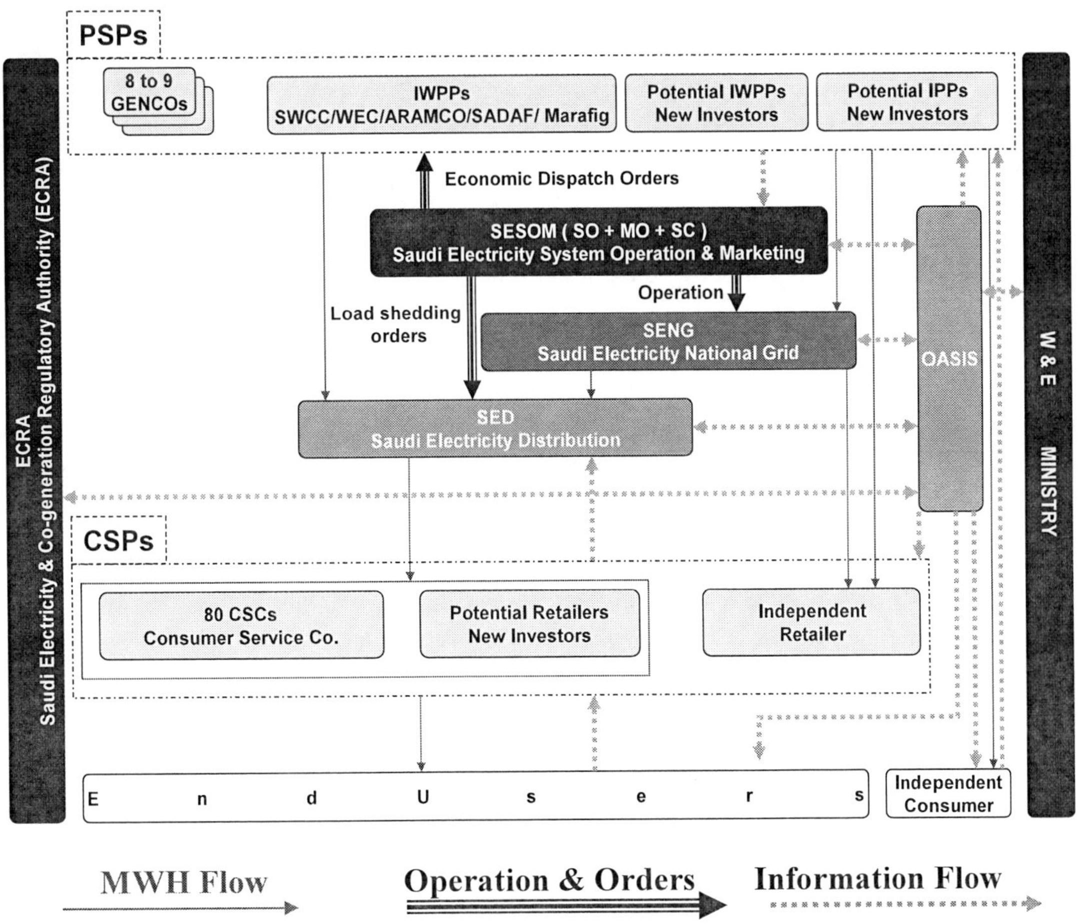

Figure 8. The proposed SESI Framework (MWH, Operation, and Info. Flow).

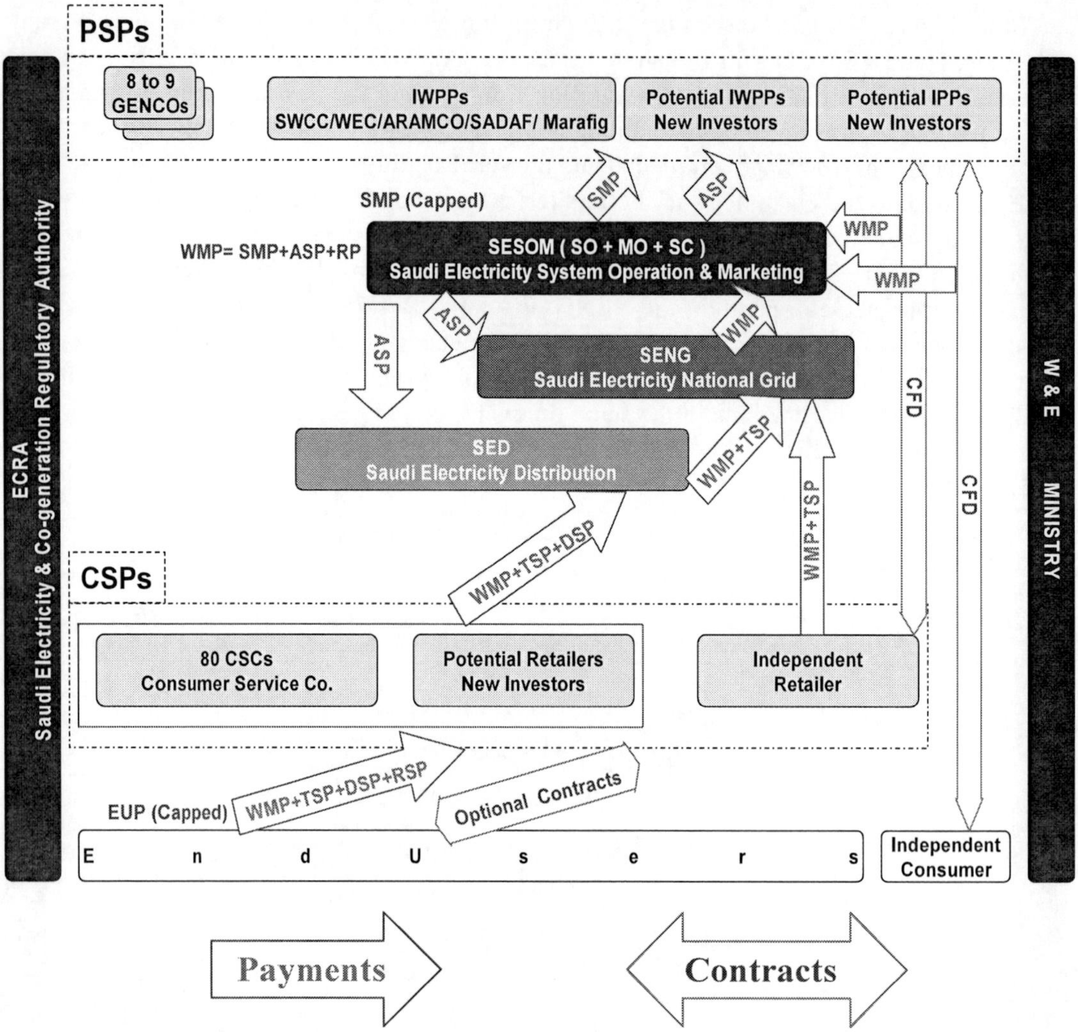

Figure 9. The proposed SESI Framework (Financial Flow).

Figure 9 illustrates the financial flow among participants which will be elaborated in the trading and pricing sections. However, it is worth to report here that all money movement is to be controlled and managed by the schedule coordinator under SESOM.

Figure 10 gives the complete proposal outlining the main interactions among participants which will be elaborated in the next sections.

ECONOMIC LOAD DISPATCH [1]

ELD is considered the basic optimization tool to dispatch the power system. Its objective is to keep the total operation cost to minimum while allocating the total load to the available generation units without breaking operation constraints and neglecting transmission constraints.

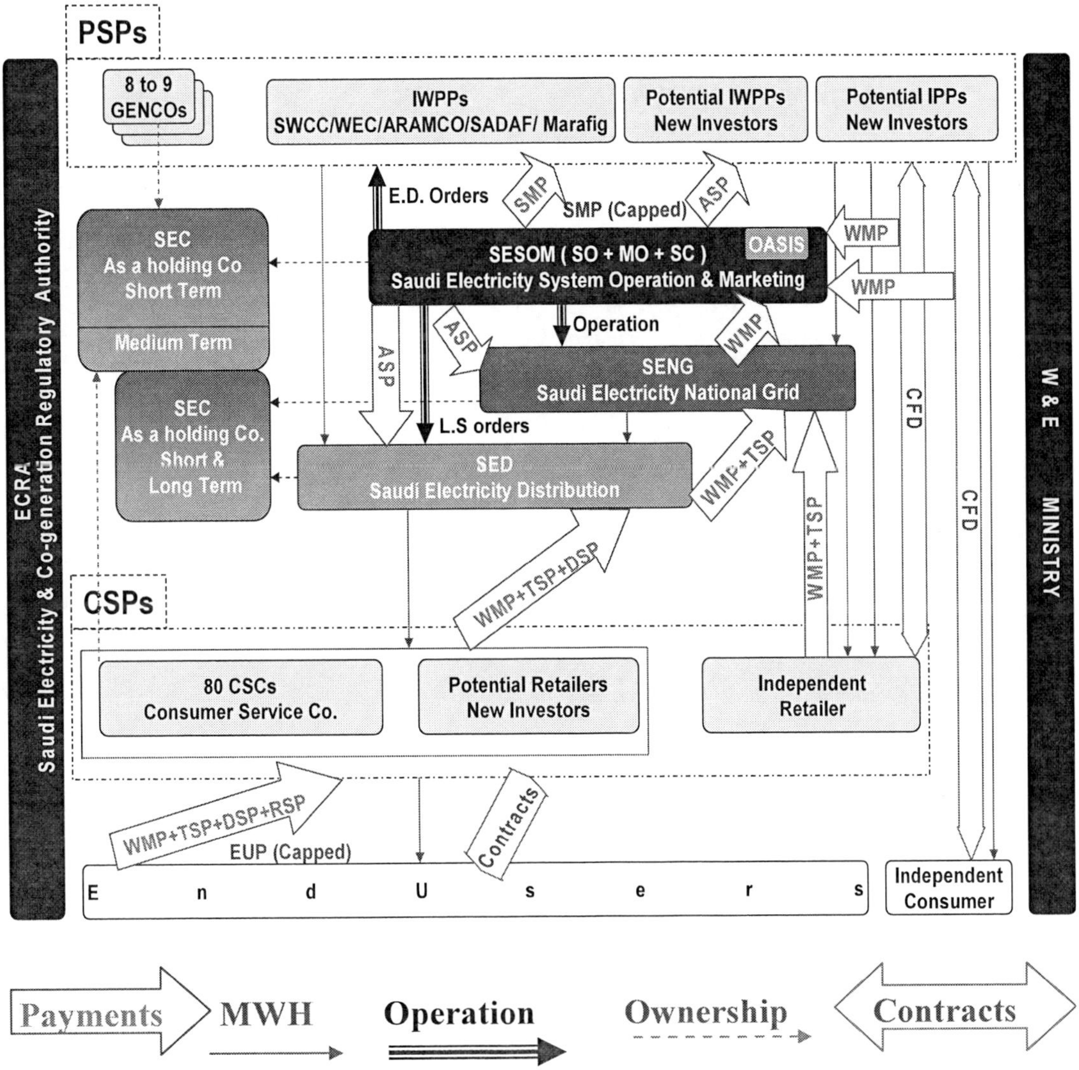

Figure 10. The proposed SESI Framework (Complete Picture).

The unit production cost can be given by the system operator or calculated as the product of the fuel cost function by the input-output function. So by knowing the heat rate characteristic to each generator, the unit cost can be calculated. Model parameters are classified as follows:-

- Decision variables are Pgi in MW where the values are the model output.
- Data to be given or calculated before running the model are:-

P_{li}, C_{gi}, n, P_{gi}^{min}, P_{gi}^{max}, and, P_{loss}, *where:*

P_{li}	The real power in MW delivered to a load from bus i.
C_{gi}	Unit Generation cost of producing one MWH of energy at bus i.
n	Set of all bus indices.

P_{gi}^{min} Minimum real power output at generation bus i.
P_{gi}^{max} Maximum real power output at generation bus i.
P_{Loss} The power loss in MW

ELD can be formulated in a simple general form with the following objective function and constraints:-

The objective function is to minimize the cost:-

$$\text{Minimize} \sum_{i=1}^{n} C_{gi} (P_{gi}) \tag{1}$$

Subject to the operating constraints which are:-

1. *System real power balance*

   As the total generated power is equal the total load added to the lost power which is represented by the following formula:-

$$\sum_{i=1}^{n} P_{gi} = \sum_{i=1}^{n} P_{li} + P_{loss} \tag{2}$$

2. *Generation unit limits*

   As the generation output level to be within the max and min of the capacity of the generation unit which can be represented as:-

$$P_{gi}^{min} \le P_{gi} \le P_{gi}^{max} \text{ for } i = 1,2,..,n \tag{3}$$

The above model can be solved by any of the advanced mathematical methods like Lagrange Relaxation or using linear programming tool either manual or by suitable software depending on the problem size. The shortcoming of this method is ignoring the transmission constraints. But still ELD can be valid for some applications like power dispatching in non-congested networks.

Optimal Power Flow

The power system is non-linear by nature which could be operated using an optimum power flow (OPF) procedure to meet all demands at minimum total cost, subject to operation and transmission line capacity constraints, meeting minimum voltage constraints. OPF is not only solving the nonlinear power balance equations, but it often determines the correct values for large number of automatic controls like capacitors and transformer tap settings, generator AVR status, phase shifting transformer, reactive compensator, and the values of FACTS. [10]

The output of OPF model is more accurate estimate than ELD outputs due to the nature of the electric system which requires the non-linearity. This output contains voltage magnitudes and phase angles at each bus in addition to the real and reactive power flow.

OPF model can be formulated in its basic general form in the same way ELD is formulated with the following differences [1]:-

Objective function is to minimize system cost with enhancement:-

- Reactive power cost is to be added to the cost function.

Subject to operating constraints which are:-

- *System power balance* as in ELD with insertion of:-
  a. Reactive power
  b. Bus voltages and angle
  c. Bus admittance and angle
- *Generation unit limits* as in ELD with real and reactive power limitations

Other constrains to be added for the OPF model are:-

- *Bus voltage limits*
- *Power flow limits in transmission network*
- *Other* constraint could be added according to the nature of the problem.

OPF model can be solved by any of the advanced mathematical methods which are very complicated processes. However computer software handling non-linear programming method will be the most suitable approach to solve OPF in addition to a linear programming approach after linearizing the non linear coefficients in the formulation process.

Study Objectives

This study has two objectives; the first one which is minor in electrical engineering field and it is to be studied in more details by economists. This objective is to give the most suitable framework of energy market and specifying the interface points among parties. This work will give a proposed framework for SEC which should encourage power system expansion in the right direction. The methodology will depend on:-

- Other experiences.
- Knowledge of the real situations and constrains.
- Knowledge of religion and culture.

Even though, the market framework will depend on the pricing method as well as the network configuration and constraints. This part will not be completely covered in this work and it is to be left to economists. The absent of realism will make it hard to economists also to ensure any conclusion related to the market power which will be a good future research after deregulation take place in Saudi Arabia.

The second objective is to establish a tool to be used to know the future generation potential areas in any utility by achieving the following:-

- Linear Programming to be used in order to get the optimal location of future generation.
- To built a MW LP-Based OPF (using Linear Interactive & Discrete Optimizer "LINDO" which is a powerful tool for solving linear programming problems).
- To test the model by applying it to a real whole eastern province system in SEC.
- To run the model to different scenarios to know the potential generation areas.
- To analyze the results.
- To give recommendations.

Beside the above listed objectives, this study will recommend future studies needed in order to enhance the electric industry.

MW LP- Based OPF (Proposed Model)

This proposed model is a Linear Program-Based OPF established to help in deciding the future generation potential areas in a deregulated industry. Its objective is to keep the total operation cost, generation and transmission, to minimum while allocating the total load to the available generation units without breaking operation constraints and transmission constraints.

Since the objective of the electrical system is to deliver real power to satisfy the loads; the model is set in a way to be extremely simplified without affecting the main limitations in the power flow. The model is limited to the following characteristics and constraints:-

- MW-only formulation (Real power only to be considered in the model).
- Generator capacity (Power plant min and max).
- Transmission lines MW limits.
- Transformers MW limits.
- The cost of generating one MWH is constant for a specific bus.
- The cost of wheeling one MWH is per KM per Circuit.
- Power reserves in the system are not considered.
- Losses are considered to be fixed.

Even though, the power system is non-linear by nature, this modified model uses a linear-programming method which will do the purpose of the optimization.

Model Formulation

Model parameters are classified as follows:-

- *Decision variables are Pgi, Pdij in MW where the values are the model output.*
- *Data to be given or calculated before running the model are:-*

P_{li}, C_{gi}, C_{ij}, n, N_{ij}, P_{gi}^{min}, P_{gi}^{max} and C_{lij}

Where most of the symbols have been defined earlier and the rest of the symbols are:

P_{gi} The real power in MW generated by a power plant at bus i.

C_{ij} Unit cost of wheeling one MWH from bus i to bus j.

N_{ij} No. of circuits between bus i and bus j.

C_{lij} Line capacity limit of MW's between bus i and bus j

The general formulation of the optimization model is as follows:-

A. Objetive Function

To minimize the generation cost and the wheeling cost without priority, so, the accuracy of the cost given for the generated unit of MWH and the transferred unit of MWH will affect the model output. So, It is not recommended to use normal costs used in other models (the operation cost based in the fuel cost & heat exchange) but the actual cost is to be used involving all parameters where GENCO's are willing to sell to outsiders and TRANSCO's are willing to be paid for utilizing their networks. The calculation of the cost will be explained in next chapter while applying the model to actual network as a case study.

$$\text{Minimize} \sum_{i=1}^{n} C_{gi}(P_{gi}) + \sum_{i=1}^{n} \sum_{j=1}^{n} N_{ij} C_{ij} P_{dij} \text{ for } i \neq j \tag{4}$$

This general form covers all busses in the system and all connections between them allowing reveres flow. So, it should be tailored well when applying it to a real system to avoid maximizing the size of the problem. *For any bus without a power plant, the cost should be set high to avoid the system to propose locations without control.*

B. Constraints

1. *Real Power Balance at each Bus*

At each bus the power flow-in from generators and from other transmission lines should be equal to power-out to load and to transmission lines at any point of time,

Subject to:-

$$P_{gi} - \sum_{j=1}^{n} N_{ij} P_{dij} + \sum_{j=1}^{n} N_{ij} P_{dji} = P_{li} \text{ where } i = 1, 2, \ldots, n \ \& \text{ for } i \neq j \tag{5}$$

Only real power is considered.

2. *System Real Power Balance*

As the total generated power equal to the total load added assuming fixed losses in the system.

Subject to:-

$$\sum_{i=1}^{n} P_{gi} = \sum_{i=1}^{n} P_{li.} + \text{Losses} \tag{6}$$

Only real power is considered.

3. *Generation Unit Limits*

As the generation output level is to be within the max. and min. of the capacity of the generation at each bus,

Subject to:-

$$P_{gi}^{min} \leq P_{gi} \leq P_{gi}^{max} \text{ for } i = 1,2,\ldots n \tag{7}$$

Only real power is considered.

4. *Limits on Power flow*

As the transmission lines are limited by its capacity which should be determined by the thermal capacity; however, it could be approximated and determined by MW capacity,

Subject to:-

$$P_{dij} \leq C_{lij} \quad \text{for all i -j pairs } \& \ i \neq j \tag{8}$$

5. *Non Negativity*

To satisfy the linear programming restrictions and the direction of the flow, non negativity is a must,

Subject to:-

$$P_{dij} \geq 0 \text{ for all i -j pairs.} \tag{9}$$

Testing the model by applying it to a real system and comparing the outputs is covered in the coming sections.

Solution Method

The model developed in the previous section is proposed by this study to be solved using LINDO optimizer software as a powerful tool for solving linear programming problems.

Figure 11 is a flow chart listing steps to be followed to make use of the model.

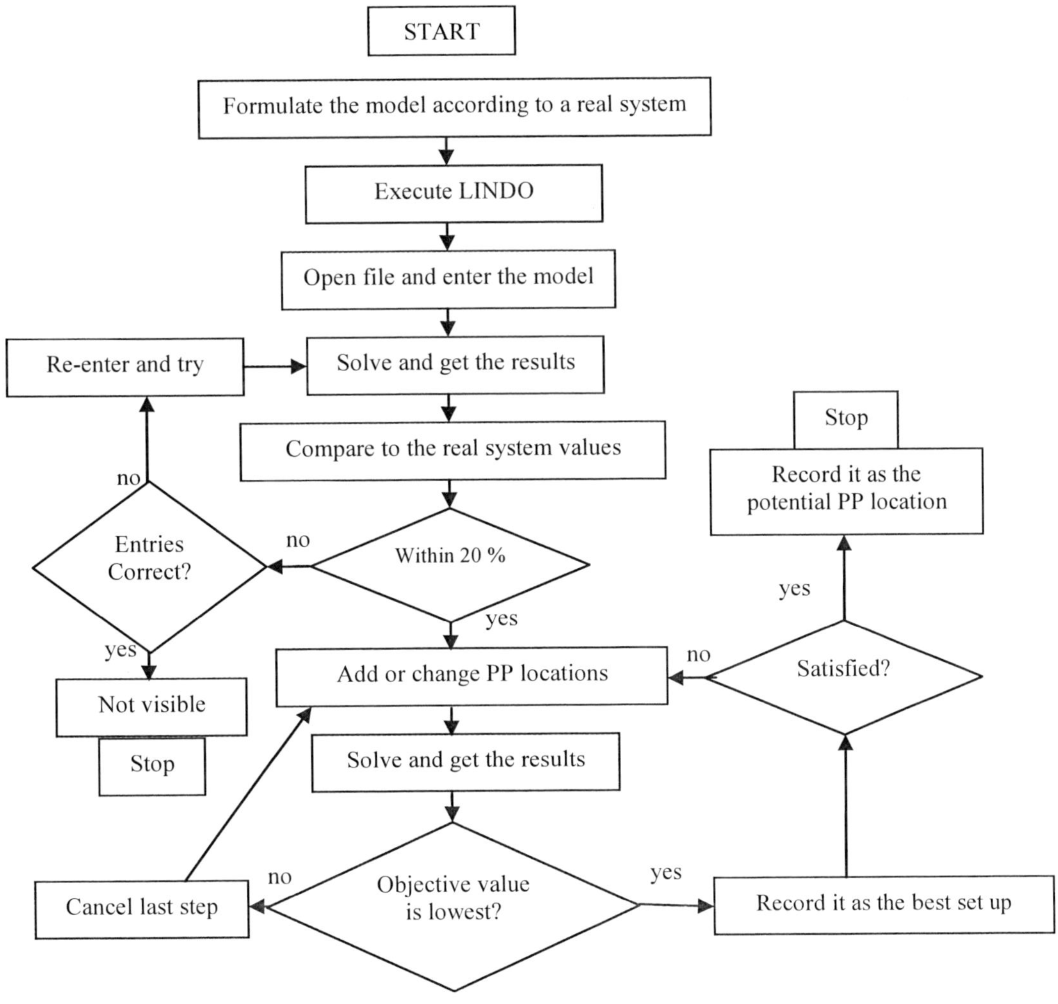

Figure 11. MW LP- Based OPF Flow Chart.

Application of the Modified Model

Application to Practical System

As per the flow chart in Figure 11; the established MW LP-Based OPF model will be applied to the real power 56-bus system in eastern province of Saudi Arabia, which is presented in Figure 12.

The model is to be formulated as follows:-

A) Objective Function

Equation (4) is the general model where values can be substituted as follows:-

- In Figure 12 the busses are numbered from 1 to 56 and each bus represents either a power plant or a grid station or a pure load and it could represent a combination of some or all of them. So, i = 1, 2, 3, up to 56 and j = 1, 2, 3, up to 56.
- C_{gi} is the generation cost per MWH in Saudi Riyals (SR)3 for each power plant as listed in Table 1. The cost calculation is one of the products of this study illustrated in the next section.
- C_{gi} is set high (1000) for all busses without generation to avoid the system to propose a power plant in every bus.
- C_{ij} is the wheeling cost per MWH per circuit in Saudi Riyals as listed in Table 2. The cost calculation is also a product of this study illustrated in Section (Unit Wheeling Cost Calculation).
- N_{ij} is the number of circuits between bus i and bus j from Table 2 and Figure 12.
- P_{gi}, Pdij are the decision variables in MW where the values are the model output.

B) Constraints

Subject to:-

I. Real power balance at each bus

Equation # (5) is the general model where it can be substituted for P_{li} from Table 1 and the other variables as illustrated in (A) above.

II. System real power balance

Equation # (6) is the general model where Figure 12 and Table 1 are used for substitutions.

III. Generation unit limits

Equation # (7) is the general model where the data of maximum and minimum output of each power plant are listed in Figure 12 and Table 1.

IV. Limits on power flow

Equation # (8) is the general model where Figure 12 and Table 2 are used for substitutions.

V. Non negativity

Equation # (9) is the general model where Lindo software is handling non negativity as a built-in feature.

* (1000) is set high for all busses without generation to avoid the system to propose a power plant in any bus.

The bus information for the whole Saudi eastern province electrical network shown in Figure 12 is summarized in Table 1. This table is to be used as the input data for the formed model which is to be solved by LINDO software. Table 1 contains the unit generation cost (C_{gi}) which is the product of Table 3. Also, other data related to busses from generation limitations including maximum and minimum power plants capacities and load information,

* 1 $ = 3.761 SR.

are listed in Table 1. This data is actual as per reference [42] except for the SR 1000 value which is set for all busses (without generation) to avoid proposing a power plant in any bus without control.

The line information for the whole Saudi eastern province electrical network (Figure 12) is summarized in Table 2. This table is to be used as the input data for the formed model which is to be solved by LINDO software. Table 2 contains the wheeling cost (C_{ij}) which is the product of Section (Unit Wheeling Cost Calculation).

Table 1. 56 Bus Information for Bus System with 16 Power Plants

Bus No.I	PP OR SS NAME	BUS VOLTAGE (kV)	Min MW P_{gi}^{min} (MW)	Max MW P_{gi}^{max} (MW)	LOAD P_{li} (MW)	Unit Generation Cost C_{gi} (SR/MWH)	Type of Fuel Used
1	GHZ380 PP	380	100	2528	120	64	Steam-Gas Crude
2	Q PP	380	200	2500	102	51	Steam-Gas Crude
3	FAD380 SS	380			--	1000*	
4	GHU380 SS	380			--	1000*	
5	SHE380 SS	380			--	1000*	
6	FAR380 SS	380			--	1000*	
7	HAW380 SS	380			--	1000*	
8	FADHLI230 SS	230			33	1000*	
9	GHZ230 PP	230	220	1600	273	63	Steam-Gas Crude
10	GHU230 SS	230			6	1000*	
11	SHE230 PP	230	100	1219	690	72	Gas
12	FAR230 PP	230	100	910	578	74	Gas
13	MAH SS	230			341	1000*	
14	RWPS SS	230			27	1000*	
15	KHURAIS SS	230			7	1000*	
16	WASIA SS	230			50	1000*	
17	HAW230 SS	230			191	1000*	
18	JAD	230			--	1000*	
19	ABQ	230			172	1000*	
20	IBK	230			490	1000*	
21	DMP PP	230	100	400	468	108	Gas-Diesel
22	DBP	230			369	1000*	
23	RAK	230			120	1000*	
24	KHO	230			380	1000*	
25	THQ	230			377	1000*	
26	AZY PP	230	100	850	--	132	Steam
27	DHC	230			139	1000*	
28	QTF	230			389	1000*	
29	RAS PP	230	20	120	148	101	Gas
30	JBSE	23			1	1000*	
31	SAF PP	230	20	75	223	100	Gas
32	IZH	230			62	1000*	
33	JBCM	230			189	1000*	
34	SADAF	230			273	1000*	
35	SHARQ	230			127	1000*	
36	PETROKEM	230			143	1000*	

Table 1. Continued

Bus No. I	PP OR SS NAME	BUS VOL-TAGE (kV)	Min MW P_{gi}^{min} (MW)	Max MW P_{gi}^{max} (MW)	LOAD P_{li} (MW)	Unit Generation CostC_{gi} (SR/MWH)	Type of Fuel Used
37	JBN	230			9	1000*	
38	JBSW	230			153	1000*	
39	SABAYEK	230			2	1000*	
40	KEMYA	230			43	1000*	
41	HAD	230			461	1000*	
42	NIGC	230			141	1000*	
43	JO1 PP	230	80	300	--	132	Steam
44	J2A PP	230	80	300	--	132	Steam
45	J2B PP	230	80	300	--	132	Steam
46	BER PP	230	80	215	500	75	Gas
47	BNGL	230			68	1000*	
48	QAIS PP	230	20	108	210	148	Diesel
49	NAR	230			103	1000*	
50	KHURAIS	230			55	1000*	
51	SIPC	230			1	1000*	
52	UTH PP	115	80	272	--	70	Gas
53	JUY PP	115	40	118	--	85	Gas
54	R9001	380			925	1000*	
55	KH9003	380			814	1000*	
56	R9000	380			158	1000*	
Total System Loads					10131	Loads + 4% (System Losses) = 10553 MW	

Table 2. Line Information for 56-Bus System with 16 Power Plants

Line No. ij	Line Name	Line Voltage (kV)	Normal Capacity CL_{ij} (MW)	Circuit Length (km)	No. of Circuits N_{ij} (No.)	Wheeling Cost Per Km Per circuit (SR/MWH)	Wheeling Cost Per Circuit C_{ij} (SR/MWH)	N_{ij} * C_{ij} (SR/MWH)
0103	GHZ2 – FAD	380	2659	89	4	0.1473	13.1057	52.42
0104	GHZ2 – GHU	380	2659	112	2	0.1473	16.4925	32.99
0204	QPP – GHU	380	2659	44	1	0.2356	10.3667	10.37
0205	QPP – SHE	380	2659	85	2	0.1473	12.5166	25.03
0206	QPP – FAR	380	2659	136	1	0.2356	32.0426	32.04
0308	FAD T/F	380 – 230	3*T750	--	4	14.0885	14.0885	56.35
0405	GHU – SHE	380	2659	74	1	0.2356	17.4349	17.43
0410	GHU T/F	380 – 230	T750	--	2	28.1771	28.1771	56.35
0506	SHE – FAR	380	2659	54	1	0.2356	12.7228	12.72
0507	SHE – HAW	380	2659	110	1	0.2356	25.9168	25.92
0511	SHE T/F	380 - 230	T750	--	2	28.1771	28.1771	56.35
0554	SHE – R9001	380	2659	313	2	0.1473	46.0907	92.18
0607	FAR – HAW	380	2659	47	1	0.2356	11.0735	11.07
0612	FAR T/F	380 - 230	T750	--	2	28.1771	28.1771	56.35
0655	FAR-KH9003	380	2659	266	2	0.1473	39.1697	78.34
0717	HAW T/	380 - 230	T750	--	2	28.1771	28.1771	56.35

Line No. ij	Line Name	Line Voltage (kV)	Normal Capacity CL_{ij} (MW)	Circuit Length (km)	No. of Circuits N_{ij} (No.)	Wheeling Cost Per Km Per circuit (SR/MWH)	Wheeling Cost Per Circuit C_{ij} (SR/MWH)	$N_{ij} * C_{ij}$ (SR/MWH)
0830	FAD – JBSE	230	741	38.2	2	0.1080	4.1251	8.25
0837	FAD – JBN	230	741	48	2	0.1080	5.1834	10.37
0838	FAD – JBSW	230	741	41	1	0.1669	6.8424	6.84
0850	FAD – KHU	230	741	42	1	0.1669	7.0093	7.01
0922	GHZI – DBP	230	741	49.4	1	0.1669	8.2443	8.24
0927	GHZI – DHC	230	741	69.2	2	0.1080	7.4727	14.95
0928	GHZI – QTF	230	741	31.8	1	0.1669	5.3071	5.31
0929	GHZI – RAS	230	741	33.3	2	0.1080	3.5960	7.19
0930	GHZI – JBSE	230	741	52.2	1	0.1669	8.7116	8.71
0943	GHZI - J01	230	741	23.9	1	0.1669	3.9886	3.99
0946	GHZI – BER	230	741	38.6	2	0.1080	4.1683	8.34
0953	GHZI- JUY+T/F	115	185/T400	7.4	2	13.7993	14.8046	29.61
1020	GHU – IBK	230	741	23	1	0.1669	3.8384	3.84
1024	GHU – KHO	230	741	25	1	0.1669	4.1722	4.17
1118	SHE – JAD	230	741	19	2	0.1080	2.0517	4.10
1213	FAR – MAH	230	741	32.5	1	0.1669	5.4239	5.42
1214	FAR – RWPS	230	741	102	1	0.1669	17.0226	17.02
1215	FAR – KHR	230	741	131	1	0.1669	21.8624	21.86
1217	FAR – HAW	230	741	52	2	0.1080	5.6153	11.23
1218	FAR – JAD	230	741	73	1	0.1669	12.1829	12.18
1252	FAR - UTH+T/F	115	370/T400	2.4	2	13.7993	14.0192	28.04
1318	MAH – JAD	230	741	40.5	1	0.1669	6.7590	6.76
1415	RWPS – KHR	230	741	30.6	1	0.1669	5.1068	5.11
1516	KHR – WASIA	230	741	52.2	2	0.1080	5.6369	11.27
1656	WASIA - R9000	380	2659	77	2	0.1473	11.3386	22.68
1819	JAD – ABQ	230	741	26	2	0.1080	2.8077	5.62
1927	ABQ – DHC	230	741	52	2	0.1080	5.6153	11.23
2021	IBK – DMP	230	741	7.8	1	0.1669	1.3017	1.30
2022	IBK – DBP	230	741	2.1	2	0.1080	0.2268	0.45
2123	DMP – RAK	230	297	9	1	0.1669	1.5020	1.50
2127	DMP – DHC	230	741	18	2	0.1080	1.9438	3.89
2228	DBP – QTF	230	741	17.6	1	0.1669	2.9372	2.94
2324	RAK – KHO	230	297	2	1	0.1669	0.3338	0.33
2425	KHO – THQ	230	297	2.9	1	0.1669	0.4840	0.48
2526	THQ – AZY	230	741	7.6	2	0.1080	0.8207	1.64
2627	AZY – DHC	230	741	15.6	2	0.1080	1.6846	3.37
2829	QTF – RAS	230	741	43.6	1	0.1669	7.2763	7.28
3032	JBSE – IZH	230	741	2	1	0.1669	0.3338	0.33
3033	JBSE – JBCM	230	741	9.3	1	0.1669	1.5521	1.55
3034	JBSE – SADAF	230	741	2	2	0.1080	0.2160	0.43
3035	JBSE – SHARQ	230	741	2	2	0.1080	0.2160	0.43
3036	JBSE – PETRO	230	370	2	2	0.1080	0.2160	0.43
3046	JBSE – BER	230	741	2	1	0.1669	0.3338	0.33
3051	JBSE – SIPC	230	741	2	2	0.1080	0.2160	0.43
3150	SAF – KHU	230	741	107.9	2	0.1080	11.6518	23.30
3237	IZH – JBN	230	741	2	1	0.1669	0.3338	0.33
3337	JBCM – JBN	230	741	7.1	1	0.1669	1.1849	1.18

Table 2. Continued

Line No. ij	Line Name	Line Voltage (kV)	Normal Capacity CL ij (MW)	Circuit Length (km)	No. of Circuits N_{ij} (No.)	Wheeling Cost Per Km Per circuit (SR/MWH)	Wheeling Cost Per Circuit C_{ij} (SR/MWH)	N_{ij} * C_{ij} (SR/MWH)
3738	JBN – JBSW	230	741	6.4	1	0.1669	1.0681	1.07
3741	JBN – HAD	230	741	2	1	0.1669	0.3338	0.33
3742	JBN – NIGC	230	741	2	1	0.1669	0.3338	0.33
3746	JBN – BER	230	741	12	1	0.1669	2.0027	2.00
3749	JBN – NAR	230	741	129.5	1	0.1669	21.6121	21.61
3750	JBN – KHU	230	741	43.5	1	0.1669	7.2597	7.26
3839	JBSW-SABAYEK	230	741	2	2	0.1080	0.2160	0.43
3840	JBSW–KEMYA	230	741	2	2	0.1080	0.2160	0.43
3841	JBSW – HAD	230	741	2	1	0.1669	0.3338	0.33
3842	JBSW – NIGC	230	741	2	1	0.1669	0.3338	0.33
3843	JBSW - J01	230	741	29.5	1	0.1669	4.9232	4.92
3846	JBSW – BER	230	741	9.4	1	0.1669	1.5688	1.57
4344	J01 – J2A	230	741	2	1	0.1669	0.3338	0.33
4445	J2A – J2B	230	741	2	1	0.1669	0.3338	0.33
4446	J2A – BER	230	741	21.5	1	0.1669	3.5881	3.59
4546	J2B – BER	230	741	21.5	1	0.1669	3.5881	3.59
4647	BER – BNG	230	741	2	2	0.1080	0.2160	0.43
4650	BER – KHU	230	741	48.5	1	0.1669	8.0941	8.09
4849	QAIS – NAR	230	741	282	2	0.1080	30.4523	60.90
4950	NAR – KHU	230	741	86	1	0.1669	14.3524	14.35

Unit Generation Cost Calculation

The cost calculation is differs from one way to another depending on two main factors; namely the information availability and other is the aim of calculating a specific cost.

Since the unit generation cost will be used in the model as explained in the previous chapter to decide potential areas for future generation expansions, it should not be limited to operation costs. But it should cover total generation costs including revenue requirements to help the investor to decide whether to build a proposed P.P. in a selected location.

Figure 13 illustrates all unit generation cost components, from which the following equation has been derived:-

$$\text{Unit Generation Cost} = [\text{Fixed Costs} + \text{Variable Costs}] \text{ X } [1+\text{Risk Factor}] \qquad (10)$$

Where:-

Fixed Costs =

Insurance Costs + Depreciation Costs + Capital Inv. Profit. + Fixed O&M Costs.

And;

Variable Costs = Fuel Costs + Variable O&M Costs.

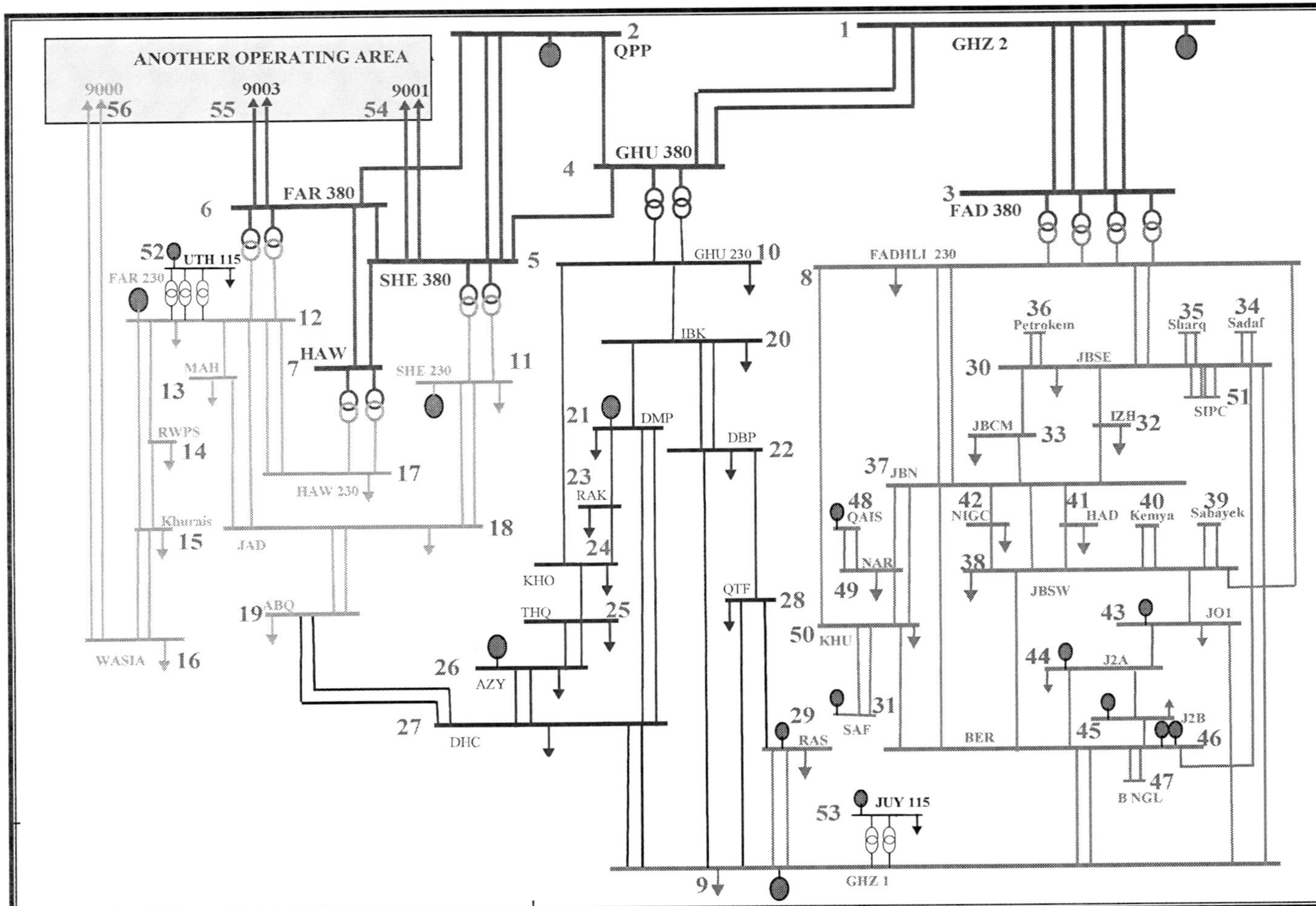

Figure 12. Case Study 1 Original 56 Bus System with 16 Power Plants.

Table 3. Unit Generation Cost SEC-EOA

PP Name	Life Cycle (Year)	Usage Factor	Asset Value (SR*1000)	Generated Energy[1] (GWH)	Fixed Costs (SR/MWH)					Variable Cost (SR/MWH)			Risk Factor[4]	Total Unit Generation Costs (SR/MWH)
					Insurance Costs[2]	Depreciation Costs	Capital Inv. Profits[3]	O&M Fixed Costs	Total Fixed Costs	Fuel Costs	O&M Variable Costs	Total Var. Costs		
QPP	30	20 %	2,710,000	17,100	0.1	5.3	12.6	2.5	20.5	26	1.8	27.8	5%	51
GHZ230	30	25 %	2,357,800	9,600	0.2	8.2	18.4	4.7	31.5	27.4	1.4	28.8	5%	63
GHQ380	30	20 %	4,902,390	17,100	0.2	9.5	22.8	2.5	35	24.7	1.4	26.1	5%	64
UTH	25	80 %	160,500	600	0.1	10,7	5.4	7.8	24	40	3	43	5%	70
SHE230	25	60 %	1,112,500	3,600	0.1	12.4	12.4	6.7	31.6	34.5	2.1	36.6	5%	72
FAR230	25	60 %	801,500	2,600	0.1	12.3	12.3	7.1	31.8	36.6	2.4	38.4	5%	74
BER	25	75 %	139,600	500	0.1	11.2	9.8	7.9	29	38.6	3.6	42.2	5%	75
JUY	25	80 %	77,100	200	0.1	15.4	7.7	8.7	32	45.7	3.3	49	5%	85
SAF	25	85 %	44,100	100	0.1	17.6	6.6	11.3	35.6	56.3	3.5	59.8	5%	100
RAS	25	60 %	80,000	200	0.2	16	16	10	42.2	50	4	54	5%	101
DMP	25	80 %	361,800	600	0.1	24.1	12.1	15.4	51.7	46.4	4.4	50.8	5%	108
AZY	25	60 %	1,200,000	1,200	0.4	40	40	7	87.4	36	2.5	38.5	5%	132
J01	25	60 %	400,000	400	0.4	40	40	7	87.4	36	2.5	38.5	5%	132
J2A	25	60 %	400,000	400	0.4	40	40	7	87.4	36	2.5	38.5	5%	132
J2B	25	60 %	400,000	400	0.4	40	40	7	87.4	36	2.5	38.5	5%	132
QAIS	25	90 %	279,400	200	0.1	55.9	14	14.5	84.5	49	51.3	56.2	5%	148

[1] Generated Energy is based on annual average energy for the last 5 years.
[2] Insurance factor is assumed to be (1/1000) of the depreciated book values.
[3] The cap. Inv. profit margin is assumed to be 10% of the depreciated book values.
[4] The risk factor is assumed to be 5% of the total unit generation cost.

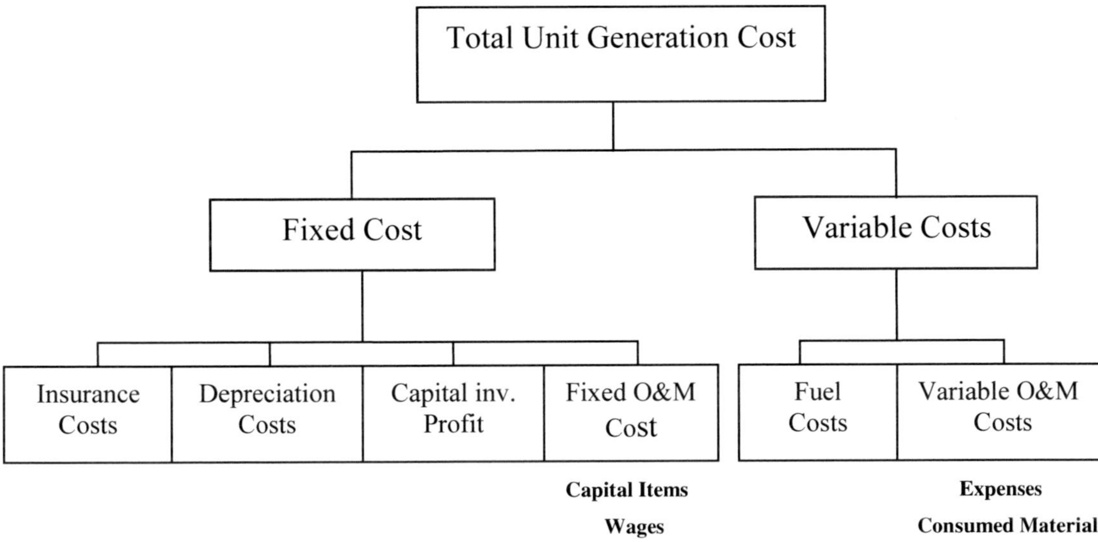

Figure 13. Unit Generation Cost Components.

Table 3 presents all required data to calculate the total unit generation cost in SR per MVA. This table is organized to facilitate any future updates to be a useful tool. Table 4 illustrates Table 3 fields with more details.

Table 4. Unit Generation Cost Fields Descriptions

Field name	Give / cal.	Field Description	Assumption
Life Cycle	Given	30 years for Qur. / Gazl. PP's and the rest are 25 years. [39]	_
Usage Factor	Given	As the average ages of P.P equipments. [39]	_
Asset Value	Given	The original asset values without depreciation. [39]	_
Generated Energy	Calculated	Calculated as the annual average generated energy for the last 5 years (2001-2005). [39]	_
Insurance	Calculated	As a percentage of the net book value after depreciation is deducted over annual generated energy = $\frac{\text{Asset Value}}{\text{Annual Generated Energy}}$ * [1-Usage F.]* [Insurance F.]	Insurance Factor = 0.1 %
Depreciation	Calculated	$\frac{\text{Asset Value}}{\text{Annual Generated Energy}} * \frac{1}{\text{Life Cycle}}$	_

Table 4. (Continued)

Field name	Give / cal.	Field Description	Assumption
Capital Inv. Profit	Calculated	As a percentage of the net book value after depreciation is deducted over annual generated energy = $\frac{\text{Asset Value}}{\text{Annual Generated Energy}}$ * [1-Usage F.]* [Profit M.]	Profit Margin = 10 %
Fixed O&M Costs	Given	Capital Items & Wages which is independent of generated energy on actual sense. [39]	_
Fuel Costs	Given	Actual costs data for the last 5 years (2001-2005). [39]	_
Variable O&M Costs	Given	Non-fuel consumable expenditures including materials, invoices, and other indirect expenditures. [39]	_
Risk Factor	Calculated	Assumed by this work	5%
Total Gen. Unit Cost	Calculated	[Total Fixed cost s + Total Variable Costs] X 1.05 in SR/MWH	_

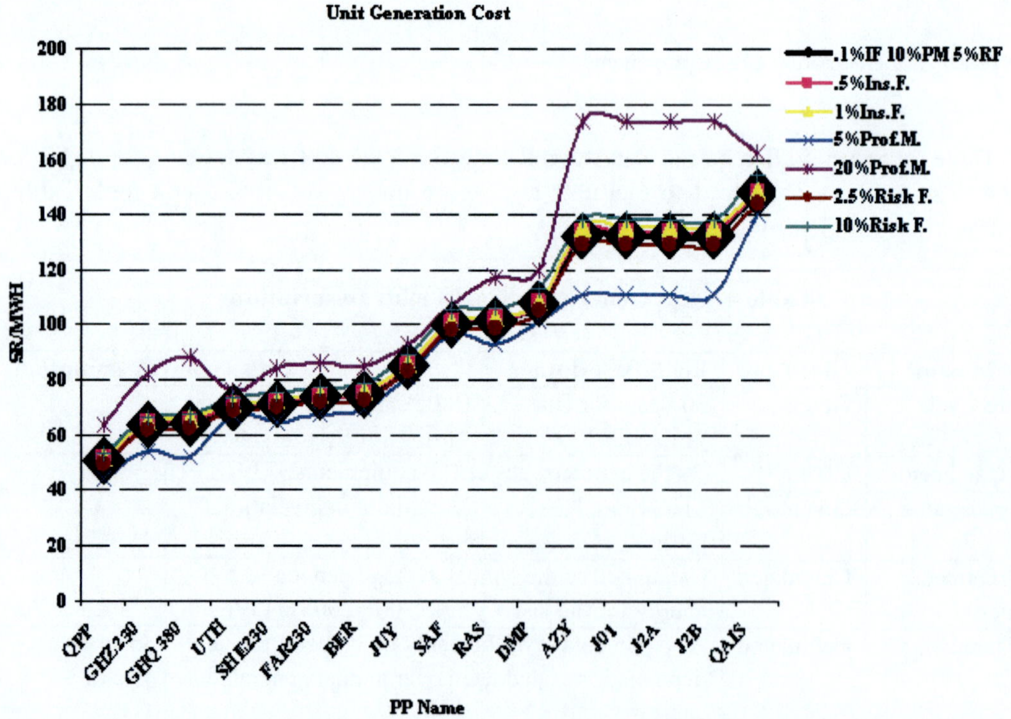

Figure 14. Unit Generation Cost Chart (Varying Ins. Factor, Risk Factor, and Profit Margin).

The basis for calculating the unit generation cost is mainly related to a given data from the utility. However, three factors are made by this study as illustrated by Table 4. These factors, which are debatable, namely: Insurance Factor, Profit Margin, and Risk Factor. So, it is needed to study and analyze the effect of changing their values on the unit generation cost for each power plant. This practice is done under a separate excel worksheet by fixing two variables and changing the third. The results are illustrated in Figure 14.

From Figure 14, it can be noticed that even though the insurance factor value changed from 0.1% to 1.0% (10 times); it has a negligible effect on the unit generation cost. In the same way, the risk factor has a minor effect on the cost even if it changed from 2.5 % to 10 %. However, the profit margin is the two lines in Figure 14 which represent the upper and the lower generation cost values while changing from 5% to 20%. So, it is worth considering this new generation costs and analyze how it could affect the model output. This activity is to be covered by Case Study 1-2.

Unit Wheeling Cost Calculation

In order to get the wheeling cost in each line of any network per km per circuit, it is needed first to calculate the total unit wheeling cost for the whole network and then proportionate each specific branch cost to the total cost. The total network unit wheeling cost could be calculated according to the following equation:-

Total unit wheeling cost for the whole network in SR/MWH = [Cap. Inv. Profit + Depreciation Cost + Insurance Cost + O&M Cost] X [1+ Risk Factor] (11)

Where:

Capital Inv. Profit (SR/MWH) = (Transmission Depreciated Book Value ÷ Average Energy Consumption) X Profit Margin.

And:

Depreciation Cost (SR/MWH) = (Transmission Asset Value ÷ Average Energy Consumption) X (1÷ Life C.).

And:

Insurance Cost (SR/MWH) = (Transmission depreciated book value ÷ Average Energy Consumption) X Insurance Factor.

And:

O&M Cost (SR/MWH) = Average annual variable & fixed O&M cost per unit of energy.

The capital investment profit for transmission is needed to be considered despite the fact that transmission is needed to be a non-profit entity. This price will be used to finance new expansions.

The transmission data related to the model of Figure 12 are summarized in Table 5.

Table 5. Unit Wheeling Cost Fields Descriptions

Field name	Source of information	Value
Transmission Network Book Value	Given in reference 39	16,000 million
Transmission Depreciated Book Value	Given in reference 39	7,000 million
Annual Energy Consumption	Average value of last five years actual energy generation	55,000,000 MWH
Profit Margin	Assumed by this work as per actual experiences in eastern province.	10%
Life Cycle	Given in reference 39	25 years
Insurance Factor	Assumed by this work as an approximation to actual	0.1%
Annual Transmission O&M Cost	Average value of last five years actual O&M costs.	2.3 SR/MWH
Risk Factor	Assumed by this work	5%

Substituting the data of Table 5 to equation (11):-

Capital Inv. Profit = 7000.000.000X0.10 ÷ 55.000.000 = 12.73 SR/MWH
Depreciation Cost = 16.000.000.000X0.04 ÷ 55.000.000 = 11.64 SR/MWH
Insurance Cost = 7.000.000.000X0.001 ÷ 55.000.000 = 0.13 SR/MWH
Average Annual O&M cost from the last 5 Years = 2.3 SR/MWH
The Risk Factor = 5%

Unit Wheeling Cost as per eq. (11) = [12.73+11.64+0.13+2.3] X 1.05 = 28.14 SR/MWH

The basis for calculating the unit wheeling cost is mainly related to a given data from the utility. However, three factors are made in this study as illustrated in Table 5. These factors, which are debatable, namely: Insurance Factor, Profit Margin, and Risk Factor. So, it is needed to study and analyze the effect of changing their values on the unit wheeling cost. This practice is done under a separate excel worksheet by fixing two variables and changing the third. The results are illustrated by Figure 15.

From Figure 15, it can be noticed that insurance factor line is almost a flat line which indicates that it has a negligible effect on the unit wheeling cost. Also, the chart indicates that the risk factor has a minor effect on the cost. However, the profit margin is the most related to the wheeling cost. The fourth line in Figure 15 represents the extreme case with the maximum values of all factors. It is worth to analyze how the extreme case values could affect the wheeling costs and the model output. This activity is to be covered by Case Study 1-3.

The wheeling cost for each type of OHL or substation could be calculated according to the following equation:-

Wheeling Cost for Z type of OHL or Substation per MWH = (Z type of OHL or Substation Cost per KM or T/F ÷ Z type Transmission Cost)

X Unit Transmission Cost per MWH X Weighing Factor. (12)

Where:

The weighing factor for OHL's = Unit Transmission Cost per MWH ÷ Cluster average wheeling cost per main links = 28.14 ÷ 1.0082 = 27.91 (Unit less)

And to avoid over weighing the substation locations, since, it will be required in future expansions which is not considered in this work formulation, the weighing factor for substations = the weighing factor ÷ No. of circuits = 27.91÷ No. of circuits.

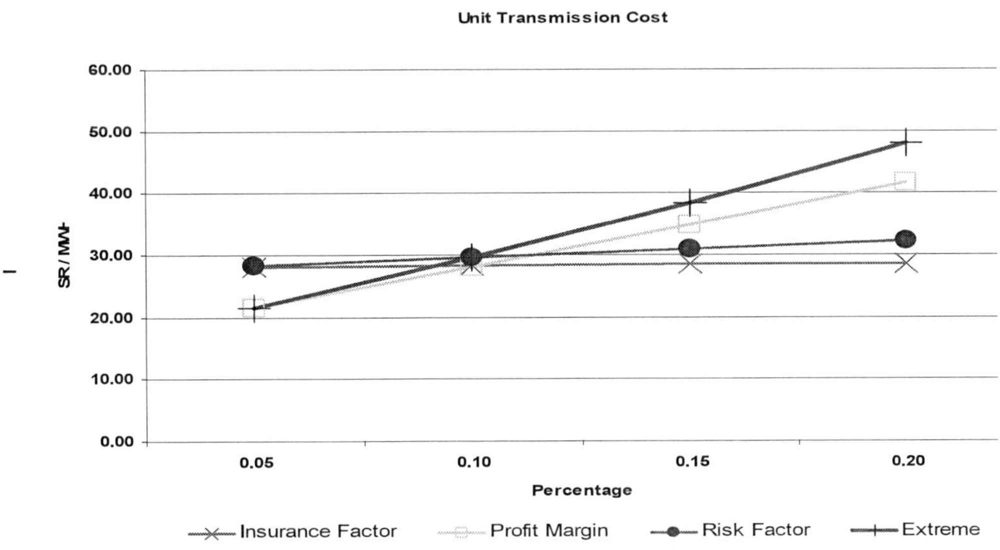

Figure 15. Unit Wheeling Cost Chart (Varying Ins. F., Risk Factor, and Profit Margin).

Table 6. Cost Data for Transmission Network

Voltage & Size	Length (km)	No. of Circuits	Cost (SR/km)	Total Line Cost (SR)
380S	465	1	1,200,000	558,000,000
380D	954	2	750000	1,431,000,000
230S	1,223	1	850000	1,039,550,000
230D	876	2	550000	963,600,000
110D	10	1	800000	8,000,000
Total OHD Lines. Cost				**4,000,150,000**
Item	**No. of T/F**	**Cost of one T/F & its breakers (SR)**		**Total Cost of T/F's & its breakers (SR)**
750 MVA	12	52,875,000		634,500,000
400 MVA	4	25,600,000		102,400,000
Total Substation Cost				**736,900,000**

The network of Figure 12 model mainly consists of Overhead Lines & Substations with different parameters listed in Table 6.

Substituting Table 6 data into equation (12); the wheeling cost for Z type of OHL per circuit per MWH will be equal:-

Wheeling cost for 380 D (single and double) = (750 ÷ 4,000,150) X 28.14 X 27.91 = 0.1473 SR/MWH/KM/Circuit

Wheeling cost for 380 S (single) = (1200 ÷ 4,000,150) X 28.14 X 27.91 = 0.2356 SR/MWH/KM/Circuit

Wheeling cost for 230 D (single and double) = (550 ÷ 4,000,150) X 28.14 X 27.91 = 0.1080 SR/MWH/KM/Circuit

Wheeling cost for 230 S (single) = (850 ÷ 4,000,150) X 28.14 X 27.91 = 0.1669 SR/MWH/KM/Circuit

Wheeling cost for 110 D (single and double) = (800 ÷ 4,000,150) X 28.14 X 27.91 = 0.1571 SR/MWH/KM/Circuit

And the wheeling cost for Z type of substation per T/F per MWH:-

Wheeling cost for 750 T/F as part of two circuits = (52,875 ÷ 736,900) X 28.14 X 27.91 ÷ 2 = 28.1771 SR/MWH/Circuit

Wheeling cost for 750 T/F as part of four circuit = (52,875 ÷ 736,900) X 28.14 X 27.91 ÷ 4 = 14.0885 SR/MWH/Circuit

Wheeling cost for 400 T/F as part of two circuit = (25,600 ÷ 736,900) X 28.14 X 27.91 ÷ 2 = 13.6422 SR/KWH/Circuit

All the above wheeling costs are listed in Table 2 to be used as a direct reference of case studies.

Case Studies

The flow chart in Figure 11 listed the steps to be followed to utilize the model in the actual business. As per the flow chart, the process is divided mainly into two parts. The first part is to check validity of the proposed model by comparing it to a real system. This part is to be covered in this section as the first case study which compares the actual system output with three of the model outputs. These three models cover the original model data calculated earlier along with changing the data according to the sensitivity analysis for both unit generation cost and unit wheeling cost calculation.

The second part of Figure 11 is to search for the potential power plant locations. This part is to be covered in this section as the second and third case studies. The second case study covers the interface locations between eastern province and central province of Saudi Arabia while the third case study is to be carried out by adding one power plant at a time in each bus location of the existing network analyzing the results.

Case Study 1

As reported earlier this case study is meant to check the validity of the proposed model by three sub-case studies.

Case Study 1-1

In this first case study; the model is applied to the actual eastern province power system in Saudi Arabia as it is without any changes as formulated by LINDO software using Tables 1 and 2 as the data source. The optimum solution is found at objective function value of 1001137 SR with unit generation cost of 71 SR/MWH as an average cost.

Case Study 1-2

In this case study; the model is applied to the actual eastern province power system in Saudi Arabia but with changes in the values of unit generation cost as recommended earlier after analyzing Figure 14. The unit generation values are the ones corresponding to a profit margin of 20 %. Case Study 1-2 is formulated by LINDO software using Tables 1 and 2 as the data source. The formulation is a copy of Case1-1 with only change in the coefficients of Pgi in the first part of the objective function. It is solved by running the software where the output and the sensitivity ranges. The optimum solution is found at objective function value of 1190010 SR with unit generation cost of 73 SR/MWH as an average cost.

Case Study 1-3

In this case study; the model is applied to the actual eastern province power system in Saudi Arabia but with changes in the values of unit wheeling cost as recommended after analyzing Figure 15. The unit wheeling values are the ones corresponding to the extreme case with 20% profit margin, 20% risk factor, and 0.4% insurance factor. Case Study 1-3 is formulated by LINDO software using Tables 1 and 2 as the data source. The formulation is a copy of Case1-1 with only change in the coefficients of Pdij in the second part of the objective function. It is solved by running the software. The optimum solution is found at an objective function value of 1181413 SR with unit generation cost of 71 SR/MWH as an average cost.

Table 7 summarizes all case studies' dispatched MW's along with the optimal values. This table is used as a worksheet to calculate the unit generation average cost and distribute the optimal values among generation costs and wheeling costs

Table 7. All Case Studies' Dispatched Power

Bus No. i	PP NAME	Act. Disp. MW	Case 1-1 Disp. MW	Case 1-2 Disp. MW	Case 1-3 Disp. MW	Case 2-1 Disp. MW	Case 2-2 Disp. MW	Case 2-3 Disp. MW	Case 2-4 Disp. MW	Case3 Bus#/ PP NAME	Optimal Value	Case3 Bus No./ PP NAME	Optimal Value	U.G. Cost Cgi	U.G. Cost case 1-2 Cgi
1	GHZ380	2718	2372	2482	2372	2195	2195	2280	2280	3/FAD	987389	27/DHC	945055	64	67
2	Q PP	2569	2500	2500	2500	1466	852	1317	1317	4/GHU	994427	28QTF	937395	51	53
9	GHZ230	1527	1600	1600	1600	1600	1600	1600	1600	5/SHE	993774	30/JBSE	917958	63	66
11	SHE230	720	1219	1219	1219	1219	1219	1219	1219	6/FAR	985006	32/IZH	948194	72	75
12	FAR230	529	910	910	910	910	910	910	910	7/HAW	990796	33/JBCM	945892	74	77
21	DMP	223	400	400	400	100	100	100	100	8/FAD	928440	34/SADAF	942684	108	113
26	AZY	832	404	294	404	100	100	100	100	10/GHU	953905	35/SHARQ	946667	132	138
29	RAS	142	120	120	120	120	120	120	120	13/MAH	969683	36/PET	966702	101	106
31	SAF	0	75	75	75	75	75	75	75	14/RWPS	980126	37/JBN	913049	100	105
43	JO1	291	80	80	80	80	80	80	80	15/KHU	979822	38/JBSW	913097	132	138
44	J2A	290	80	80	80	80	80	80	80	16/WAS	979202	39/SAB	950737	132	138
45	J2B	290	80	80	80	80	80	80	80	17/HAW	976950	40/KEMIA	949605	132	138
46	BER	138	215	215	215	215	215	215	215	18/JAD	947102	41/HAD	938310	75	78
48	QAIS	14	108	108	108	20	20	20	20	19/ABQ	947111	42/NIGC	946239	148	155
52	UTH	197	272	272	272	272	272	272	272	20IBK	912238	47/BNGL	915022	70	74
53	JUY	73	118	118	118	40	40	40	40	22DMP	916516	49/NAR	961718	85	89
54	R9001	Prop.	0	0	0	925	1539	925	925	23/RAK	967808	50/KHU	927491		
55	KH9003	Prop.	0	0	0	814	814	875	814	24/KHO	946666	51/SIPC	950198		
56	R9000	Prop.	0	0	0	242	242	242	306	25/THQ	946104				
			CASE 1			**CASE 2**				**CASE 3**					
Average Cost		75	**71**	73	71	69	62	66	69						
Optimal Value			1001137	1190010	1181413	903766	853555	871170	901100						
Gen. Cost SR/H			744223	769516	744223	726818	657300	695442	726818						
Wheel. Cost SR/H			256914	420494	437190	176948	196255	175728	174282						

Case Study 1. Observations

The values of the decision variables for the three sub-cases are compared with actual system dispatching values as illustrated in Figure 16.

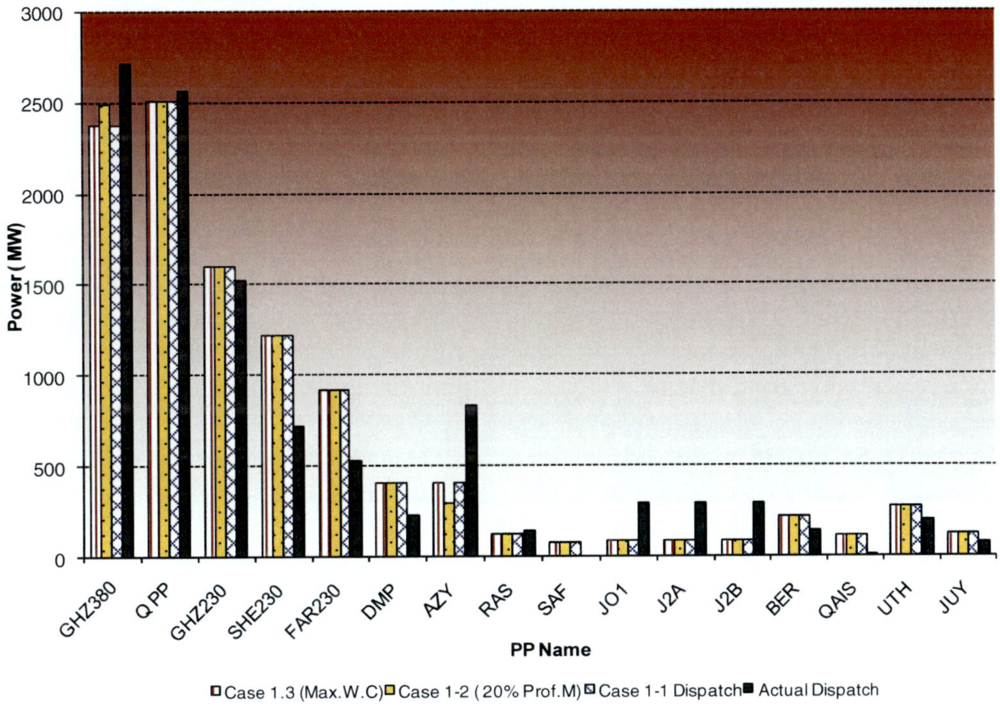

Figure 16. Actual vs. Model Dispach Results (case 1).

The general shape of Figure 16 validates that the model is working in the same direction of the actual dispatch process. However, the actual dispatch is affected by other aspects like maintenance schedules and existing contracts.

The clear differences from the graph is in Aziziyah (AZY) power plant and the three Jubail power plants (JO1, J2A, and J2B) which are actually separate entities owned by SWCC where the energy is contracted with SEC in fair prices. This explains why the actual is much more than the model. The model selected the more economical power plants based on the unit generation cost calculation (not as per contract) which explains the increase in dispatched power from Shadgam, Faras, Dammam, Barri, and Uthmaniyah power plants. Qaisomah (QAIS) & Safaniyah (SAF) power plants were under maintenance which explains why the actual is much less than the model.

Other observations are listed as follows:-

- From Table 7 the percentage of the wheeling cost in Case 1-1 to the generation cost is almost 35%. This percentage is not far from 38% representing the percentage of the unit wheeling cost calculated earlier (28.14 SR/MWH) over the average unit generation cost (75 SR/MWH). This indicates that the model is supporting the calculation accuracy and keeping the correct weighing between the generation and the transmission costs.

- The optimal value of Case 1-1 is lesser than each of cas1-2 and Case1-3 which is expected due to the selection of higher values of unit generation cost in Case 1-2 and higher unit wheeling cost in Case 1-3.
- The higher average cost of 73 SR/MWH in Case 1-2 indicates the effect of raising the profit margin from 10% to 20% in the generation industry.

The above observations in Case Study 1 prove the validity of the model.

Case Study 2

This case study is conducted to help in taking a decision whether to continue sending power from EOA to COA or to build one or more power plants in the interface locations and, if so at what cost. The costs are to be encouraging for investors to build a new plant in COA, otherwise, increasing the capacity of EOA and transmit to COA would be the option to choose.

Busses 54, 55, and 56 are the existing loads in COA and fed from EOA as per Figure 12. So, to conduct this study, four sub-case studies were carried out.

Case Study 2-1

In this case study, three power plants are proposed at R9001, KH9003, and R9000 corresponding to busses 54, 55, and 56 respectively. The power plants are to be proposed at the same time with unit generation costs equal to the average cost on original model (71 SR/MWH of Case Study 1-1) keeping other power plants without changes.

To consider busses 54, 55, and 56 as generating units in our original model; the objective function is modified by assigning a cost of 71 SR/MWH instead of the high cost of 1000 SR/MWH that was assigned to these busses earlier. Also it is required to add the MW generated by the selected busses "Pg54, Pg55, and Pg56" to the real power balancing constraint.

The maximum and minimum of generator limits were not imposed to have more flexibility.

Case Study 2-1 is formulated by LINDO software using Tables 1 and 2 as the data source. The formulation is a copy of Case1-1 with only changes mentioned in the last two paragraphs. It is solved by running the software where the output and the sensitivity ranges are listed in Appendix 7. The optimum solution is found at objective function value of 903766 SR with unit generation cost of 69 SR/MWH as an average cost. This information and the dispatched power values are listed in Table 7. It can be noticed that, the proposed power plants are dispatched only to cover their own loads except for R9000 (bus 56) which feeds 84 MW more than its loads.

Case Study 2-2

The aim of this case study is to report at which unit generation cost the proposed power plant at R9001 bus 54 is going to be dispatched for loads from other busses.

Running a sensitive analysis by LINDO software for Case Study 2-1 can help to specify the required value reducing the trial and error iterations. From the obtained results, the allowable decrease value corresponding to the variable Pg54 is 54. Then the recommended cost will be (71 -54 = 17 SR/MWH).

In Case Study 2-2, three power plants are proposed at bus 54, 55, and 56 at the same time with unit generation cost equal to the average cost of 71 SR/MWH for busses 55 and 56 while bus 54 with cost of 17 SR/MWH keeping other power plants without changes.

Case Study 2-2 is formulated using LINDO software. The formulation is a copy of Case 2-1 with only one change, replacing 71Pg54 by 17Pg54 in the first part of the objective function. It is solved by running the LINDO software. The optimum solution is found at objective function value of 853555 SR with unit generation cost of 62 SR/MWH as an average cost. This information and the dispatched power values are listed in Table 7. It can be noticed that, the proposed power plant at bus 54 is dispatched to cover 1539 MW of loads while the bus loads are 925. The selection of 17 SR/MWH as the border value for R9001 (bus 54) can be tested by running the software with 18Pg54 instead of 17Pg54, then the dispatched power should go back to 925 MW.

Case Study 2-3

The aim of this case study is to report at which unit generation cost the proposed power plant at KH9003 bus 55 is going to be dispatched for loads from other busses.

Running a sensitivity analysis by LINDO software for Case Study 2-1 more than one time to reach to the border value of 31 SR/MWH.

In Case Study 2-3, three power plants are proposed at bus 54, 55, and 56 at the same time with unit generation cost equal to the average cost of 71 SR/MWH for busses 54 and 56 while bus 55 with cost of 31 SR/MWH keeping other power plants without changes.

Case Study 2-3 is formulated using LINDO software. The formulation is a copy of Case 2-1 with only one change of replacing 71Pg55 by 31Pg55 in the first part of the objective function. It is solved by running the LINDO software. The optimum solution found at objective function value of 871170 SR with unit generation cost of 66 SR/MWH as an average cost. This information and the dispatched power values are listed in Table 7. It can be noticed that, the proposed power plant at bus 55 is dispatched to cover 875 MW of loads while the bus loads are 814. The selection of 31 SR/MWH as the border value for KH9003 (bus 55) can be tested by running the software with 32Pg55 instead of 31Pg55, then the dispatched power should go back to 814 MW.

Case Study 2-4

The aim of this case study is to report at which unit generation cost the proposed power plant at R9000 bus 56 is going to be dispatched for loads from other busses.

Running a sensitive analysis by LINDO software for Case Study 2-1 more than one time to reach to the border value of 60 SR/MWH.

In Case Study 2-4, three power plants are proposed at bus 54, 55, and 56 at the same time with unit generation cost equal to the average cost of 71 SR/MWH for busses 54 and 55 while bus 56 with cost of 60 SR/MWH keeping other power plants without changes.

Case Study 2-4 is formulated using LINDO software. The formulation is a copy of Case 2-1 with only one change of replacing 71Pg56 by 60Pg56 in the first part of the objective function. It is solved by running the LINDO software. The optimum solution found at objective function value of 901100 SR with unit generation cost of 69 SR/MWH as an average cost. This information and the dispatched power values are listed in Table 7. It can be noticed that, the proposed power plant at bus 56 is dispatched to cover 306 MW of loads while the bus loads are 158.

Case Study 2 Observations

Figure 17 compares the four sub-cases with each other and with the original model in terms of optimal value, system generation cost, and system wheeling cost. The figure also illustrates which sub-case is with low system average unit generation cost and at which cost the investor is forced not to exceed.

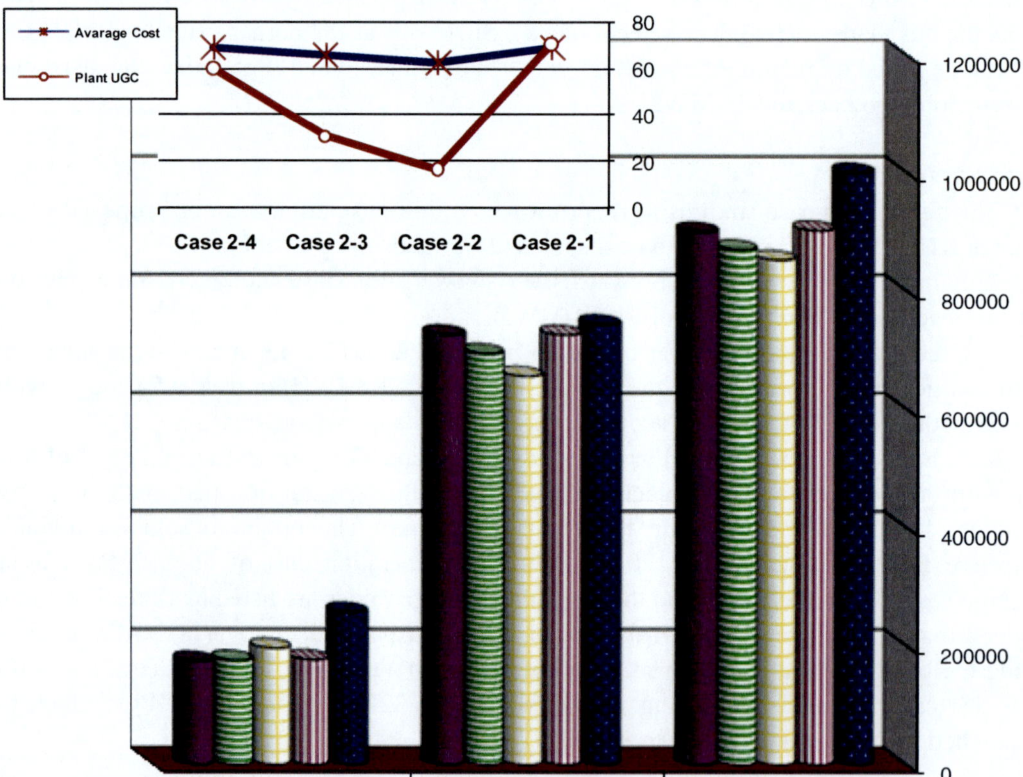

	Wheel. Cost SR/MWH	Gen. Cost SR/MWH	Optimal Value
Original	256914	744223	1001137
Case2-1	176947.5278	726818.4722	903766
Case2-2	196255.3343	657299.6657	853555
Case2-3	175727.6468	695442.3532	871170
Case2-4	174282	726818	901100

Figure 17. Case 2 Analysis.

Reading Figure 17 reveals that the optimal values for any of the sub-cases is less than the optimal value for the original model. This means that the system could run with less cost. However, the challenge is at what price will the system allow each power plant to dispatch the stated amount of power?

Other observations could be summarized from Figure 17 and Table 7 as follows:-

- In Case Study 2-1 (71 SR/MWH for the three locations) the system runs with less cost than the original. However, the proposed power plants are dispatching only to cover their bus loads. This is true even for bus 56 where the difference between the dispatched power (242 MW) and the bus loads (158 MW) is only to feed (as can be seen from Figure 12 and Table 1) Wasia, Khurais, and RWPS which are still in the borders of COA.
- In Case Study 2-2 (17 SR/MWH for R90001 bus 54) the system runs with less cost than all other cases, however, the mandatory cost of 17 SR/MWH is a very low cost and it is not encouraging to any investor to built a power plant with low purchasing prices.
- In Case Study 2-3 (31 SR/MWH for KH90003 bus 55) the system runs with less cost than other cases except for Case 2-2. However, the mandatory cost of 31 SR/MWH is still very low cost compared to the lowest existing Qurayah power plant with unit generation cost of 51 SR/MWH. So, it is not visible to investors to build with these limitations.
- In Case Study 2-4 (60 SR/MWH for R9000 bus 56) the system runs with less cost than the original and the allowable unit generation cost is good enough to grab investors. However, the loads on the bus are quite low, so, investors should check first the feasibility study with the existing limitations of load quantity and related prices if a load increases.

Case Study 3

This case study is conducted to help in taking a decision whether to build a power plant on the existing busses' locations and in which priority?

This third case study is to be carried out by adding one power plant with the original model average cost of 71 SR/MWH at a time in each bus location of the existing network and analyze the results.

To consider bus i as a generating unit in our original model; the objective function is modified by assigning a cost of 71 SR/MWH instead of the high cost of 1000 SR/MWH that was assigned to the selected bus earlier. Also it is needed to add the MW generated by the selected bus "P_{gi}" to the real power balancing constraint. This step is repeated for each bus in Figure 12 without power plant where i = 3, 4, 5, 6, 7, 8, 10, 13, 14, 15, 16, 17, 18, 19, 20, 22, 23, 24, 25, 27, 28, 30, 32, 33, 34, 35, 36, 37, 38, 39, 40, 41, 42, 47, 49, 50, and 51.

The maximum and minimum of generator limits were not imposed to have more flexibility.

Case Study 3 is formulated by LINDO software for bus 3 as a sample for all busses. The formulation is a copy of Case1-1 with only changes mentioned in the above two paragraphs.

It is solved by running the software where the optimum solution found at objective function value listed in Table 7.

Case Study 3 Observations

Table 8 lists all selected busses in a priority order starting with the least optimal value. The table also reports the power dispatched by each proposed power plant to be a supportive information in making a decision to build a new power plant in the selected location or not.

It could be noticed from Table 8 that all optimal values are less than the original model. This is expected, since, each proposed power plant has average cost of the model and will take its adjacent loads with less wheeling cost.

The first five priorities selected by the model are IBK (bus20), JBN (bus37), JBSW (bus38), BNGL (bus47), and DMP (bus22). These selected locations are actually the center of loads in eastern province of Saudi Arabia. The outcome of Case 3 approves two main points:-

- Validity of the model
- No congestion on the exiting transmission network which allows the model to select the center locations.

Table 8. Proposed Power Plants Priority

Priority	Bus No. i	PP NAME According to Priority	Case 3 Dispatched MW	Optimal Value
1	20	IBK	3454	912238
2	37	JBN	3449	913049
3	38	JBSW	3602	913097
4	47	BNGL	3449	915022
5	22	DMP	3333	916516
6	30	JBSE	3122	917958
7	50	KHU	3177	927491
8	8	FAD	3314	928440
9	28	QTF	2612	937395
10	41	HAD	1943	938310
11	34	SADAF	1755	942684
12	27	DHC	1632	945055
13	33	JBCM	1671	945892
14	25	THQ	1632	946104
15	42	NIGC	1623	946239
16	24	KHO	1715	946666
17	35	SHARQ	1609	946667
18	18	JAD	2961	947102
19	19	ABQ	2961	947111
20	32	IZH	1544	948194
21	40	KEMYA	1525	949605

Priority	Bus No. i	PP NAME According to Priority	Case 3 Dispatched MW	Optimal Value
22	51	SIPC	1483	950198
23	39	SABAYEK	1484	950737
24	10	GHU	1488	953905
25	49	NAR	2533	961718
26	36	PETROKEMIA	883	966702
27	23	RAK	714	967808
28	13	MAH	1557	969683
29	17	HAW	1543	976950
30	16	WASIA	306	979202
31	15	KHURAIS	306	979822
32	14	RWPS	340	980126
33	6	FAR	1120	985006
34	3	FAD	2252	987389
35	7	HAW	304	990796
36	5	SHE	304	993774
37	4	GHU	304	994427

COMMENTS

Proper location selection for a new power plant to be built is a complete process with many factors like: land ownership, type of fuel, load distribution, environment limitations, water and other resources accessibility. The proposed model in this chapter is handling the major factor affecting the decision which is the cost related part along with load distribution.

OPF model is a very sophisticated method in the optimization problems but most of the times we need to do approximations and reach to our objectives from the short paths since there are always better ways to hit different targets.

Beside the main usage of the modified model which is to dispatch the power system for the purpose of localizing the potential power plant; the model could be used for other purposes with some modifications like:-

- Addition of power loss and power reserve parameters.
- Addition of a binary equation to the model to facilitate the possibility of the units to be "on" or "of", where it is needed in this case to count for the start-up cost and the minimum time before in or out the unit. And LINDO has this tool in a very simple form.

However, the model cannot be used to optimize the power flow for operation purposes due to the lack of the following parameters:-

- The impedance
- The reactive power
- The time index
- The bus represents a power plant (not each unit).

- The model uses a constant cost for any bus (Power plant).

Despite the shortcomings of the model, it is simple, fast, efficient, and effective. The model can be considered as a unique model in solving both operation and power flow equations by the flow of active power only. The model is more effective by putting more emphasis to the costs which effect the decision.

Accuracy is expensive in terms of both money and time; however, in the competitive market a balance between accuracy and cost is needed.

Conclusion

This study was started in year 2003, which is the same year SEC had started functioning. This was considered the first practical stage in Saudi Electricity Supply Industry deregulating process. In the middle of this study, the second stage of deregulating SESI was carried out by SEC. The third stage is already planned by SEC. However, SEC has its own timetable in going through some or all of the restructuring stages. So, it is hoped that the timing of this study is supportive to add additional analysis learning from the worldwide experiences. The aim is to find out the best way to provide reliable and low-cost electricity suiting the uniqueness of SESI.

Summary of Findings

It can be concluded that each country has its own ESI characteristics which may call for deregulating the industry or not. And each ESI, if to be deregulated, needs to have its own framework suitable for its needs but not necessary for others.

Restructuring is recommended in Saudi Arabia due to its geographical large area, the availability of investment power, high demand growth rate, political stability, and for the fact that power generation is thermal (no nuclear power plants).

SEC efforts in the ongoing deregulating process explained earlier indicates that the direction toward deregulating the industry is handled seriously by SEC and pushed by the ministry of W&E. Despite that, the data are not showing any financial support from outside SEC to recover the restructuring costs. This could affect other programs which could lead, in the short run, to customer dissatisfaction even if the intention is toward enhancing the quality of customer service.

By reviewing the most popular market models in, it can be concluded that it is not practical to design an ideal market model. This is due to the complexity and uniqueness of any ESI. However, the aim should be to have near ideal model to fit most needs with the existing limitations.

SESOM is to be established to act as SO, MO, and SC. It would administer and operate the wholesale market where supply and demand are to be balanced at the real time through a centralized mechanism to be efficient and secured. SESOM is required to dispatch generators according to economic merits order. The last generator being dispatched to meet the required demands will be considered the price leader. This uniform price is the SMP which is proposed to be capped.

The end user price will be a combination of wholesale price (includes the ancillary prices and the regulatory prices) and wires services prices for both transmission and distribution in addition to the retailer price. The end user price is also to be capped by the regulator.

In the recommended structures, customers can choose their supplier who can provide additional value-added services.

Through the conclusion same comments on the ongoing SEC restructuring process can be summarized as follow:-

- Inefficient generation units are not to be released by SEC to private investors.
- To reduce the share of WEC by SEC.
- CCC is to be a priced optional service.
- Existing bilateral contracts by transmission BL's are to be converted to CFD by SESOM.
- SLA actual implementation is to be expedited.
- To put hold on the action of actual separation of TRANSCO and instead financial separation of operation activities from transmission wires.
- To group SEC GENCO's into 8 to 9 groups.
- To put hold on the idea of dividing DISTCO's and instead pursue financial separation of customer services from distribution wires.

Although the market rules (to be issued by ECRA) are mandatory to all market participants, competitive participants (PSP's & CSP's) will have some room of freedom to set their own plans. This freedom could be in terms of operation schedules by PSP's or additional services by CSP's. Moreover, it could be in terms of contracts agreements and profit margins for both PSP & CSP.

PSP's system planners are willing to use the most suitable optimization tool to know the future generation potential areas. Those tools including economic load dispatch, unit commitment, optimal power flow, & DC optimal power flow are briefed earlier. This brief links the available tools together and compares them in terms of complexity, advantages, limitations and the proper usages.

The above learning process enables researcher to set the characteristics of the required model. The modified model is a linear program MW-based OPF established by this study to help in deciding the future generation potential areas in a deregulated industry. The same model with slight modifications can be used for additional purposes.

The proposed model is a unique model in solving all operations and power flow equations by the flow of active power only for simplicity and to give more emphasis to the costs which effect the decision.

LINDO is the optimization software used as a tool to solve the MW LP-Based OPF model. The flow chart in Figure 11 illustrates all steps to be followed to make use of the model.

The proposed model is tested in earlier by applying it to the real power 56-bus system in the eastern province of Saudi Arabia.

All input data required to run the program are mainly direct data except two costs are to be calculated namely: Unit generation cost and unit wheeling cost.

Unit generation costs calculation proposed scheme is one of the products of this Study. Since this cost is to be used in the model to decide potential areas for future generation, it should not be limited to operation costs. So, it is proposed to cover total generation costs including revenue requirements.

Unit wheeling cost calculation scheme is also proposed in the same chapter. The technique used to calculate the unit wheeling cost in SR/MWH/km is based on the unit transmission cost for the whole network. Then, this cost is to be proportionate to each specific branch cost to the total cost.

The validity of the two costs calculation scheme depends on two factors. The first factor is the availability and accuracy of the data used in the calculation process. The second factor is the correctness and the acceptance of the assumptions made for the profit margin, the insurance factor, and the risk factor. The profit margin in transmission is only to finance new expansions. Moreover, sensitivity analysis is carried out for all assumptions to answer (what if?) question for each case.

Case studies are carried out on the modified optimization model and the first case study approves the validity of the model. The second case study covers the interface locations between eastern province and central province of Saudi Arabia while the third case study is carried out by adding one power plant at a time in each bus location of the existing network and recommended all possible locations according to priority.

Challenges

In the short and medium terms, ECRA, SESOM and other market participants are expected to face a lot of challenges, some of which are summarized in the next paragraphs.

- Reverse flow characteristics of electricity limit the pricing choices and complicate the methodology. Moreover, up to nowadays technology, energy can not be stored which requires real time balancing of supply and demand which is critical and no shortcomings are allowed to occur.
- Major investment requirements by SEC, in the short term, to establish SESOM as well as funding transmission expansion for the intended interconnected network. More investment is needed for funding generation expansion to maintain enough medium term capacity and to deplete the old units. Finally, some funds are required for distribution infrastructure.
- Even if the wholesale prices went down, in the start of real competition stage, retail prices and end user prices may not be decreased to the desired level. Such expectation is due to the need of recovering capital investment costs especially for the interconnection projects.
- Market power is difficult to be avoided in the short and medium terms of moving toward competitive market. The expected market power is due to the fact that SEC is providing almost 85% of the whole energy consumptions.

- Unavailability of reasonable priced smart energy meters. Such meters are to be used for households as well as for other customers. This is needed to let the end users sense the price signals of wholesale market.

Expected Benefits

The proposed framework satisfies most of the criteria needed to have a stable competitive market. This could drive the industry to have more security, more reliability, less fuel consumption, less pollution emissions, more innovations, new renewable energy technologies, and more value-added services to end users. Such improvements will lead to customer satisfaction and government satisfaction for hitting the targets without subsidies.

Another expected benefit is having more reliability and at the same time lesser reserve capacity which is an incentive to reduce generation cost. When SESOM start operating the system, reserve capacity is expected to be lower than when the industry is under regulation. However, the competition is expected to improve the operating efficiency which will lead to more reliability.

At retailing level and after promoting real competition among various CSP's, the customer has the right to choose his CSP. Therefore, CSP would compete to grab more customers by avoiding disconnections for non-payment bill, cost hedging, lowering contracted prices under the regulated price, and providing other value-added services. Value-added services could include one or a combination of related services such as: desalinated water, cable TV services, internet services, and energy conservation consultations.

In the long term, end users are expected to participate in the demand side management. This is to be accomplished if the price signals passes to them. So, they could reduce their loads in the peak time. In the medium term, this participation could also be accomplished with smart retailing business moves.

Recommendations

The following summary of main recommendations covers all the Study key actions advised to be implemented by SESI participants or any other outside interested readers. Some of these recommendations are general which is applicable to most of ESI's and some of them are specifically related to SESI structure and uniqueness.

The recommendations are:-

5. ECRA is advised to form various function groups to be membered by a well selected expertise from ECRA, W&E ministry, SEC expertise, and legal consultant companies. The groups are required to develop all rules and regulations. This includes codes, contracts, efficient pricing criteria, and all other issues needed to drive the industry to the right direction.
6. At this ongoing reform stage in SESI, all participants are advised to shorten the period of any new contract to be signed. The reason behind that is to give a chance to

different kinds of contracts to be carried out like CFD when SESOM is established and the wholesale market started.

7. It is recommended to transfer to ECRA all required regulatory authority from existing concerned agencies namely: Deputy Ministry of Electricity, S.A. General Investment Authority, Ministry of Oil and Natural resources, Ministry of Finance, and Supreme Economic Council. In addition to that, a clear determination of duties and responsibilities of each concerned agency is required. The aim of this recommendation is to avoid overlapping of authority which could discourage new investors or induce high regulatory in the intended competitive market.
8. ECRA and SEC are advised to seek one time subsidy from government to cover infrastructure costs in deregulation process to enhance the industry and drive the process to success.
9. ECRA is advised to review the existing connection fees and consumption tariff. Chapter 4 proposes brief ideas for more suitable prices.
10. It is recommended to ensure 100% independency of SESOM and ECRA. Ensuring fairness among all market participants is behind this recommendation.
11. Transmission wires and distribution wires are to be regulated due to their nature of monopoly. While generation and retail businesses are to be competitive with no particular large scale power plant or retailer. Also, distinct separation between the regulated entities and the competitive entities is a mandatory action in the long term structure.
12. OASIS is to be sophisticated enough to meet the complexity of the industry and the intended increase of participants.
13. SESOM is advised to learn from other industries incidences. High protection for the system is required to avoid any chance of cascaded blackouts that have happened in other countries like: California, France, Canada, and New Zealand.
14. New investors are to be supported in the beginning by SEC allowing SESOM to give them priority to run their capacity. Later, when a competitive market is well established, SESOM should dispatch with priority to the power plants of lower prices regardless of being SEC property or other's property.
15. SEC has to encourage investors by all means to invest in the generation as well as retailing businesses to have large number of competitors in the industry. Otherwise, market power could be induced affecting the market and the industry in a negative way.
16. No particular PSP or CSP is to be allowed for supply monopoly over any specific area, city, town, village, or any government or non government authority. But rural areas could be allowed for supply monopoly under specific conditions, since it is mostly not economical to compete unless the network is available.
17. The proposed MW LP-based OPF tool can be used by new investors to decide the correct location of their generators. This would be beneficial in case of SESOM support of giving detailed and accurate data.

POSSIBLE FUTURE RESEARCHERS

Problems have been reported during deregulating electrical industries and after the deregulation had taken place along with problems raised by researchers in the field, make this area a good land for economist and electrical engineer researchers.

Due to complexity of electricity commodity, economists could seek a help from electrical engineers even if the problem is related to pricing or trading. The opposite is true also, due to the need of cost calculation criteria in most of the newly raised problems. So, the output of a researcher with electrical engineering background and master in economics is expected to be more comprehensive output. This could be accomplished by joint work also.

One of the Study products is the calculation criteria for the wheeling cost per energy per KM of electrical lines. However, the criterion used has shortcoming of requiring a lot of data in order to establish the cost. So, economists are called to develop a formula to calculate cost per MWH per KM without long calculation.

There is a need for future research in co-generation of water and electricity process, its viability and how far it should go.

Enhancement of the proposed MW LP-based OPE, is another subject for future research. The key issue here is to select a wide range of usage in addition to the localization of potential power plants. Then modify the model without losing the advantage of the approximation by using active power only in the formula (No reactive power and no line characteristics). This is meant to let the cost drive the decision.

Still more problems are expected to be raised when SESI is completing the final stage of reform process and SESOM starts to run the competitive market. At that time, immediate researches would be called for and they will be based on realism.

REFERENCES

[1] M. Shahidehpour & M. Alomoush, *"Restructured Electrical Power Systems: Operation, Trading, and Volatility"*, Marcel Dekkar, Inc., 2001, New York.

[2] G. Robertson, *"Regulators Urged to Demand Generation"*, Calgary Herald (www.canada.com), October 18, 2002.

[3] L. Philipson and H. Willis, *"Understanding Electric Utilities and De-regulation"*, Marcel Dekkar, Inc., 1999, New York.

[4] M. Takahashi, "The Current Status of Electric Power Industry Deregulation in Japan and the Influence of the California Crisis", *Japan and the World Economy,* Vol. 14, Iss. 3, August 2002, pp. 341-345.

[5] J. Hein, *"An Essential Industry at the Crossroads: Deregulation, Restructuring, and Anew Model for the United States' Bulk Power System"*, Master Thesis, University of Colorado, 2003, Denver.

[6] T. Overbye, X. Cheng, and Y. Sun, *"A Comparison of the AC and DC Power Flow Models for LMP Calculations"*, 37th Hawaii International Conference on System Sciences, 2004, USA.

[7] L. Philipson and H. Willis, *"Understanding Electric Utilities and De-regulation"*, Marcel Dekkar, Inc., 1999, New York.

[8] J. Casazza and G. Loehr, *"The Evolution of Electric Power Transmission under Deregulation"*, IEEE, Inc., 2000, USA.
[9] L. Lai, "Power System Restructuring and Deregulation: *Trading, Performance and Information Technology",* City University, 2002, London.
[10] K. Bhattacharya, B. Math, and J. Daalder, *"Operation of Restructured Power Systems"*, Kluwer Academic Publishers, 2001, Portland.
[11] G. Robertson, "Regulators Urged to Demand Generation", *Calgary Herald* (www.canada.com), October 18, 2002.
[12] P. Meyer, "City Finds Electric Deregulation Causes Problems", *Plano Star Courier* (www. planostar.com), September 25, 2002, Pg. 461.
[13] W. Stueck, "B.C. Warned Against Power Deregulation", *The Globe and Mail,* September 26, 2002, Pg. B7.
[14] D. Gillmor, "State Needs Intelligent Energy Deregulation", *San Jose Mercury News* (www.siliconvalley.com), September 25, 2002, Pg. C1.
[15] J. Nisse and L. Wadhwani, "The Future of Electricity", *Independent News* (news.independent.co.uk), October 20, 2002.
[16] J. Glater, "Dynegy Says It Will Exit the Energy Trading Business", *The New York Times* (www.nytimes.com), October 17, 2002, Pg. C5.
[17] D. Ackman, *"Dynegy Says it's New and Improved",* Forbes (forbes.com), October 17, 2002, Pg. 1017.
[18] C. Rose, *"Bogotá's Christmas Lights to Shine Again",* The Paris News (web.theparisnews.com), October 17, 2002.
[19] J. Kbal, "Nothing Gained By Deregulation" *Brookings Register* (www .brookingsregister.com), October 16, 2002.
[20] R. D'Errico, "Assemblyman's Report: N. Y. Power Industry in Crisis", *The Business Review* (Albany.bizjournals.com, October 18, 2002, Pg. 77.
[21] D. Lunsford,"Electric Deregulation has City Toying with Strategy", *My West Texas* (www.mywesttexas.com), September 29, 2002, Pg. 461.
[22] I. Al-Shaikh, "Gulf Needs 8 Billion Dollar Annually to Raise Electric Production", *Alsharg Alawsat News Paper* (www.asharqalawsat.com), November 02, 2002.
[23] G. Lee, G. Heydt, "An Interactive-Dynamic Mechanism Conceptualizing the Cost and Benefit of Electric Power Quality", *Science Direct-Electric Power Systems Research,* 2004, pp. 69-75.
[24] M. Al-Mutairi, "SEC New Hierarchy at Next Year", *Alsharg Alawsat News Paper* (www.asharqalawsat.com), November 02, 2002.
[25] M. Al-Mutairi, "Gulf Interconnecting Project is already a head", *Alsharg alawsat News Paper* (www.asharqalawsat.com), November 02, 2002.
[26] P. Navarro and M. Shames, "Electricity Deregulation: Lessons Learned from California", *Energy Law Jornal, Vol.* 24, Iss. 1, 2003, Washington, Pg 33.
[27] S. Chen, T. Zhan and M. Tsay, "Generation Expansion Planning of the Utility with Refined Immune Algorithm", *Science Direct-Electric Power Systems Research,* Vol. 76, Iss. 4, January 2006, pp. 251-258.
[28] B. Lu and M. Shahidehpour, "Unit Commitment with Flexible Generating Units", *IEEE Transaction on Power Systems,* Vol. 20, Iss. 2, May 2005.

[29] H. Bjornsson, R. Crow, and H. Huntington, *"International Comparisons of Electricity Restructuring: Consideration for Japan"*, Center for Integrated Facility Engineering, *Technical Report #150,* February 2004, Stanford.

[30] D. Lee, "Intermediary Report on Comparative Analysis of Electricity Reform in OECD Pacific Countries", IEA Report, 2004.

[31] A. Ritschel and Greg P. Smestad, "Energy Subsidies in California's Electricity Market Deregulation", *Energy Policy,* October 2002, pp. 01-13.

[32] M. Takahashi, "The Current Status of Electric Power Industry Deregulation in Japan and the Influence of the California Crisis", *Japan and the World Economy,* Vol. 14, Iss. 3, August 2002, pp. 341-345.

[33] L. Geerli, S. Niioka and R. Yokoyama, "Electricity Pricing and Load Dispatching in Deregulated Electricity Market", *International Journal of Electrical Power & Energy Systems,* March 2002, pp. 01-08.

[34] L. Geerli, L. Chen and R. Yokoyama, "Pricing and Operation in Deregulated Electricity Market by Non cooperative games", *Electric Power Systems Research,* Vol. 57, Iss. 2, March 2001, pp. 133-139.

[35] R. Schuler, "Analytic and Experimentally Derived Estimates of Market Power in Deregulated Electricity Systems: Policy Implications for the Management and Institutional Evaluation of the Industry", *Decision Support Systems,* Vol. 30, Iss. 3, January 2001, pp. 341-355.

[36] W. Hall II, "Securitization and Stranded Cost Recovery", *Energy Low Journal,* Vol. 25, Iss. 1, 2004, Washington, pp. 41-173.

[37] W. Blacconiere, M. Johnson and M. Johnson, "Market Valuation and Deregulation of Electric Utilities", *Journal of Accounting and Economics,* Vol. 29, Iss. 2, April 2000, pp. 231-260.

[38] C. Chompoo, C. Yingvivatanapong, P. Fuangfoo, and W. Lee, *"Transmission Congestion Management during Transition Period of Electricity Deregulation in Thailand",* Energy System Research Center, University of Texas, 2005, Arlington, pp. 2665-2671.

[39] A. Meliopoulos, G. Cokkinides and A. Bakirtzis, "Load-frequency Control Service in a Deregulated Environment", *Decision Support Systems,* Vol. 24, Iss. 3-4, January 1999, pp. 243-250.

[40] G. Doorman and B. Nygreen, "Market Price Calculations in Restructured Electricity Markets", Annals of Operations Research, Basel, *Vol. 124, Iss. 1,* November 2003, *Pg. 49.*

[41] S. Pickle, C. Marnay and F. Olken, "Information Systems Requirements for a Deregulated Electric Power Industry: Assessment and Perspectives from the 1996 Informatics Workshop", *Utilities Policy,* Vol. 6, Iss. 2, June 1997, pp. 163-176.

[42] Planning Department, "Annual Reports", "Official Releases", Saudi Electricity Company, Years 2000- 2005, Saudi Arabia.

[43] T. Al-Betairi, "SNG & IWPP's: Opportunities & Challenges", *Presentation,* AUPTDE, May 2005, Bahrain.

[44] A. Al-Swaha, "الشركة السعودية للكهرباء- آخر المستجدات والتوجهات" *Presentation,* SEC, March 2005, Saudi Arabia.

[45] I. El-Amin, "A Tutorial on Electricity Deregulation and its Application to Saudi Arabia", *Presentation,* KFUPM, May 2005, Saudi Arabia.

[46] T. A. Al-Muhawesh and I. S. Qamber, "The established Mega Watt Linear Programming-Based Optimal Power Flow model applied to the real power 56-bus system in eastern province of Saudi Arabia", *International Journal of Energy,* Vol. 33, No. 1, 2008, pp. 12-21.

[47] Tareq Al-Muhawesh and Isa Qamber, "The Prerequisite for Competition in Restructured Wholesale Saudi Electricity Market", *International Journal of Energy Policy, V*ol. 36 No. 1, 2008, pp. 477-484.

In: Energy Policy: Economic Effects, Security Aspects... ISBN 978-1-60692-894-3
Editor: Noah B. Jacobs

Chapter 3

RENEWABLE ENERGY IN A WARMING WORLD

Patrick Moriarty and Damon Honnery
Department of Mechanical and Aerospace Engineering, Monash University, Australia

ABSTRACT

Many researchers argue that renewable energy (RE) will eventually have to replace the present fossil fuel dominated global energy supply system, for two reasons. First, resource depletion will increasingly limit annual availability of oil, and later natural gas and even coal. Second, the European Union is calling for a 2 °C cap on allowable temperature increases compared with the pre-industrial era, in order to avert dangerous anthropogenic climate change. Keeping within this 2 °C limit will mean cutting present world emissions of greenhouse gases to as little as 15 % of their 2000 value. Some researchers also argue that other solutions, such as nuclear energy and carbon sequestration and storage, will merely solve some problems but introduce others. This leaves RE, which can be conveniently divided into non-intermittent and intermittent sources. The first category includes hydro, biomass, and geothermal. Using an earth system science approach, we analyse each of these in turn, and demonstrate that their likely future potential is small, because of limited availability, competition for scarce inputs (especially biomass), or because their potential will be adversely affected by unavoidable climate change. The second category includes solar and wind energy, each with much higher global technical potential, and the various forms of ocean energy. However, the technical potential for both wind and solar—particularly solar thermal electricity with its need for direct insolation—are unevenly distributed among the earth's regions. For this reason, large-scale very long-distance energy transmission will be increasingly needed as RE penetration rises. Further, these intermittent sources will require conversion of electricity to some other energy carrier, possibly hydrogen, and storage. The need for international transmission of energy, for example from the North African Sahara to the countries of Western Europe, raises security issues similar to those for oil. The energy losses and capital expenditure associated with energy conversion, storage and transmission will both raise costs of delivered energy and delay their implementation. Overall, energy security, together with economic and environmental cost considerations, point to a lower energy future than most energy analysts forecast.

GLOSSARY

ASPO:	Association for the Study of Peak Oil
CO_2:	carbon dioxide
EC:	European Commission
EGS:	enhanced geothermal systems
EJ:	exajoule (10^{18} joule)
EWEA:	European Wind Energy Association
GJ:	gigajoule (10^9 joule)
GW:	gigawatt (10^9 watt)
HDR:	hot dry rock
IEA:	International Energy Agency
IPCC:	Intergovernmental Panel on Climate Change
MW:	megawatt (10^6 watt)
MWh:	megawatt-hours
NPP:	Net Primary Production
OECD:	Organisation for Economic Cooperation and Development
OTEC:	Ocean Thermal Energy Conversion
PV:	photovoltaic
RE:	renewable energy
SPS:	satellite power systems
STEC:	solar thermal electricity conversion
THC:	thermo-haline current
WEC:	World Energy Council

INTRODUCTION

In 2006 world primary energy use was around 493 EJ, including non-commercial biomass [Moriarty and Honnery 2009]. For comparison, authoritative estimates in the literature for global primary energy use for various scenarios in 2050 are 800-940 EJ [European Commission (EC) 2006], 690-930 EJ [International Energy Agency (IEA) 2007a] and 798-1173 EJ [Riahi et al. 2007]. A number of other researchers have produced scenarios giving 1000 EJ or more primary energy demand in 2050 [Moriarty and Honnery 2009].

There are vast differences between countries in average energy use per capita, particularly for commercial energy. For electricity use, for example, three orders of magnitude separate Iceland and Norway (27 and 24.5 MWh per capita respectively) from a group of tropical African countries with around 0.01-0.05 MWh per capita [BP 2008]. Given this large inequality, and projected economic growth, the 1000 EJ estimates for 2050 primary energy demand seem reasonable for a ‘business-as usual’ world. Furthermore, primary energy needs could rise if, as expected, the energy requirements for a given level of food or fresh water provision increased, or if the ratio of final energy demand (secondary energy) to primary energy supply continues to rise, as it has in recent decades [IEA 2007b, Moriarty and Honnery 2009]. The number of people anticipated to experience water stress will rise in future [Parry et al. 2007], so energy needs for ocean desalination, inter-basin water transfer schemes, and deep-aquifer pumping could all rise. Similarly, extra energy will be needed for fertiliser to replenish eroded soils, to counter rising energy costs for delivered final energy and refined minerals, for carbon capture and sequestration—if attempted—and for electrolysis of water for hydrogen production. (On the other hand, with present energy accounting methods, primary energy needs will be lowered if primary electricity sources (hydro, wind, PV) dominate energy production.)

The IEA have also projected the proportion of total primary energy that would be supplied from renewable energy (RE) sources. Their reference scenario forecasts that RE will increase its share of overall primary energy from 12.7 % in 2004 to only 13.8 % in 2030. Their alternative scenario gives a slightly higher RE share (16.2 %), mainly the result of reduced coal consumption [IEA 2007b]. The EC study [2006] in its 'carbon constraint scenario', the most favourable for RE, projects the share of RE in global energy use to rise from 14.5 % in 2030 and 21 % by 2050. For electricity only, the IEA [2007a] gives a range from 17-38 % RE for 2050, compared with 18 % in 2004.

In contrast to the modest shares for RE in these forecasts for actual energy use, some recent studies see the global technical potential for RE in 2050 as enormous. Although an earlier study by Hafele [1981] gave RE technical potential as only about 470 EJ (in terms of displaced fossil fuels), Johansson et al. [2004] report an annual technical potential for RE as greater than 7500 EJ, with 5000 EJ for geothermal potential alone. A study by de Vries et al. [2007] gave even higher values. They totally excluded farmland (except for wind), forests, nature reservations and urban areas, together with most grasslands and even deserts, in calculating global potential for each RE source. Even so, the separate technical potentials in 2050 for solar-PV, wind, and biomass electricity production were 14,778, 220 and 212 EJ respectively. Particularly for solar PV and biomass, their values were several times the estimated year 2000 technical potential, because of assumed technical progress.

The world will soon need to shift from its present reliance on fossil fuels for two reasons. The first reason is global climate change. The European Union is calling for a 2 °C cap on allowable temperature increases compared with the pre-industrial era, in order to avert dangerous anthropogenic climate change. Keeping within this 2 °C limit will mean cutting present world emissions of greenhouse gases to as little as 15 % of their 2000 value by 2050 [Moriarty and Honnery 2008a, 2009]. Climatologist James Hansen [2008] has gone further, arguing that in order to avert catastrophic sea level rise, atmospheric CO_2 levels will need to be reduced from their present level of 385 ppm to 350 ppm. Second, it is becoming increasingly clear that annual growth in oil and even natural gas consumption cannot continue much longer [Leggett 2006, Association for the Study of Peak Oil and Gas (ASPO) 2008]. Some researchers believe that growth in coal will also not continue for long after oil's peak [Energy Watch Group 2007, Strahan 2007], necessitating a shift to non-fossil energy sources.

This chapter analyses the prospects for the various sources of RE—specifically hydro, biomass, geothermal, wind, solar and ocean energy—out to 2050, mainly in the light of the urgent need to cut greenhouse gas emissions. We use an earth system science approach to these global energy flows, and evaluate the consequences—for the environment, energy security, economies, and societies—of attempts to provide 1000 EJ of primary energy from these sources. ('Earth system science takes the major components of planet earth—the atmosphere, oceans, fresh water, rocks, soils and biosphere, and seeks to understand major patterns and processes in their dynamics' [Lawton 2001].) Our main findings are that RE sources that can deliver non-intermittent power—hydro, biomass and geothermal—will be able to supply at most 50-100 EJ in 2050. Only the intermittent energy sources—wind, solar and ocean energy—have the potential to together supply 1000 EJ annually. Attempting to do so will involve orders of magnitude scale-up from their present low output (together they are less than 1 EJ), and will require conversion and storage of energy to overcome their intermittent nature. Further, they face unknown costs, whether monetary or environmental. We conclude that major reductions in global primary energy use are desirable—and probably even necessary.

RENEWABLE ENERGY RESOURCE BASE AND PRESENT USE

Potential RE resources for earth can tap into three possible energy flows. By far the largest energy flow comes directly from the sun, with some 5.6 million EJ arriving at the upper atmosphere annually. About 30% of this insolation is reflected by the atmosphere and surface land and water; the remaining 70% (3.9 million EJ) is absorbed by land and oceans [Hafele 1981]. At the top of the atmosphere insolation intensity is 1367 W/m^2 [Bala et al. 2008]. A small proportion of this radiant energy is diverted to wind energy, mainly as high-speed 'jet stream' winds at some 12 km altitude. Most researchers estimate that about 2% of total insolation is converted to wind, or about 115,000 EJ [e.g. Hafele 1981, Hoogwijk et al. 2004]. The share over the inhabited continents (i.e. excluding Antarctica) we take as about 26% (their proportion of the earth's surface), or around 30,000 EJ.

The terrestrial Net Primary Production (NPP) of mainly natural vegetation can be used as a proxy for biomass theoretical potential. NPP measures the net conversion of atmospheric CO_2 by photosynthesis into plant biomass over a given time period. To obtain NPP, autotrophic (self) respiration by plants is subtracted from gross biomass production. Photosynthesis produces a total global NPP of roughly 3000 EJ. Total NPP on land is presently estimated by a number of researchers to be about 120 billion tonnes dry matter annually or around 63% of the global total [e.g. Haberl 2006, Running et al. 2004]. This corresponds to about 1900 EJ, or less than four times world primary energy use [BP 2008]. Other forms of indirect solar energy include the energy in ocean waves and currents, and hydro. Ocean waves/currents account for a further 630 EJ. Even though one-third of Earth's absorbed insolation presently drives the hydrological cycle, the total potential energy of all terrestrial run-off is only 130-160 EJ [Hafele 1981].

The second source of energy to the earth's surface is internal. Geothermal energy flows in small amounts from the Earth's interior as a result of heat generated during the Earth's formation; the temperature of the earth's outer core boundary is estimated at around 4000 °C. For the continents, this heat flow is supplemented by the heat from decaying radioactive isotopes (nearly all from U_{235}, U_{238}, Th_{232} and K_{40}) [Gupta and Roy 2007]. Nevertheless, the heat flow rate to the surface is less for the continents than for the oceans. For the continental crust, oceans and globe overall, Barbier [2002] gives 0.057, 0.099 and 0.082 W/m^2 respectively, some four orders of magnitude less than globally averaged solar energy. (Nordell [2003] has somewhat higher values: 0.065, 0.101, and 0.087 W/m^2 respectively.) Global geothermal energy flowing to the surface is thus about 1320-1400 EJ annually, with the inhabited continents receiving 240-275 EJ of this. This annual energy flow is gradually declining over the millenia as the earth's core cools and the radioactive elements in the crust decay.

The third source of earth's energy, tidal energy, results from the gravitational attraction of the moon, and to a lesser extent, the sun and the other planets. Of the 2500 GW total power input from the Moon-Earth system, nearly all (2400 GW or 76 EJ annually) is dissipated in the oceans, mainly at coastlines [Munk 1997]. Like geothermal energy, this energy input is gradually decreasing, and will eventually decline to zero.

Table 1 presents the total resource base for each of these renewable energy sources. The values refer to present energy flows. Since climate is already changing, and further change can be expected both because of committed temperature rises [Hansen 2008] and inertia in the fossil fuel-dominated energy supply system [Riahi et al. 2007], the partitioning of incoming solar energy between the various RE types shown in Table 1 will also change to some extent in future. (Note that energy sources like wind that are derived from solar energy should strictly be subtracted from the insolation figure to avoid double-counting. However they are all very small compared with insolation.) Further, as will be detailed below, only a small fraction of the terrestrial energy flows can usually be diverted for human energy use. The table shows that only three RE sources—direct solar, wind, and biomass—have annual terrestrial energy flows greater than the 1000 EJ level. The others, including hydro, geothermal, and the various individual forms of ocean energy (tidal, wave etc), each have annual energy flows even lower than present global use.

Table 1. Earth global and terrestrial energy flows (EJ/yr)

Energy type	Global energy flows	Terrestrial energy flows
Solar (direct)	3 900 000	780 000
Wind	115 000	30 000
Biomass	3000	1900
Ocean wave/current	630	NA
Hydro	130-160	130-160
Geothermal	1320-1400	240-275
Tidal	76	NA

NA: not applicable.
Source: Moriarty and Honnery 2005a, Hafele 1981.

With the exception of biomass and direct geothermal and solar heat energy, all these RE sources as presently used generate only electricity. Global biomass use, at perhaps 50 EJ for all uses, is by far the largest RE source, but since most is non-commercial fuelwood in low-income countries, the exact figure is uncertain. Table 2 shows that despite the declared support for RE, particularly in the Organisation for Economic Cooperation and Development (OECD) countries, its share of global electricity production is falling, as it has done for many decades. Although wind and solar energy (both photovoltaic (PV) cells and now solar thermal plants) are growing much faster than overall electricity growth, their output has been too small to offset the decline in hydro's share. Most of the growth in new RE sources in recent decades has been in OECD countries; conversely, nearly all hydro growth has come from industrialising countries. In brief, the experience of recent decades suggests that dramatic increases in RE share of global energy demand will be difficult to achieve.

Table 2. Global RE electric output, 1980-2005 (EJ)

RE source	1980	1990	2000	2005
Hydro	6.20	7.74	9.53	10.70
Wind	*	0.01	0.13	0.45
Biomass	0.06	0.32	0.60	0.70
Geothermal	0.05	0.13	0.18	0.20
Solar	*	*	*	0.01
Ocean	*	*	*	*
Total RE	6.31	8.20	10.44	12.06
% total elec	*23.7*	*21.6*	*21.3*	*20.0*

* less than 0.01 EJ

Source: BP 2008, IEA 2007b.

EVALUATING RENEWABLE ENERGY: ENERGY RATIOS

On what basis do we decide to use one energy source in preference to another? No single energy source today dominates worldwide, as was the case in the past for wood, then coal. True, oil still provides over 95 % of transport energy, but for electricity generation, coal, natural gas, hydro, nuclear and oil presently all have a greater than five percent share. In individual countries, however, a single energy source can dominate for electricity production, with hydro in a number of countries providing over 80 % [IEA 2007b]. Given that most countries regard energy security (and the local employment that use of local energy resources brings) as important, many rely on indigenous energy sources for electricity, even if imported electricity inputs would be cheaper. Cost is another obvious criterion for energy selection, but here we concentrate on energy return on energy invested.

Energy analysis of conventional and alternative fuels was popular in the 1970s and 1980s, but much less so today. Yet energy analysis is vital in assessing the prospects for energy sources. Too often, annual potentials for RE sources are given without any indication of their likely energy ratio (the ratio of gross energy output over the life of the energy generating system to total energy inputs, both usually measured in primary energy units). Energy ratios vary, not only for different existing or proposed RE systems, but also with total annual output of each RE source, since, in general, the most favourable sites/resources are used first for economic reasons.

Clearly, if the energy ratio is one or less, not only is no net energy generated, but also additional money and environmental costs are imposed by the need to construct and operate the energy conversion equipment. Some energy researchers have felt that ratios much larger than one are needed for a proposed energy source to be viable. Hall [2004] estimates that a minimum ratio of 5.0 may be needed. Here, the input energy costs include (as usual) an allowance for the indirect energy used, for example, to build PV cell fabrication plants. Dynamic energy analysis, which looks at the net annual energy delivered to the economy after subtracting the energy costs of energy infrastructure construction for that year, is another approach for estimating the minimum energy ratio. Weinberg [1979] has argued that for energy payback periods of >5 years, the required energy subsidy from the rest of the economic system will probably impose a constraint on the rate of introduction of the new energy system. For a proposed RE plant with a typical operating life of 25 years, the energy payback period will be 5 years if the energy ratio is 5.0, i.e. the two approaches of Hall and Weinberg give the same answer.

If this cut-off ratio of five is accepted, it should in principle be easy to assess the present and future potential for each proposed RE source. However, there are serious obstacles to performing such assessments. First, as will be shown below, intermittent sources of electrical energy, such as from wind turbines or solar PV plants, will eventually require conversion and storage in some other energy form. This could lead to a halving of the energy ratio [Moriarty and Honnery 2007a]. Second, published energy analyses for particular RE sources show large variation in resulting values [Weinberg 1979, Mortimer 1991, Moriarty and Honnery 2005b, 2007a]. Some of the variation is caused by incomplete assessment of input energy costs. However, Moriarty and Honnery [2007b] showed that for short-rotation coppiced energy plantations in NW Europe, the energy ratio for two studies varied from 14.4 to 64.8, even with the same set of input items. Indeed, order of magnitude differences in energy ratios for a given RE source are quite common [Lenzen 1999].

Increasingly, the environmental effects of different energy sources are seen as important, particularly air pollution and global climate change effects. In this chapter, although the effect on global climate change is the main criterion used for RE evaluation, any new energy source must also deliver net energy. If the energy inputs for a new RE source are largely from fossil fuels, as is the case today, an energy analysis will give similar results to a greenhouse gas analysis [Mason and Zweibel 2007]. However the results can diverge if (as is the case for hydro and even geothermal), greenhouse gas emissions can arise from operation of the plant. Further, climate change effects can also occur independently of greenhouse gas emissions, as may be the case if some biomass energy plantations directly affect local albedo. (The albedo measures the share of incoming radiation that is directly reflected.)

The following sections use the period from the present to 2050 as the relevant time period for evaluating RE potential, first, because a number of energy forecasts use the 2050 as their base year, and second, because climate scientists are increasingly stressing the need for major reductions in global emissions in a few decades, rather than by the end of the 21st century [e.g. Hansen et al. 2008].

BIOMASS ENERGY

Table 1 showed that terrestrial NPP is around 1900 EJ/year. This NPP has to support all the world's heterotrophic species, including humans with their huge requirements, both direct and indirect, for food, fodder and fibre from agricultural land, and timber and wood pulp from managed forests. In effect, we are already heavily exploiting the biomass energy flow for non-energy uses. The potential for land use conflict between these expanding existing uses for land and any increase in bioenergy production is evident. A recent World Bank report found that growth in biofuels was responsible for 75 % of recent food price surges [Ngo 2008]. Pimentel et al. [2007] point out that in 1950, an estimated 500 million people (20 % of world population) were malnourished; today 3.7 billion (nearly 60 % of world population) suffer some degree of malnutrition, which higher food prices can only worsen. Naylor et al. [2007] have detailed the mechanisms connecting biofuel expansion and food prices.

It might be thought that technology can now liberate us from constraints on NPP, with intensive agriculture producing higher annual biomass yields/hectare than natural systems. After all, high-input industrial agriculture produces much higher output/hectare than low-input traditional agriculture. However, a satellite-based study of NPP of all the earth's vegetated surface at 1.0 km^2 resolution shows that this is not so [Running et al. 2004]. Average biomass yield for cropland is similar to that for 'needleleaf forest', but far below those for 'savanna' or 'evergreen broadleaf forest' categories.

One reason, of course, is that natural systems do not need the network of roads or farm structures (small dams, buildings) necessary for agricultural production, which reduces output on a km^2 basis. Also, natural ecosystems mostly recycle their nutrients and carbon, whereas the whole point of agriculture and forestry is to remove crops and timber, and unavoidably, the plant nutrients and carbon they contain. What intensive agriculture has done is to increase the harvestable fraction of the above-ground biomass, particularly for cereals. Consistent with this argument, Haberl [2006] reports NPP before agriculture as higher than today. Further, future application of biotechnology to crops does not seem to offer much hope for substantial yield increases [Sinclair et al. 2004].

At present, worldwide growth in biomass energy is mainly for liquid transport fuels—ethanol from grain or sugarcane, and to a much smaller extent, biodiesel from vegetable oils. Much controversy exists in the US as to whether or not ethanol from corn gives a net return on energy invested. All agree that net output, if it exists, is small. Indeed, Scharlemann and Laurance [2008] report on recent research claiming that the environmental costs of all present liquid biofuels, including Brazilian sugarcane ethanol, are higher than for petroleum fuels. Cellulosic biomass is already used for electricity production, particularly in Finland and Sweden, where forestry wastes are used. However, some researchers argue that most or all wastes, whether forestry or agricultural, should be left in place to avoid erosion and loss of soil carbon [Patzek 2004, Field et al. 2008].

In general, zero or near-zero input energy costs only arise for feedstocks such as municipal garbage, sewage gas, sawmill wastes, and bagasse, resources that are often already fully used [Moriarty and Honnery 2007b]. For biomass plantations, including food crops such as cereals, sugar cane or oilseeds as well as short rotation coppice (SRC) plantations, the input energy for the cultivation, fertilisation, protection, harvesting, transport, storage and preparation of feedstock must be considered. Total energy inputs vary with soil fertility and rainfall. Nevertheless, our analysis of the literature shows that, excluding the need for irrigation or forced drying of crops, input energy costs for feedstocks—whether trees, grasses or food crops—lie mainly in the range of 0.5-1.0 GJ primary energy per tonne of above-ground oven dry biomass [Moriarty and Honnery 2005a, and references therein]. However, somewhat higher values are reported for studies based on commercially grown crops than for studies based on experimental plots or modelled energy crops. Oven-dry biomass in turn produces an output energy (lower heating value) of about 18 GJ/tonne [Moriarty and Honnery 2007b].

But if irrigation is needed, input energy costs can rise sharply. Already, about 40 % of global grain output is grown using irrigation, up from only about 5 % in 1900. Given the shortage of unused rain-fed land, this proportion can only rise, which suggests that if biomass is not to displace food production, water will be an increasing constraint [Moriarty and Honnery 2005a, 2007b]. In many important food-producing regions of the world, water tables are falling and 'fossil' water is being mined—as much as 200 km^3 annually [Brown 2008].

Recently, the authors published an analysis for energy crops which showed that for pumping energy costs alone, a pumping depth greater than 167 metres would not produce net energy; all the resulting biomass would be needed to fuel the power stations which provided input electrical energy for the pumps. On the other hand, in humid areas, irrigation is not needed, but forced drying may be, again with high energy costs [Moriarty and Honnery 2007b].

Even if biomass energy gave an adequate return on energy inputs, it may still have serious climate change effects. For most RE sources, GHG emissions are largely CO_2, and energy analysis and GHG analysis produce similar results. But for biomass, results may greatly diverge, for two reasons. First, bioenergy plantations may change soil carbon storage—either increasing or decreasing it, depending on previous land use. Converting forests to bioenergy plantations will lead to soil carbon losses; conversely, conversion from agricultural crops will usually increase it [Field et al. 2007]. An extreme case of soil carbon loss occurs on the island of Borneo, where peat bogs are being drained for palm oil plantations. Drainage causes rapid oxidation of the peat, resulting in high CO_2 emissions [Pearce 2007].

Second, biomass energy plantations can affect albedo, again with effects depending on the circumstances. In tropical regions, energy plantations will tend to increase albedo, which in turn promotes local cooling. But in high latitude regions, energy plantations may well decrease albedo by replacing reflective snow with more absorbtive vegetation, thus offsetting to some extent the benefits of a carbon-neutral energy source [Field et al. 2007, Bonan 2008]. Unfortunately, energy plantations are far more likely in high latitude regions than in the tropics, where net deforestation is occurring [Moriarty and Honnery 2007b].

Biomass is unique among RE sources in that a combustible solid is produced, rather than electricity or heat. It can therefore be readily stored and transported, and is a much better fit to our existing energy system. Genuine biomass wastes should therefore be regarded as a premium energy source. However, such wastes are of modest extent globally, and attempts to remedy this shortage by bioenergy plantations will either give low energy returns on invested energy on marginal lands—or displace crops for food on good farmland and raise food prices. Indeed, Field and colleagues [2008] estimate that only about 27 EJ of biomass energy could be harvested annually without threatening food supplies or worsening climate change. This is less than today's use, because much of the approximately 50 EJ of present biomass energy use, largely fuelwood in industrialising countries, is being produced unsustainably.

HYDROPOWER

Hydroelectricity in 2005 provided about 10.7 EJ (Table 2), or 17.1 % of global electricity production, but its share has steadily fallen over the past century. Hydroelectricity is desirable in that it can be very quickly brought on-line, and so used for both base- and peak-load power. Although at good sites its electricity costs are competitive with fossil fuels, the EIA [2007a] forecast hydro providing only about 20 EJ in 2050, even in their most favourable scenario. The total technical potential for large hydro schemes is only about 50 EJ/year, about one third of the resource base [World Energy Council (WEC) 2007]. Unusually for an RE source, there is little disagreement in the literature on its technical potential. Much of the total gravitational energy can't be tapped because of frictional losses from downstream water flow, and because a minimum flow rate is needed if rivers are to deliver their various environmental services [Hafele 1981]. Further, much of the gravitational energy is too diffuse to be collected without excessive energy expenditure.

Large hydro dams have other problems. In the devastating May 2008 earthquake in Sichuan Province, China, hundreds of dam structures were weakened, including some large hydro dams, although fortunately at that time of the year, water levels are low [Pearce 2008]. It is now becoming clear that dams should be built neither in earthquake-prone areas, nor mountainous areas with slope stability problems. Large dams in turn can cause micro-seismic tremors. They can also lead to major population displacement, adverse impacts on riverine ecosystems and fisheries decline [Sims et al. 2007]. Indeed, Abbasi and Abbasi [2000] stress than one reason why we now know so much about the environmental and social problems of large hydro schemes is the wisdom of hindsight—we have built many such schemes.

Past river flow rates and especially seasonal flow rates can no longer be assumed to be stable over the long life of hydro schemes. As Milly el al. [2008] put it in the context of water management, 'stationarity is dead'. There is evidence that annual river flow rates are already changing, with increases for rivers flowing into the Arctic Sea, for example, but decreases in some other regions [Parry et al. 2007]. On the other hand, any further temperature rise means an increasing share of total precipitation falling as rain rather than snow, as well as earlier melting of snow, leading to river flows much more temporally skewed than formerly. Extreme rainfall events are also expected to be more common, increasing soil erosion and thus dam sedimentation [Prichard 2002, Thacker 2004]. Unlike regional precipitation, increased regional temperatures are predicted with high confidence [Barrett et al. 2005] and higher temperatures will increase potential evaporation from soils and dam reservoirs, lowering hydro output.

All these changes—as well as land-use changes unrelated to climate change—can have significant implications for the economic viability of hydro power. Harrison et al. [2006] point out that for existing plants operation will become non-optimal under altered flow conditions. For a proposed central African project, they found that the modelled financial performance of the scheme (internal rate of return, for example) was very sensitive to changes in catchment rainfall patterns and temperature. They conclude that because of hydropower's reliance on climatic conditions, any changes can adversely affect hydro project financial viability. Similarly, Lehner et al [2005] have shown that projected decreased precipitation for southern Europe will entail disproportionately larger decreases in annual hydro production.

But just as ongoing climate change will affect the output and cost of both existing and proposed hydro schemes, so can hydro schemes affect climate. They do this by their emissions from the impounded water, particularly in tropical regions, where most of the remaining hydro potential is located [Moriarty and Honnery 2005b]. If tropical forests are inundated, the decaying vegetation produces CO_2, and methane if the decay is anaerobic. The magnitude of the climate change effects of hydro dams are controversial [Giles 2006, Cullenward and Victor 2006], but there is agreement that at least in the early years after reservoir filling, CO_2-equivalent emissions can exceed those from a gas-fired power station of similar output.

Baines [1987] showed that for New Zealand, the energy ratio for hydro schemes rose steadily until 1980, reaching a value of 18.8, but was then projected to fall to 16.8 by 1995 as opportunities for good sites became limited. Presumably hydro projects in other countries would show a similar pattern. Even an energy ratio of 16.8 is quite satisfactory, since no conversion and storage of energy in another form is needed. However, as the negative impacts of hydro schemes are increasingly recognised, the remedial measures needed will lower the calculated energy ratio.

Geothermal Energy

Geothermal energy is unusual in that although the flow to the land surface is very small compared to insolation (Table 1), there are local hot spots where heat energy can be 'mined' in a manner similar to fossil fuel energy. These localised heat sources can therefore be depleted in much the same way as fossil fuels, and are therefore not strictly renewable. But geothermal heat can provide power on a continuous basis over the plant's life, and like tidal energy, its potential can only be indirectly affected by climate change.

Geothermal globally provided only about 0.2 EJ primary electricity in 2005 (Table 2). Bertani [2003] compared a number of estimates for electricity technical potential from geothermal energy, reporting a range from 1.2 to 151 EJ/year. For conventional hydrothermal systems, but with improved technology such as permeability enhancement, another study, using a country-by-country analysis, found that globally, geothermal systems could only sustain the generation of 3.9 EJ of electricity annually [Gawell 1999], although the potential for low temperature direct heat is much higher. This modest value is consistent with a more recent IEA assessment [IEA 2007a], which states that 'world potential capacity for geothermal power generation is estimated at 85 GW over the next 30 years.' With the year 2005 capacity factor of 75 %, annual production would thus only be 2.0 EJ. Even this estimate may prove optimistic: installed capacity only grew by 12 % between 2000 and 2005 [Bertani 2005].

These low estimates for geothermal electricity potential are supported by the recent experience of OECD countries. Unlike the situation in many industrialising countries, OECD geothermal electricity production is constrained neither by lack of relevant skills nor investment capital, and at suitable locations geothermal electric power can be delivered at competive cost [IEA 2007a]. Particularly for dry steam plants, high energy ratios are also possible [Mortimer 1991]. Nevertheless, several leading OECD producers (US, NZ, Japan) have experienced absolute declines in geothermal electricity output in recent years, as has the OECD overall [Moriarty and Honnery 2005b]. In the US, for example, net electrical output fell from 16.8 TWh in 1993 to 15.1 TWh in 2005.

An earlier review of hot dry rock systems (HDR) in the journal Geothermics [Abe et al. 1999] pointed out the likely difficulties of this untried energy source. First, the authors stress that there will be no 'typical' HDR site; each potential site will need to be thoroughly explored to understand the local geology. Second, power output varies with both temperature of the water when brought to the surface, and its mass flow rate. At low water temperatures, the pumping power required to circulate the water through the target rock mass will consume a larger fraction of the output power. Lower temperatures will also mean lower electrical conversion efficiency. If, to raise flow rate, pumping pressures are increased too much, there is risk of induced microseismicity and increased water consumption as water is lost. Pipe friction power losses are also a function of depth.

A study by Massachusetts Institute of Technology (MIT) researchers found that the resource base in the US alone for HDR, (or enhanced geothermal systems (EGS) as they call it) is in the millions of EJ, orders of magnitude higher than estimates of global fossil fuel reserves. But the vital question, as for all RE, is whether net energy can be obtained from this resource. To our knowledge, no detailed energy analysis has ever been done for EGS. Such studies are urgently needed to evaluate what fraction of the resource can ever be exploited. Given that for the US, all the source rock at 200 °C or greater lies below four km depth [MIT 2006], we suspect that only modest additions to geothermal electric potential can be expected from EGS/HDR, mainly adjacent to existing hydrothermal systems.

For direct use of geothermal heat, the review by Bertani [2003] also gave a wide range of estimates—from 2.4 to 600,000 EJ/year. Because it is not technically feasible to distribute low-grade heat over more than about 8 km, only a small fraction of the resource can be used for applications such as domestic heating or hot water. For the US, a study by Lienau and Ross [1996] found that for 10 western states, only 7.4 million people (13.4 % of their population) lived in areas that were within 8 km of a geothermal resource with a temperature of at least 50 °C. Since geothermal energy in the US is heavily concentrated in these states, only about 2.5 % of total US population would have access. In summary, low temperature geothermal energy is unevenly distributed both within and among the nations of the world, so that the total resource figure is of little value: most communities will have little or none, a few will have more than they can use, but cannot export to other areas.

Geothermal has some adverse environmental effects, and its operation can even produce CO_2 in excess of that released annually before the geothermal plant was built. According to an Icelandic study [Kristmannsdottir and Armannsson 2003], geothermal electrical energy can release up to 400 gm of CO_2-e per kWh, compared with around 850-1300 gm/kWh for a coal-fired power plant. However, most of this CO_2 is from the geothermal fluids, and would eventually be released naturally. Not only GHG emissions, but other pollutants vary greatly from field to field, depending on their geological and hydrological properties. Land subsidence and micro-seismic events can also occur.

Wind Energy

Wind energy worldwide is growing rapidly, with installed capacity averaging over 27 % annual growth in the 2000-2007 period, but in 2005 it still only provided 0.45 EJ (Table 2), less than 0.1 % of global primary energy use. Although global growth is still strong, in the European Union (EU) countries—formerly leaders in wind implementation—growth has decreased in recent years, with only 11 % growth in installed capacity in 2007 [Global Wind Energy Council 2008]. Denmark, the country with the highest wind penetration, has experienced almost no growth since 2003. This slowdown has been compensated to some extent by rapid growth in recent years in other formerly lagging EU countries, including France, Italy, and the UK.

The European Wind Energy Association (EWEA) [2006a] reported that global wind electricity could grow to 3.2 EJ by 2030, and to 5.5 EJ by 2050 in their base case scenario. Their most optimistic scenario envisages 18.6 EJ in 2030 and 28.5 EJ in 2050. The latter figures would require an annual 20 % growth rate to continue until 2010, then falling thereafter, with only 1.0 % annual growth 2040-2050 [EWEA 2006a]. Mainly because of anticipated increasing hub heights, capacity factors are assumed to rise from 24 % in 2005 to 30 % in all scenarios, despite their fall in recent years [WEC 2007]. The upper value of 28.5 EJ for 2050 is similar to the EC report highest figure of 26.5 EJ by the same year [EC 2006]. Clearly, these projections fall a long way short of the 1000 EJ projection from all RE sources.

These projections for wind energy production must be compared with potential production. Evaluation of the global potential for wind energy is much more advanced than for other new RE sources. For example, significant research has been done on criteria for exclusion of areas from wind development, whereas for solar and even hydro, no such criteria are usually applied. Here, the global potential for wind energy and its distribution is analysed using a standard turbine with properties shown in Table 3, based on information in Vestas Wind Systems [2005] and Lenzen and Munksgaard [2002]. Figure 1 is a global map of energy ratios that would be obtained by placement of a 2 MW turbine on each square kilometre of land [Honnery and Moriarty 2009]. The potential is clearly very unevenly distributed globally, with higher class winds (for example, classes 6 and 7) even more so. Canada, US, Russia, north-west Europe and southern and eastern South America are well-favoured, but the world's tropical and sub-tropical regions usually have low wind potential, a result also found by Hoogwijk et al. [2004]. There is thus a global imbalance between regions giving good energy returns on invested energy and the most densely populated regions such as Asia.

Figure 2 shows the potential cumulative global annual electricity output from land-based wind turbines plotted against energy ratio. The upper curve assumes that there are no constraints at all on location of turbines, while in the lower curve, a set of constraints similar to those of de Vries et al [2007] is applied. These constraints totally exclude all forests, nature reserves and urban areas, and assume that only 60 % of agricultural land can be utilised, for example. Total potential output for an energy ratio >1.0 in the unconstrained case is 534 EJ, comparable with present primary energy use. In the constrained case, potential output falls to only 103 EJ. Were the data available, similar graphs could be drawn for other RE sources.

Table 3. Turbine parameters

Rated power	2 MW
Diameter	80 m
Hub height	70 m
Cut-in wind speed	4 m s^{-1}
Rated wind speed	15 m s^{-1}
Cut-out wind speed	25 m s^{-1}
Turbine density	1.0 km^{-2}
Energy input	720 GJ

Source: Honnery and Moriarty 2009.

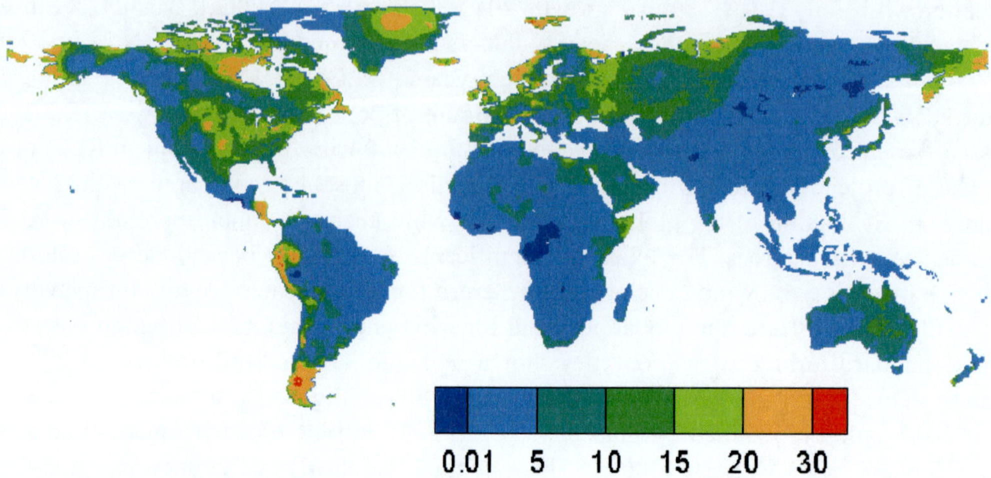

Figure 1. Global distribution of energy ratio for an infinite array of 2 MW wind turbines, Table 3.

Germany is the world leader in wind energy output, with an end-2005 installed capacity of 18,400 MW. Installed capacity grew by about 12 % in 2005, but output rose by only 1.5 %, because of below-average wind speeds. The electricity grid operator E.ON Netz controlled 7600 MW, or 41.3 % of total German capacity [E.ON Netz 2006]. Its experience with wind integration can thus provide important clues to a future with wind providing a major share of global energy. In 2005, the average wind in-feed was 1300 MW, with an average capacity factor of 18 %. But the instantaneous in-feed varied from 6300 MW (an instantaneous capacity factor of over 82 %) down to just 8 MW, or a little over 0.1 %. Also, the output in January 2005 was four times that of September 2005, despite a slightly lower installed capacity.

It is easy to show, using the E.ON 2005 data, that when the installed rated wind power is 122 % (1/0.82) of average power demand the maximum instantaneous in-feed will equal this average demand. Higher values of installed wind power will either require ever-increasing discarding of wind electricity or expensive conversion and storage, unless electricity demand can be modified. Yet at 122 % of average demand, wind would be providing little more than 20 % of total electricity, with a capacity factor of 18 %. For regions with higher capacity factors, for example 30 % as in the UK, the maximum infeed will also be higher, as much as 99 % of the installed capacity [Oswald 2006]. Without storage, discarding would occur at an installed rated wind power of only 101 % of average power demand, but wind would then be providing 30 % of total power. Drake and Hubacek [2007] show that for the UK, having four widely dispersed wind farms would reduce wind power unreliability by 36 % compared with a single wind farm, but at some increase in transmission energy losses. Energy conversion and storage can entirely eliminate the need for discarding wind electricity, but will both increase wind power costs and lower its energy ratio [Moriarty and Honnery 2007a].

Even though wind power capacity world wide in 2007 was only 94 GW, public opposition has been significant in many countries, leading to the cancellation of some projects. Local opposition for expansion has been particularly strong in North America, Australia and some European countries. Reasons for the opposition include local concerns effects such as visual intrusion, noise and radio and television interference, but effects on wildlife, especially birds and bats, have also been of concern. However, the number of bird kills annually is tiny compared to the other hazards for birds, such as vehicles, buildings, domestic pets and hunting [EWEA 2006b, Moriarty and Honnery 2007a]. Wind energy will probably need to expand beyond the constrained regions in Figure 2 if all RE sources are to deliver 1000 EJ of primary energy by 2050. The problem is that we have little idea whether local residential opposition will fade or intensify with such a large scale-up. On the one hand, visual obtrusion from access roads, turbines and transmission links will be far more prevalent, and effects on wildlife (and, with it, public opposition) may show non-linear increases with installed capacity, as opportunities for wind farms to avoid more sensitive areas decrease. On the other hand, the public might accept wind energy growth as necessary to combat global warming and fossil fuel depletion.

Opposition from great expansion of wind farms can also be expected from existing wind farm owners. As more and more turbines are installed, interference effects between wind farms will become more common. Already, a court case in Germany concerns owners of an established wind farm suing the owners of one planned nearby for potentially reducing its turbines' output by 15 %. Adjacent wind farm can also pass on more irregular flows, increasing turbine rotor vibrations [IndWindActionGroup 2007]. Such legal challenges, which could become more common as installed capacity rises greatly, can be expected to slow the rate of wind penetration.

Placing wind turbines off-shore can sometimes (but not always) lessen environmental opposition. Off-shore winds tend to be stronger and steadier than on-shore winds. However construction costs are much higher, particularly as sea depths increase. Operation and maintenance costs are also much higher than for land-based turbines. Placement is limited to shallow seas (such as the North Sea) and not too far from shore. The global potential for off-shore wind has been estimated as more than 3.6 EJ for sea depths of 15 metres or less [Pelc and Fujita 2002], and as high as 133 EJ for depths up to 50 metres [Hoogwijk et al. 2004]. At the end of 2005, 679 MW of off-shore wind capacity had been installed—only about 1% of the world total [Sims et al. 2007].

One high-tech solution proposed for dealing with intermittency (and ground-based environmental conflicts) is to place turbines in the jet stream, high above the earth's surface. Roberts et al. [2007] report on a plan to tether turbines 4.5 km from the surface, feeding the electric power developed back in an aluminium conducting core of the tether cable. The main advantages claimed are the much higher wind speeds and their relative constancy compared with near-surface locations. The tethered turbines could also be shifted to new locations. However, as with surface winds, the low latitude regions (roughly 30 °S to 30 °N) would not benefit from the jet stream—only higher latitude countries would, which as just shown, usually have good surface winds. Further, the jet stream is also intermittent, with speeds varying throughout the year, being very low in the northern hemisphere in the June-August period. At five km height, the wind power is only 10 W/m^2 at 30 °N, rising to 200-500 W/m^2 at 50 °N. Again at 50 °N, the turbines would need to be positioned at 10 km height to enjoy 1000 W/m^2 winds during these months [Sky WindPower Corporation n.d.].

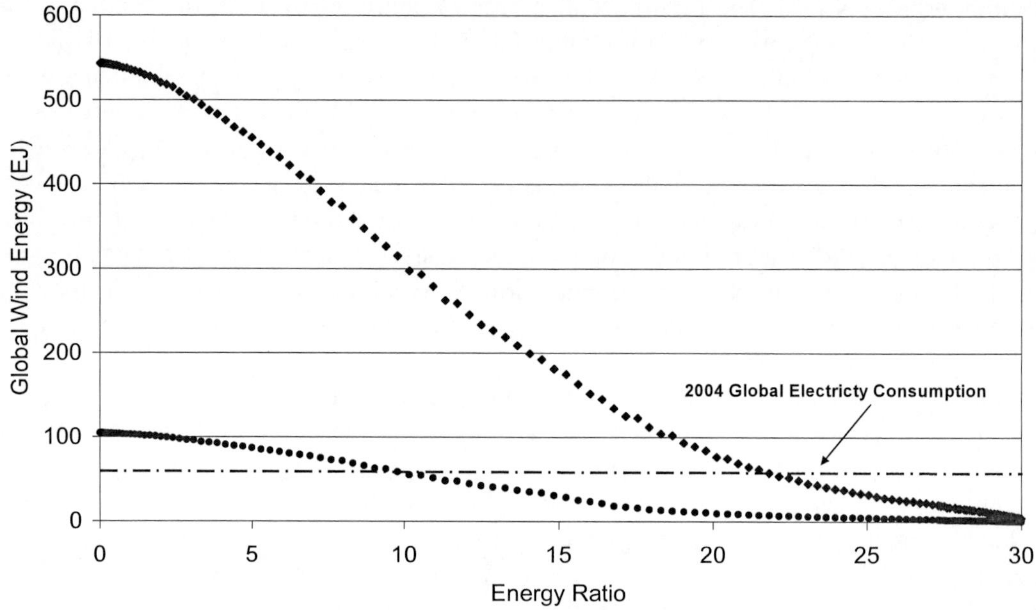

Figure 2. Global wind potential (EJ/year) available from all locations with energy ratios greater than or equal to the energy ratio values shown. ◆ Unconstrained case; ● Constrained case (based on de Vries et al., 2007). Also shown is the 2004 global electricity consumption.

Implementation of wind energy on a large scale appears to have negligible adverse effects on global climate [Keith et al. 2004]. A few studies have used General Circulation Models to show that wind energy will in turn be little affected by on-going climate change [e.g. Breslow and Sailor 2002, Pryor 2006]. However, most models do not represent the stratosphere well; those that do conclude that westerly winds would weaken outside the tropics in winter and spring, with evident effects on the extent and location of favourable surface winds [Baldwin et al. 2007]. Since turbines have a design life of up to 30 years, actual wind conditions could differ significantly from design values over the turbine's working life.

Solar Energy

Solar energy, along with wind, are the only two global energy flows which are many times present energy use (Table 1). An important advantage of solar energy over wind energy in that, although intermittent, it is more predictable. Given that there is no night-time insolation, an obvious response if solar energy use becomes more important in future is to reduce night-time electrical demand by increasing rather than decreasing costs of night-time power use. Solar PV sales are expanding even faster than for wind turbines, but total solar electricity production in 2005 was a negligible 0.01 EJ, nearly all of it from PV cells.

A number of solar thermal electricity conversion (STEC) plants were built near Barstow, California in the 1980s and are still operating, but it is only recently that new plants are being built in the southern areas of the US and Spain. The key advantage of STEC compared with PV systems is that thermal energy storage is possible, since the sun's energy is collected as heat rather than electricity, and heat can be stored in systems such as nitrate salt mixtures [Blankinship 2006, Shinnar and Citro 2007].

Although PV cells can use diffuse radiation, solar thermal technologies such as parabolic troughs which focus sunlight need direct radiation, and thus the clear skies found in arid or semi-arid climates [IEA 2007a]. This limitation may become even more critical in future since cloud cover is on average expected to increase in a warmer world. Both PV and especially solar thermal plant effectiveness will also be reduced in regions with significant atmospheric pollution, such as large regions of Asia, where sulphate aerosols and dust are important [Solomon et al. 2007]. Since biomass burning also causes air pollution, much would remain even if fossil fuel use was curbed. In short, local insolation values today in many regions may change in the coming decades. Dust deposition on PV cells or solar thermal collectors will also degrade their output [El-Shobokshy and Hussein 1993, Trainer 2007]. Fathi and Salem [2007] have also pointed out the deterioration in PV cell performance at the high temperatures encountered in interior North Africa. Also, limited reserves of gallium and indium could constrain production of high-efficiency PV cells using these materials [Cohen 2007].

For both PV and solar thermal systems, researchers have sought ways of overcoming the lack of insolation in the high latitude regions—which usually also have high-energy use. European researchers in particular have looked to the deserts of North Africa for locating huge solar farms to supply Europe with energy [Czisch et al. 2003, Czisch 2004]. The electric power would be sent to Europe using high voltage DC transmission. In addition to AC/DC conversion losses, power resistance losses are estimated, perhaps optimistically, as about 4 % per 1000 km. The roughly 5000 km distance to central Europe from North Africa could still make economic sense, given the latter region's much higher insolation—and presently lower labour costs—but it is important also to check that such ambitious projects give an adequate energy return on energy invested.

As with other RE sources, there is a large variation in reported energy payback periods and energy ratios for solar energy, with much of this resulting from insolation variations in different regions [Quashning 2004, Moriarty and Honnery 2005b]. The energy ratio for STEC systems are consistently reported as much higher than for equivalent PV systems [Mortimer 1991, Pehnt 2006]. Electricity costs are also lower [Trieb et al. 1997, Shinnar and Citro 2007]. The advantages of STEC over PV systems (except for off-grid applications), both from a cost and energy return viewpoint, would be even greater if energy conversion and storage were needed.

Even transmission of energy from North Africa to the EU would not avoid the seasonal variation in insolation, which is a problem since peak energy demand in Europe is in the winter months. Typically, at latitudes 35° N and S, insolation in winter months is about half that for summer months. An even more ambitious proposal thus envisages a global network of seven huge solar energy plants, placed in the desert areas both north and south of the equator, to avoid the need for much storage [Seboldt 2004]. They would be linked to load centres by high-voltage transmission lines or undersea cables. To supply the world in 2020 with 2630 GW (82.9 EJ/year) of continuous solar electric power, Seboldt estimates a total cost of US$ 198 trillion, with nearly two-thirds of this massive cost for transmission lines. Even this huge monetary outlay would only provide electric power, not the rest of the world's energy needs.

Seasonal and daily insolation variation can also be solved in principle by locating solar energy PV arrays in space. Unlike surface solar energy, all countries potentially have equal access to space solar energy. Satellite power systems (SPS) were first proposed in the US in 1968, and the idea has recently been revived by several researchers [Hempsell 2006, Seboldt 2004, Sasaki 2006]. However, apart from basic feasibility and very high energy and money costs, the proposals face other serious problems. Ruppe [2003] stresses that four processes are involved (solar to PV electricity, electricity to microwave energy, transmission of microwave energy to earth receiving station, reconversion to electricity) with energy losses and money costs at each stage. Solar satellite lifetime in a space environment, and reliability and maintenance, will be added problems. The costs for the electricity he estimates as anywhere from US$ 1.0 to 20 per kWh, 1-2 orders of magnitude higher than present electricity prices.

Shapiro [2002] points out that safety concerns will limit siting possibilities for microwave ground receiving stations. Not only is the area required similar to that for conventional solar systems of the same power output, but a 'safety zone' would be needed around the receiving area because of radiation concerns. With SPS, solar energy could rapidly lose its present widespread public acceptance. In summary, the remaining technical challenges, the high costs of the electricity produced, and possibly, public opposition, all make SPS irrelevant over the next half century at least, and possibly forever.

In contrast to these high-tech proposals, passive solar energy has been used for millenia for heat and light, but is not included in official statistics for primary or secondary energy. There is large potential for using passive solar energy to reduce energy needs for both building lighting and climate control. Levine et al. [2007] estimate that, globally, residential and commercial buildings in 2004 were responsible for 8.6 billion tonne CO_2, or nearly a quarter of global CO_2 emissions. In the US in 2005, and China in 2000, well over half of all energy used in buildings was for space heating and cooling, water heating and lighting. In China, Lam et al. [2008] have modelled commercial building energy for different climate regions, and have found that passive solar design can significantly reduce building energy needs. In Harbin, in the north, a reduction of a third in the heating load alone was found possible.

Ocean Energy

The only form of ocean energy that is presently deployed is tidal energy, and nearly all is from the Rance estuary installation in France. Tidal energy is only intermittently available, but at predictable times. Total electrical output is presently only 0.003 EJ annually. Based on an analysis of suitable locations for all continents, the WEC estimate the technical potential for tidal energy as about 1.4 EJ annual electrical output [WEC 2001]. Pelc and Fujita [2002] give a higher value of 1.8-3.6 EJ, but note that the economic potential is much less. The IEA [2007a] forecast negligible output of ocean energy of any type in 2050, even in their scenario projecting optimistic technical advances for RE.

A number of small wave generators are operating, but total annual electricity output is negligible. Over a thousand competing designs are available for this intermittent power source, although no design is clearly superior [Pelc and Fujita 2002]. The key problem for wave energy is that systems must be designed to operate at low wave heights, but must also withstand intense storms without being damaged or breaking loose from their moorings. The conceptual wave energy schemes analysed in Mortimer [1991] suggest very low values of energy return on energy investment.

Ocean Thermal Energy Conversion (OTEC) generates power by exploiting the temperature difference between warm tropical ocean surface waters and cold deep ocean currents that originate from the polar regions. The roughly 20 °C temperature difference is maintained by the thermo-haline currents (THC) which operate in the world's oceans. In this sense, OTEC works in a 'reverse' way to geothermal, where temperature increases with depth. No commercial plants presently operate, although small prototypes have operated intermittently. Small land-based OTEC plants could produce a steady year-round power output, and the high costs could be partly offset by joint production of desalinated water.

However, if more than a tiny fraction of OTEC potential was to be realised, the plants would have to move to the open tropical oceans. One plan is for OTEC ships to move continually and 'graze' the oceans [Avery and Berl 1997/8]. Since they would be far from land, they could no longer economically produce by-product desalinated water, and being mobile, the electric power could no longer be sent ashore by cable, as with offshore wind farms. Instead the electricity would all have to be converted to hydrogen or ammonia, stored on-board, then periodically shipped back to shore.

Costs could be very high, as the thermodynamic efficiency is very low, of the order of 3 per cent [Nihous 2005]. But there are doubts as to whether OTEC gives a net energy return, since construction energy, operation and maintenance energy (for water pumping and movement of the plant ship with its 1000 metre long cold water pipe), and energy losses in conversion and transport of the produced energy will all be high. Although large plants (200-400 MW) in general give better economies than small ones (10-50 MW), large plants lose the advantage of possible shore-based operation, and continuous energy production. Very large-scale production of energy would thus convert OTEC to an intermittent energy source, and raise its energy payback period.

Estimates of OTEC's technical potential range as high as 150,000 GW (4730 EJ/year) of electricity. Nihous [2005] argues that this figure is far high, and calculates that at most 2700 GW (85 EJ/year) of electricity could be extracted without the huge flow rates of seawater required perturbing 'the thermal structure of the ocean'. Any weakening of the THC brought about by global climate change would lower this value [Nihous 2007]. Abbasi and Abbasi [2000] also point the potentially damaging environmental consequences of the pumping and entraining of the enormous quantities of sea water needed if OTEC is to be a major future energy source. Overall, although some tidal current schemes may be built, high costs and long energy payback periods mean that ocean energy is likely at best to provide no more than a few EJ by 2050.

Implications for Energy Use

The uneven distribution of RE has important consequences for energy security. Although fossil fuels, particulaly oil, are widely traded internationally (in 2006, 66.2 % of oil production crossed international boundaries [BP 2008]), such is not the case for electricity. In 2005 only 3.4 % of electricity did so, and for most importing countries, net imports were very small [IEA 2007b]. Most trade is similar to the Canadian-US case, with neighbouring countries both importing and exporting electricity along their common border. There could be immense political problems in whole regions (such as Western Europe) being totally dependent on electric power imported from thousands of kilometres away in North Africa. Security of energy supply could be even more serious than for today's oil, since unlike coal, petroleum or even nuclear fuels, electricity cannot be stockpiled. Even after conversion to hydrogen, storing the very large volumes needed to improve security is impracticable.

With the shift to RE, real energy costs can be expected to rise, for several reasons. First fossil fuel energy is still in general cheaper than RE if the costs of GHG emissions are ignored, even with the recent fossil fuel cost increases. Second, as we have shown above, if the world needs 1000 EJ primary energy from RE, nearly all will need to come from intermittent sources, mainly wind and solar. As these sources increase their share, they will increasingly need to be converted from electricity to another energy form (probably hydrogen) and stored. Depending on the final energy mix demanded, reconversion back to electricity may also be needed. It is possible that only 50 % of the primary electricity produced from wind, for example, will be delivered to consumers, in which case the cost of the electricity will double, even if money costs for electrolysis of water and storage of hydrogen were ignored. The latter costs are large: Shinnar and Citro (2007) state that direct use of electricity is three times cheaper than hydrogen from electrolysis. Bossel [2006] demonstrates the high energy costs involved in producing, storing and transporting hydrogen—and the large quantities of de-mineralised water needed for electrolysis. Hultman and Koomey [2007] make the important point, based on a retrospective analysis of new energy sources, that actual costs and times needed for implementation of new energy technologies are always far greater than originally estimated.

A third reason for future real cost rises for energy is the need to move to progressively lower-quality RE sources as annual outputs rise. We have already shown how the energy ratio for wind falls as output rises. Figure 2 shows that for an energy ratio of 20, annual output will only be a quarter of that at an energy ratio of 10. Such a drop might not matter if potential output far exceeded any possible human demand, but, as we have seen, the maximum possible wind output is only 534 EJ, and feasible location and cost constraints might bring this down to only 103 EJ, with perhaps only half these values available to consumers.

Similar declines in energy ratio with output would apply to biomass, hydro, geothermal and ocean power. Also, hydro and biomass output will probably fall with on-going climate change, and, further, have adverse climate change effects in operation. It follows that solar energy will have to provide nearly all the 1000 EJ that a business-as-usual world economy might require in 2050. Although the potential for solar energy far exceeds that of other RE sources, and is several orders of magnitude larger than any conceivable global energy demand, in general it appears to be a low-quality energy source compared with the others. Trainer [2007] produced evidence to show that solar energy electricity production could be at least 10 times as costly as coal-fired thermal electricity. Such a high cost suggests that energy payback periods could also be high. Mason [2007], on the other hand gives a far more optimistic picture of PV systems, with an energy ratio of nearly 10, even with conversion of PV electricity to hydrogen. Such differences are common between results based on experience with existing systems (Trainer) and on future modelled systems (Mason).

The above arguments suggest that provision of large quantities of environmentally sustainable RE will be an expensive process which will take many decades to implement, particularly for grandiose schemes involving inter-continental transfers of hundreds of EJ of energy. Some of these proposals may also give a low energy return on input energy. For many of the proposals, the monetary and environmental costs are far from fully known, and rely on unproven technology. Hence an obvious response is to abandon the target of 1000 EJ primary energy and implement energy conservation on a large scale. Energy reductions can occur in two ways, either by improving the efficiency of energy-using equipment such as power stations or transport vehicles, or by reducing use (and probably numbers) of energy-using devices.

Energy analyist Amory Lovins has argued that there is a large potential for energy efficiency improvements in all sectors, particularly buildings, industry, and transport vehicles [Lovins 2005, Weizacker et al. 1997]. However, the global economic system is far better at delivering energy efficiency than overall energy reductions. Thus recent decades have shown an overall rise in economy-wide energy efficiency, as measured by primary energy use per unit of real global Gross National Income [World Bank 2008], but at the same time, a steady rise in global primary energy use. As shown in the Introduction, the world has very large inequalities in energy use, even when measured on a national basis. Given this inequality, and continued income growth even in the wealthy countries (as assumed, for example, in all IPCC scenarios [Barker et al. 2007], absolute energy reductions through energy efficiency improvements alone will likely improve impossible.

So, if we do have to drastically cut global energy use, we may also need to the abandon the primary aim of continued economic growth, starting with the already-wealthy OECD countries. If we do not do so, any short-term conventionally-measured economic growth is likely to be ultimately at the expense of the conditions for sustainable living on this planet. We need to adopt a human needs approach, providing for essential needs such as food, water, health, education, and sociality for the global population in a manner that minimises environmental damage [Moriarty and Honnery 2008a,b]. As conventionally-measured, resulting economic growth could be either positive or negative: the point is that it should no longer drive our economies. Nevertheless we need to remember that two important energy sources in such an economy are presently not included in energy accounting: passive solar energy, already discussed, and more use of human energy, for example in personal travel.

CONCLUSION

Renewable energy (RE) will eventually have to replace the present fossil fuel-dominated global energy supply system because of both resource depletion of fossil fuels, and the environmental effects, particularly on global climate, of fossil fuel combustion. However, in shifting from reliance on fossil fuels to RE, we are entering unknown territory, as a large scale-up of RE, together with adoption of technology unproven in the field, will be necessary. In a business-as-usual world with continued economic growth, and reduced energy and income inequality, the world could well need 1000 EJ primary energy by 2050. We argue that because of the limited technical potential of non-intermittent RE sources, such as hydro, biomass and geothermal energy, and their non-negligible global climate change effects, future RE will mainly have to come intermittent sources, particularly solar energy. As intermittent sources increase their share of total energy, they will need to be converted to an energy form that can be stored and transported, probably hydrogen. The actual amount of non-solar RE that can be produced is uncertain, as wind potential, the largest, depends heavily on what land has to be excluded. In the extreme case no more than about 50-100 EJ may be available, or perhaps half this after conversion and transmission. Solar energy, presently produced in tiny amounts, would need to be scaled up by four to five orders of magnitude from its 2005 level.

Further, the present technical potential for both wind and solar—particularly solar thermal electricity with its need for direct insolation, are unevenly distributed among the earth's regions. For this reason, large-scale intra- and even inter-contintenal scale transmission of electricity or hydrogen will be increasingly needed as RE penetration rises. Advanced Western countries are even investigating the possibility of overcoming energy resource limitations by going to the open oceans, deep into the earth, or even into space. Clearly, many decades will be needed for these to happen, if they ever do.

We conclude that it will be very difficult to provide 1000 EJ of RE primary energy by 2050. It is more likely in our view, that the difficulties, costs and risks of such an ambitious project will make energy conservation the more likely option. Energy conservation can occur by energy efficiency or by less use of energy-consuming devices. Given the global unmet demand for primary energy, it is probable that any further economic growth will continue to yield both energy efficiency gains—and absolute energy growth. Future energy will have to come from renewable sources, but we may have to trim both our energy needs and even the primary aim of continued economic growth if we are to live sustainably.

Acknowledgments

Patrick Moriarty would like to acknowledge the financial support of the Australasian Centre for the Governance and Management of Urban Transport (GAMUT) in the preparation of this paper. This Centre is in turn funded through a Volvo Research and Education Foundations grant.

References

Abbasi S.A., Abbasi N., 2000. The likely adverse environmental impacts of renewable energy sources. *Applied Energy* 65, 121-144.

Abe H., Duchane D.V., Parker R.H., et al., 1999. Present status and remaining problems of HDR/HWR system design. *Geothermics* 28, 573-590.

Association for the Study of Peak Oil and Gas (ASPO), 2008. *ASPO Newsletter No. 91 July* (Also earlier newsletters). Available at (www.peakoil.net).

Avery W.H., Berl W.G., 1997/8. Solar energy from the tropical oceans. *Issues in Science and Technology* 14 (2), 41-44.

Bala G., Duffy P.B., Taylor K.E., 2008. Impact of geoengineering schemes on the global hydrological cycle. *PNAS* 105 (22), 7664-7669.

Barbier E., 2002. Geothermal energy technology and current status: an overview. *Renewable and Sustainable Energy Reviews* 6, 3-65.

Baines J.T., 1987. Resources, reserves and their economic potential or how scarce is scarce? *Technological Forecasting & Social Change* 32, 295-310.

Baldwin M.P., Dameris M., Shepherd T.G., 2007. How will the stratosphere affect climate change? *Science* 316, 1576-1577.

Barker T., Bashmakov I., Bernstein L., et al., 2007. Technical Summary In: Metz B, Davidson O.R., Bosch P.R., et al., editors. *Climate Change 2007*: Mitigation. Cambridge, UK, CUP, pp. 25-93.

Barnett T.P., Adam J.C., Lettenmaier D.P., 2005. Potential impacts of a warming climate on water availability in snow-dominated regions. *Nature* 438, 303-309.

Bertani R., 2003 What is geothermal potential? *IGA News* No. 53, 1-3.

Bertani R., 2005. World geothermal power generation in the period 2001-2005. *Geothermics* 34, 651-690.

Blankinship S., 2006. Solar thermal's day in the sun could be at hand. *Power Engineering* 110 (3), 11-12.

Bonan G., 2008. Forests and climate change: forcings, feedbacks, and the climate benefits of forests, *Science* 320, 1444-1449.

Bossel U., 2006. Does a hydrogen economy make sense? *Proceedings of the IEEE* 94 (10), 1826-1837.

BP, 2008. *BP Statistical Review of World Energy* 2008. London, BP.

Breslow P.B., Sailor D.J., 2002. Vulnerability of wind power resources to climate change in the continental United States. *Renewable Energy* 27, 585-598.

Brown L.R., 2008. Plan B 3.0: *Mobilizing to Save Civilization.* NY, WW Norton & Co.

Cohen D., 2007. Earth audit. *New Scientist* 26 May, 34-41.

Cullenward D., Victor D.G., 2006. The dam debate and its discontents. *Climatic Change* 75 (1/2), 81-86.

Czisch G., Giebel G., Mortensen N.G., 2003. *Effects of large-scale distribution of wind energy in and around Europe.* Available at (http://www.saharawind.com/dedicatedpapers.php).

Czisch G., 2004. *Least-cost European/transeuropean electricity supply entirely with renewable energies.* Available at (http://www.saharawind.com/dedicatedpapers.php).

de Vries B.J.M., van Vuuren D.P., Hoogwijk M.M., 2007. Renewable energy sources: Their potential for the first-half of the 21st century at a global level: An integrated approach. *Energy Policy* 35, 2590-2610.

Drake B., Hubacek K., 2007. What to expect from a greater geographic dispersion of wind power?—A risk portfolio approach. *Energy Policy* 35, 3999-4008.

Energy Watch Group (EWG) 2007. *Coal: resources and future production.* EWG Series No 1/2007.

E.ON Netz., 2006. *Data and facts relating to wind power in Germany: supplement 2006 to the E.ON Netz wind report.* Bayreuth, Germany, E.ON Netz GmbH.

El-Shobokshy M.S., Hussein F.M., 1993. Effect of dust with different physical properties on the performance of photovoltaic cells. *Solar Energy* 51 (6), 505-511.

European Commission (EC), 2006. *World Energy Technology Outlook-WETO H_2. Directorate General for Research*; EUR 22038. Available at (http://ec.europa.eu.research/energy/pdf/weto-h2_en.pdf).

European Wind Energy Association (EWEA), 2006a. Plugging the gap: the prospects for wind. *Wind Directions* September, 31-33. Available at (www.ewea.org/index.php?id=14).

EWEA, 2006b. Wind power and the environment—benefits and challenges. *Wind Directions* July/August, 26-40. Available at (www.ewea.org/index.php?id=14).

Fathi N.Y., Salem A.A., 2007. The reliability of the photovoltaic utilization in southern cities of Libya. *Desalination* 209, 86-90.

Field C.B., Campbell J.E., Lobell D.B., 2008. Biomass energy: the scale of the potential resource. *Trends in Ecology and Evolution* 23 (2), 65-72.

Field C.B., Lobell D.B., Peters H.A., et al., 2007. Feedbacks of terrestrial ecosystems to climate change. *Annual Review of Environment and Resources* 32, 1-29.

Gawell K., Reed M., Wright P.M., 1999. *Geothermal Energy: The Potential for Clean Power from the Earth.* Available at (www.geo-energy.org/PotentialReport.htm).

Giles J., 2006. Methane quashes green credentials of hydropower. *Nature* 444, 524-525.

Global Wind Energy Council, 2008. *Global Wind 2007 Report.* Available at (http://www.gwec.net/uploads/media/Global_Wind_2007_Report_final.pdf).

Gupta H., Roy S., 2007. Geothermal Energy; *An Alternative Resource for the 21st Century.* Amsterdam, NL, Elsevier.

Haberl H., 2006. The global socioeconomic energetic metabolism as a sustainability problem. *Energy* 30 (1), 87-99.

Hafele W., 1981. *Energy in a Finite World: A Global Systems Analysis.* Cambridge, MA, Ballinger.

Hall C.A.S., 2004. The myth of sustainable development: personal reflections on energy, its relation to neoclassical economics, and Stanley Jevons. *Journal of Energy Resources Technology* 126, 85-89.

Hansen J., Sato M., Kharecha P., et al., 2008. *Target atmospheric CO_2: where should humanity aim?* Available at (www.arxiv.org/abs/0804.1126).

Harrison G.P., Whittington H.W., Wallace A.R., 2006. Sensitivity of hydropower performance to climate change. *International Journal of Power and Energy Systems* 26 (1), 42-48.

Hempsell M., 2006. Space power as a response to global catastrophes. *Acta Astronautica* 59, 524-530.

Honnery D., Moriarty P., 2009. Estimating global hydrogen production from wind. *International Journal of Hydrogen Energy.* doi:10.1016j.ihydene.2008.11.001.

Hoogwijk M., de Vries B., Turkenburg W., 2004. Assessment of the global and regional geographical, technical and economic potential of onshore wind energy. *Energy Economics* 26 (5), 889-919.

Hultman N.E., Koomey J.G., 2007. The risk of surprise in energy technology costs. Environmental *Research Letters 2, 1-6.*

Industrial Wind Action Group, 2007. Available at (http://www.windaction.org.news/9430).

International Energy Agency (IEA), 2007a. *Renewables in Global Energy Supply: An IEA Fact Sheet*. Paris, IEA/OECD.

IEA, 2007b. *Key World Energy Statistics.* Paris, IEA/OECD.

Johansson T.B., McCormick K., Neij L., et al., 2004. The potentials of renewable energy. *International Conference for Renewable Energies,* Bonn. Available at (www.renewables2004.de).

Keith D., DeCarolis J.F., Denkenberger D.C., et al., 2004. The influence of large-scale wind power on global climate. *PNAS* 101 (46), 16115-16120.

Kristmannsdottir H., Armannsson H., 2003. Environmental aspects of geothermal energy utilization. *Geothermics* 32, 451-461.

Lam, J.C., Wan, K.K.W., Tsang, C.L., 2008. Building energy efficiency in different climates. *Energy Conversion and Management* 49, 2354-2366.

Lawton J., 2001. Earth system science. *Science* 292, 1965.

Leggett J., 2006. *The Empty Tank.* NY, Random House.

Lehner B., Czisch G., Vassolo S., 2005. The impact of global change on the hydropower potential of Europe: a model-based analysis. *Energy Policy* 33, 839-855.

Lenzen M., Munksgaard J., 2002. Energy and CO_2 life-cycle analyses of wind turbines—review and applications. *Renewable Energy* 26, 339-362.

Lenzen M., 1999. Greenhouse gas analysis of solar-thermal electricity generation. *Solar Energy* 65 (6), 353-368.

Levine M.D., Urge-Vorsatz D., Blok K., et al., 2007. Residential and commercial buildings. In: Metz B., Davidson O.R., Bosch P.R., et al., editors. *Climate Change 2007: Mitigation.* Cambridge, UK, CUP, pp. 387-446.

Lienau PJ, Ross H. *Final Report: Low-Temperature Resource Assessment Program,* 1996. Available at (http://geoheat.oit.edu/pdf/db7.pdf).

Lovins, A.B., 2005. More profit with less carbon. *Scientific American* September, 52-61.

Mason J.E., Zweibel K., 2007. Baseline model of a centralized pv electrolytic hydrogen system. *International Journal of Hydrogen Energy* 32, 2743-2763.

Massachusetts Institute of Technology (MIT), 2006. The future of geothermal energy: Impact of enhanced geothermal systems (EGS) on the United States in the 21st century. *DOE Contract* DE-AC07-051D14517. Available at (http://geothermal.nrel.gov).

Milly P.C.D., Betancourt J., Falkenmark M., et al., 2008. Stationarity is dead: whither water management? *Science* 319, 573-574.

Moriarty P., Honnery D., 2005a. A comparison of bioenergy with other renewables. *Bioenergy Australia 2005 Conference,* Melbourne, Australia.

Moriarty P., Honnery D., 2005b. Can renewable energy avert global climate change? *Proceedings of the 17th International Clean Air & Environment Conference,* Hobart, Australia.

Moriarty P., Honnery D., 2007a. Intermittent Renewable Energy: The Only Future Source of Hydrogen? *International Journal of Hydrogen Energy* 32, 1616-1624.

Moriarty P., Honnery D., 2007b. Global bioenergy: problems and prospects. *International Journal of Global Energy Issues* 27 (2), 231-249.

Moriarty P., Honnery D., 2008a. Mitigating greenhouse: limited time, limited options. *Energy Policy* 36, 1251-1256.

Moriarty P., Honnery D., 2008b. Low mobility: the future of transport. *Futures* 40 (10), 865-872.

Moriarty P., Honnery D., 2009. Hydrogen's role in an uncertain energy future. *International Journal of Hydrogen Energy* 34, 31-39.

Mortimer N.D., 1991. Energy analysis of renewable energy sources. *Energy Policy* 19 (4), 374-385.

Munk W., 1997. Once again: once again—tidal friction. *Progress in Oceanography* 40, 7-35.

Naylor R.L., Liska A.J., Burke M.B., et al., 2007. The ripple effect: biofuels, food security and the environment. *Environment* 49 (9), 31-43.

Ngo P., 2008. World Bank report to claim biofuels caused 75 % of food price surge. *Ethanol & Biodiesel News* 20 (28).

Nilhous G.C., 2005. An order-of-magnitude estimate of ocean thermal energy resources. *Transactions of the ASME* 127, 328-333.

Nilhous G.C., 2007. An estimate of Atlantic Ocean thermal energy conversion (OTEC) resources. *Ocean Engineering* 34, 2210-2221.

Nordell, B., 2003. Thermal pollution causes global warming. *Global and Planetary Change* 38, 305-312.

Oswald Consultancy Ltd., 2006. *25GW of distributed wind on the UK electricity grid.* Available at (www.ref.org.uk/images/pdfs/ref.wind.smoothing.08.12.06.pdf).

Parry M.L., Canziani O.F., Palutikof J.P., et al., 2007. Technical Summary. In: Parry M.L., Canziani O.F., Palutikof J.P., et al., editors. *Climate Change 2007: Impacts, Adaptation and Vulnerability.* Cambridge, UK, CUP, pp.23-78.

Patzek, T.W., 2004. Thermodynamics of corn bioethanol biofuel cycle. *Critical Reviews in Plant Sciences* 23, 519-567.

Pearce F., 2007. The bog barons. *New Scientist* 1 December, 50-53.

Pearce F., 2008. Climate flood threat looms after quake. *New Scientist* 7 June, 8-9.

Pehnt M., 2006. Dynamic life cycle assessment (LCA) of renewable energy technologies. *Renewable Energy* 31, 55-71.

Pelc R., Fujita RM., 2002. Renewable energy from the ocean. *Marine Policy* 26, 471-479.

Pimentel D., Cooperstein S., Randell H., et al., 2007. Ecology of increasing diseases: population growth and environmental degradation. *Human Ecology* 35, 653-668.

Pritchard S., 2002. Overloaded. *International Water Power & Dam Construction* 54 (7), 18-22.

Pryor S.C., Schoof J.T., Barthelmie R.J., 2006. Winds of change? Projections of near-surface winds under climate change scenarios. *Geophysical Research Letters* 33, L11702.

Quaschning V., 2004. Technical and economic system comparison of photovoltaic and concentrating solar thermal power systems depending on annual global irradiation. *Solar Energy* 77, 171-178.

Riahi K., Grubler A., Nakicenovic N., 2007. Scenarios of long-term socio-economic and environmental development under climate stabilization. *Technological Forecasting & Social Change* 74, 887-935.

Roberts B.W., Shepard D.H., Caldeira K., et al., 2007. Harnessing high-altitude wind power, *IEEE Transactions on Energy Conversion* 22 (1), 136-144.

Running S.W., Nemani R.R., Heinsch F.A., et al., 2004. A continuous satellite-derived measure of global terrestrial primary production. *Bioscience* 54 (6), 547-560.

Ruppe H.O., 2003. Remarks concerning the feasibility of space-based solar-electric energy production for Earth usage ("Powersat"). *Acta Astronautica* 53, 71-73.

Sasaki S., Tanaka K., Higuchi K., et al., 2006. A new concept of solar powered satellite: tethered SDS. *Acta Astronautica* 60, 153-165.

Seboldt W., 2004. Space- and earth-based power for growing energy needs of future generations. *Acta Astronautica* 55, 389-399.

Shapiro F.R., 2002. Utilities in the sky? Comparison of space-based and terrestrial solar power systems. *Refocus* Nov/Dec, 54-57.

Scharlemann J.P.W., Laurance W.F., 2008. How green are biofuels? *Science* 315, 43-44.

Shinnar R., Citro F., 2006. A road map to U.S. decarbonization. *Science* 313, 1243-1244.

Sims R.E.H., Schock R.N., Adegbululgbe A., et al., 2007. Energy Supply. In: Metz B., Davidson O.R., Bosch P.R., et al., editors. *Climate Change 2007: Mitigation.* Cambridge, UK, CUP, pp.251-322.

Sinclair T.R., Purcell L.C., Sneller C.H., 2004. Crop transformation and the challenge to increase yield potential. *Trends in Plant Science* 9, 70-75.

Sky WindPower Corporation, (n.d.). *High altitude wind power.* Available at (http://www.skywindpower.com/ww/index.htm).

Solomon S., Qin D., Manning M., et al., 2007. Technical Summary. In: Solomon S., Qin D., Manning M., et al., editors. *Climate Change 2007: The Physical Science Basis.* Cambridge, UK, CUP, pp. 19-91.

Strahan D., 2008. The great coal hole. *New Scientist* 19 January, 38-41.

Thacker P.D., 2004. As the planet heats up, will the topsoil melt away? *Journal of Soil and Water Conservation* 59 (1), 12A-15A.

Trainer T., 2007. *Renewable Energy Cannot Sustain a Consumer Society.* Dordrecht, NL, Springer.

Trieb F., Langnib O., Klaib H., 1997. Solar electricity generation—a comparative view of technologies, costs and environmental impact. *Solar Energy* 59 (1-3), 88-99.

Vestas Wind Systems A/S, 2005. *Life cycle assessment of offshore and onshore sited wind power plants based on Vestas V90-3.0 MW turbines.* Available at (www.vestas.com/vestas/global/en/Downloads/Downloads/).

Weinberg A.M., 1979. Are the alternative energy strategies achievable? *Energy* 4, 941-951.

Weizsacker E. von, Lovins A.B., Lovins L.H., 1997. *Factor 4: Doubling Wealth—Halving Resource Use.* London, Allen & Unwin.

World Bank, 2008. *Key Development Data and Statistics.* Available at (http://web.worldbank.org).

World Commission on Dams, 2000. *Dams and Development: A New Framework for Decision-making.* World Commission on Dams.

World Energy Council (WEC), 2001. *2001 Survey of Energy Resources.* London WEC.

WEC, 2007. *2007 Survey of Energy Resources.* Available at (www.worldenergy.org/publications/).

Reviewed by Dr Geoff Lacey, Honorary Research Fellow, Department of Civil and Environmental Engineering, University of Melbourne, Australia.

In: Energy Policy: Economic Effects, Security Aspects... ISBN 978-1-60692-894-3
Editor: Noah B. Jacobs

Chapter 4

Energy Future - Global Energy Reserves, Influence of Greenhouse Gas and Hydrogen Emissions on the Environment and Quantification of Emission Sources

Lars Rose
University of British Columbia,
Department of Materials Engineering, Vancouver, Canada

Abstract

This chapter aims at giving a brief overview of the current knowledge about global CO_2 emissions and related effects on our climate. Additionally, some of the contributors of CO_2 emissions are quantified. The chapter is divided into four separate parts that aim at investigating topics related to CO_2 emissions and climate change. All topics presented here directly impact on both short and long term future of humankind. The article combines an overview over the current state of affairs concerning our oil reserves, the potential future of a hydrogen economy, explains how mass media try to create a controversy in the discussion about global warming that does not exist in the scientific community, and finally calculates different sector contributions to green house gas emissions, outlining how largely varying these numbers are between different sources.

Keywords: Energy policy, proven fossil fuel reserves, alternative energies, hydrogen economy, hydrogen environmental influence, greenhouse gas emissions, CO_2

Introduction

While the global average temperature progression might have raised little interest some decades past, it has become one of the key concerns in everyday life. The report on climate change impact on society compiled by Stern in 2006 gives a very detailed outline of not only the actual physical effects but also the minimum cost our society faces due to the effects of human induced climate changes [Stern 2006]. The price tag that comes along with the direct impacts of the changes that have been set in motion earlier than all of us today have been alive are equivalent to at least 5% of the global Gross Domestic Product (GDP) each year from now on. And considering a wider range of not only impacts but also total risks, the damage could easily rise to more than 20% of the global GDP. This cost could be limited by taking immediate actions to limit green house gas (GHG) emissions and energy waste, with an estimated annual cost of 1% of the global GDP.

Consequently, strong action has to be taken to prevent damaging effects on the environment. These effects are not only permanently costly to society, but likely irreversible within a fathomable time scale. It will later be shown how pre-historic CO_2 levels varied and what that potentially means to our immediate as well as far distant future.

It remains unknown what energy carrier could potentially replace fossil fuels. The increasingly vigorous discussion about peak oil shows that even with increased findings in traditional fossil fuels, the absolute reservoir of total fossil fuels on earth is finite and approaching its limits. Consequently, alternatives such as oil shale with much more severe detrimental impacts on the environment than even coal are already being tapped and developed for the foreseeable future. The ramifications are clear; the emission of GHG will significantly increase just to produce usable fuels. Alternatives are available, but on too small scales. Atomic reactors, apart from radiation waste with a half-life of several millennia, need uranium, which is not infinitely available either. Battery storage technologies, hydrogen as an energy carrier, and energy conversion devices such as fuel cells or solar cell research programs are too under-funded to have any significant impact on changes in the near future with the current political mindset in most states, especially on the basis of total cuts of more than 70% of the funding for alternative energies in Northern America between 1985 and 2005 [Kammen 2006].

A combination of political, social, technological, and lifestyle changes are required to alleviate the global energy crisis and to reduce the environmental damage that has already been caused by humans, especially during the past 150 years. This, however, must not be seen as a negative cost factor. In fact, it contains a vast growth potential for research and development as well as business in 'green' technologies. While the protection of ecology and nature is often rightfully correlated to a cost factor, this is an opportunity to financially gain from an increase in efforts in green technology. An impact, however, can only be achieved by wide application of available green technologies. If these measures are not taken, our society will reach a point where it can no longer sustain any kind of growth, economically or otherwise.

This chapter will focus on a variety of aspects related to climate change by asking the following key questions.

- We will probably not run out of fossil fuels, but the price tag for energy might increase several orders of magnitude. But how long will that take? What political and societal pressure will a decrease in oil, gas and coal put on developments in severely inefficient hydrocarbon fuel extraction such as tar sands or oil shale? Our arguably wasteful lifestyle is possible because petrol products are so cheap. Some basic calculations will project the time until most major oil wells run dry around the globe under certain conditions. But if we run out of fossil fuels or act on the realization that the global climate change is caused by the wasteful burning of fossil fuels, what are the alternatives? Hydrogen, ethanol and methanol have seen a significant increase in applications over recent years. But can it be shown that the alternatives available to date have the necessary volume to impact on the climate change?

- How do media reports on climate change and scientific opinion correlate? While peer-reviewed scientific findings portrait in unison a picture of severe climate changes to come, the media prefers a version in which there is much unrest and debate in the scientific community about whether we do have any human-induced climate changes or not.
- Finally, this chapter will analyze and quantify contributions to GHG emission from several sources. This will give an overview of the current scale of these emissions, and the uncertainty introduced by the different emission data sources.

Although on the first view the issues addressed here seem unrelated, all of the presented points have to be addressed if we are to slowly convert our society away from hydrocarbon fuels to a cleaner, economically and ecologically sustainable future.

1. Vanishing Cheap Resources

1.1. Introduction

New discoveries of oil reserves, for example in equatorial Africa, lead to an increase in projections of the time that fossil fuels will last [Tügel, 2000]. However, this increase only delays the date by which fossil fuel reserves will be depleted to an extent that it will have become economically unfavourable to extract the remaining residue [Hoel 1996 and Shell 2001]. Long before that point, the price of fossil fuel products, already strongly related to world politics, will reach levels that are largely out of the reach of most inhabitants of this planet [e.g. Rose 2008, Hansen 2003, BP 2005, Coimbra 2004, and API 2006]. There is some evidence that humankind has peaked new crude oil reservoir discoveries and that possible exploitation of the known reservoirs will decrease in the future, whereas total demand for energy, as well as energy demand per capita, will increase [Vuuren 2003]. From publications by global players in the oil business, one can derive certain conclusions:

Estimates of proven traditional oil reserves have increased steadily over the past years [e.g. BP 2005]. This chapter focuses on estimates of proven traditional oil reserves, and does not include ultimately recoverable resources or probable expected reserves, which vary much more widely between different studies [Ogele 2006].

1.2. A Changing World Energy Resource Base

Oil production has increased significantly over the past years, including an increase of more than three million barrels per day (MBD), almost 3% of total daily production, in 2004 alone [BP 2005]. Together, these two trends resulted in a decrease in reserves-to-production ratio in 2004, as production rose faster than new traditional reserves were discovered for the first time since 1998 [BP 2005]. Reserves-to-production ratios are an indication of the number of years that proven reserves will last. It is assumed that production rates remain at their current level. Although the worldwide ratio has changed little in the past 20 years, the reserves-to-production ratio may experience more of a decline in coming years, as rising production and demand combine with a decrease in the rate of discovery of new proven oil reserves [Campbell 2000]. Meanwhile, oil consumption has increased from 58 MBD in 1984 to 82 MBD in 2004, with the United States leading total resource consumption with 21 MBD, about 25% of the global consumption [BP 2005, NM 2006].

This chapter focuses on today's proven reserves only. These, together with the ultimately recoverable carbon reservoirs, only make up <25% of all global carbon reservoirs as shown by Kvenvolden et al. [Kvenvolden 1998]. The majority consists of gas hydrates, mostly in the form of methane crystals (clathrates) on the bottom of the oceans. These clathrates could potentially have a tremendous influence on future climate development, as they bind more greenhouse gas warming capacity than humans can ever produce by burning currently usable fossil fuels, as shown by Kvenvolden [Kvenvolden 1993]. An increase in water temperature might lead to the release and subsequent significant enrichment of green house gases in the atmosphere [Krajick 1997, Haq 1998, Econ 2003, Kvenvolden 1988]. The outcome of such an occurrence is unknown. The impact of human-made carbon dioxide on the environment will be discussed in further detail in section 3.

The World Resources Institute has published regional projections of the years to exhaust proven reserves, all of which are less than one century for oil and gas [WRI 2006]. One must keep in mind that all numbers are estimates and that oil reserves vary in quality, level of accessibility, and difficulty of extraction. This difficulty is a large factor for non traditional sources including Canada's tar sands, oil from coal, bitumen, heavy and extra heavy oil, deep water oil, polar oil, natural gas condensates, and oil shale [GeoCan 2006]. A good example of sudden changes in estimates is the raising of Canada's proven resources in 2003 by the Oil and Gas Journal from 4.9 to 180 billion barrels [OGNews 2003, O'Neill 2003]. This catapulted the North American country to second place in total proven accessible oil reserves, just after Saudi Arabia [Natmast 2006]. Based on [BP 2005], North America's proven oil reserves in 2004 were approximately 60 billion barrels. While many states possess large quantities of resources, only an increase in oil prices above a certain threshold will make extraction of certain types of resources cost-effective. These resources can then be considered (proven) reserves.

Also, significant findings of possibly more than 2 trillion barrels of oil contained in oil shale under the Rocky Mountains could, if extractable, make the US the country with the single largest supply of oil, and offer huge business opportunities within the US [Badiali 2006a-b, DOE 2006]. But oil shale is widely disputed, as it produces about four times as much greenhouse gas as ground oil when it is converted into usable petrol products [GP 2006]. Developments of future trends are additionally largely dependent on political as well as demographic developments, with the wars in the Middle East being examples of possible influences [Markussen 2004].

It should be noted that once regions are depleted of oil, the production level will have to be maintained by the other remaining regions, thus accelerating their depletion. The calculations presented in Tables 1-2 give a rough estimate of the worldwide years to complete oil depletion, without breaking the estimates down into regions. They are based on projected data from British Petroleum [BP 2005]. Table 1 shows development under steady-state conditions, assuming no new discoveries of oil reserves and a constant production rate. The oil in that scenario is depleted after less than 40 years.

The data shown in Table 2 are based on a significant increase in total oil reserves and a rapid increase in production and consumption, following the trends over the past five years. It can be seen that even with twice today's reserves, the oil would be depleted in 2050.

Calculations for other fossil fuel sources in the literature also yield estimated times to complete depletion of less than 100 years, with natural gas having an estimated time to depletion of approximately 70 years [BP 2005].

Table 1. Assumptions for calculating years to complete oil depletion, projected to 2006, followed by a steady-state (no change with time)

Proven Oil Reserves (Trillion barrels)	1.2 (no increase with time)
Daily Oil Production (Million barrels/day)	85 (no change with time)
Reserves-Production Ratio (years)	~39

Table 2. Assumptions for calculating years to complete oil depletion, projected to 2006, with trends from 2002-2004 projected forward and with an assumed doubling of proven reserves, due to new oil findings and heavy oil shale extraction (dynamic changes with time)

Total Oil Reserves (Trillion barrels)	2.5 (about twice the volume available today)
Daily Oil Production (Million barrels/day)	85 (in 2006) (projected 3 MBD increase per year until depletion)
Daily Oil Production (Million barrels/day)	220 (in 2050)
Years to complete oil depletion (years)	~44

1.3. Conclusion

The estimate of oil depletion by 2050 is based on the assumption of significant new reserve discoveries (or conversions of resources to reserves), and a steady increase in production and consumption that is based on the 2002-2004 increase and that is independent of remaining oil reserves or other factors. As more and more oil wells near depletion, production will likely drop as the remaining areas are unlikely to keep up or even increase the level of production. Alternatively, increasing use of more polluting alternatives such as tar sands and oil shale might be made. Also, as the easily available sources ebb, more costly developments with a lower production rate may predominate, thus further reducing production rate. Additionally, with the decrease in available crude oil, its price will rise and buyers will shift to other more available sources, which will likely be other fossil fuels initially. This substitution will elongate the time to depletion of the fuel source it replaces, in this case oil. This, however, only shifts the problem from one fossil fuel to another, and accelerates the depletion rates of other fossil fuel reserves. The problem is aggravated by the trend toward increasing total global population and worldwide energy consumption.

2. Potential Influences of Hydrogen Gas on the Environment

2.1. Introduction

With the advantage of zero local emissions of exhaust fumes, fuel cells promise to be a very healthy future alternative to combustion engines [Rose 2006, Rose 2007a, Rose 2007b]. Using biomass-derived fuels such as ethanol creates CO_2 exhaust in fuel cells, although at a lower level than current internal combustion engines create [DC 2004]. And even for zero local air pollution emissions using hydrogen as fuel, off-site emissions of CO_2 occur during the creation of hydrogen, unless it is produced by hydrolysis with renewable energy sources such as solar cells (excluding the CO_2 produced in the production and transport of these devices), as described by Asendorpf et al [Asendorpf 2004].

2.2. Hydrogen in our Future

In order to understand what role hydrogen will play in our future, we need to have an idea of which sources of energy will likely be chosen by our society. Shell Dutch regularly publishes projections of possible energy source distributions, e.g. [Shell 2001]. Driven by demands of energy security and cleanliness, this study assumes that choice of energy carriers changes from high-carbon fuels that will be depleted (as calculated in section 1 of this chapter) to alternative low-carbon fuels. In that scenario, electricity becomes the most important energy carrier being produced, at increasingly delocalised sources.

Hydrogen has the potential to become one of the mentioned new renewables [Shell 2001]. Hydrogen, however, has to be produced, as it is not naturally abundantly available in its gaseous form. Depending on the technology used, up to two thirds of the energy invested to create hydrogen could be lost during production, storage, transport, and energy conversion, compared to approximately 10% when electrical energy is directly fed into the power grids [Clark 2006, Ewe 2004, Grotelueschen 2004, Witthuhn 2004]. However, technological developments of still novel alternative energy devices including hydrogen storage, fuel cell and solar cell technologies have the potential to significantly reduce these losses and to create the basis for a widespread use of these high efficiency, low-emission devices. This is especially true for fuel cell technologies which can more efficiently use any type of fuel available today, including the use of carbon monoxide (CO) as a fuel in solid oxide fuel cells (SOFCs) [Van Herle 2004, Inui 2006]. Energy losses in power grid transmission have been reduced only after more than a century of developments in that area. There are several possible methods of large-scale hydrogen production. Firstly, coal can be gasified, producing hydrogen as a by-product. However, although coal is currently fairly cheap, this method does not reduce dependence on fossil fuels. Furthermore, any power plant could produce hydrogen electrolytically from water even today. This approach to produce, compress, store, and transport the hydrogen is not energetically favourable and demands for grid electricity are often high thus not allowing much of the power to be diverted to hydrolysis. The ideal procedure could be to use decentralised hydrogen production stations that utilize renewable, low emission energy, such as solar cells. However, although solar cells produce more energy than was required for their fabrication, they are still expensive, and the question of whether the produced electricity should not rather be used in the grid, or stored in batteries, remains [Shiga 1998]. However, many promising novel materials and semiconductor combinations are being investigated for their feasibility in solar cell technology [NREL 2006]. Finally, the collection and reformation of sewage and garbage biogas remains an effective way to produce hydrogen. However, the scale of this production might not be sufficient to yield enough hydrogen for a possible future with hydrogen as a major energy carrier, but it is a niche market already today.

2.3. Impact of Hydrogen on the Ecosystem

"Humanity is conducting an unintended, uncontrollable, globally pervasive experiment whose ultimate consequences could be second only to global nuclear war" [Toronto 1988, SCC 2006]. This statement refers to the impact of today's technology on the ecosystem, and possible consequences such as global warming. There is little public discussion about consequences resulting from an increased amount of hydrogen in the earth's atmosphere. Recent findings indicate that there might be some impacts varying from changes in soil bacteria with unknown effects on all ecosystems, to changes in the ozone layer [Witthuhn 2004]. As most of these predictions are based on models, the predictions vary from a significant widening of the gap in the ozone layer, as more hydrogen leaks into the atmosphere [Tromp 2003], to no increased detrimental effects or improvements of the ozone layer integrity [Schultz 2003]. Model predictions today will have to evaluate what impact imaginable replacement fuels might have on the environment in order to help in the decision making in selecting our future energy basis. According to most scenarios, hydrogen digesting bacteria in the soil will play a crucial role in dealing with an increase in atmospheric hydrogen. It is yet unclear to what extent the affected ecosystems could cope with this increase.

2.4. "New" Renewables

Alternative renewable energy carriers could be plant products that are converted into fuels. Since up to 10% of ethanol can be blended with gasoline without significantly changing the mixture's burning properties, this is a method which is increasingly used to stretch the remaining oil supplies [Pool 2006]. Beyond this level, problems emerge related to the affinity of alcohol to water. Hygroscopic additions of water will significantly change the overall combustion enthalpy of the fuel mix, changing its properties during combustion to a point where the efficiency of the motor tuned for pure alcohols is detrimentally affected. In addition, alcohols can be corrosive, and thus all materials in contact with this fuel, from transport vehicles to tanks, pumps, and fuel injection systems, must be corrosion resistant, often increasing the cost of these systems. Also, the plant stock has to grow initially and requires energy in terms of fertilisers, pumped water, labour, machinery, etc. Depending on which factors are taken into account, this process is considered to create a net energy gain or loss. The US Department of Energy has calculated a net gain of more than 35% energy for ethanol from various plant types [DOE 2006B, Allen 2006].

One of the advantages of ethanol as a fuel is its greenhouse gas neutrality. All CO_2 emitted during combustion has been incorporated into the plant from the air during growth, although additional energy from fossil fuels used for growing and producing the ethanol are not included in this consideration of CO_2 neutrality. And, even though the 85% ethanol blend with gasoline, known as E85, used for flex-fuel vehicles has slightly lower energy content than pure gasoline, the octane rating is higher for pure alcohols. This potentiates a higher compression ratio and consequently higher thermodynamic efficiency. This advantage, however, can only be fully utilised in cars that no longer have the option to switch back to gasoline.

The Renewable Fuel Association has estimated that the annual production of ethanol in the US by crop refineries is on the order of 24 billion litres [RFA 2006]. This amount is only approximately 3% of the annual US consumption of gasoline. Clearly, there is a margin for increase. However, assuming that 4000 m^2 of corn plantations can produce about 1100 litres of ethanol per season it would take 71% of all currently farmed land in the US to replace all of the gasoline used by ethanol [RFA 2006]. Land use is clearly a limiting factor. However, concerns about food being used as fuel even today can be mitigated when taking into account that ethanol is produced from the starch portion of the kernel of mostly livestock-grade corn. Large portions of the remaining fibre, protein, minerals, and vitamins can be reused to produce (starch-reduced) animal food. Alternative sources for ethanol production such as waste products from the beverage industry, rice straw, sugar cane trash, corn stalks, municipal solid waste, switch grass, and all-year-round wood products are being investigated for usability [Zerbe 1991].

Methanol, with large potential for small scale mobile applications in direct methanol fuel cells (DMFCs), is unlikely to play a large role in the massive global automotive market. Methanol is mostly produced from methane. This makes it dependent on natural gas, a finite natural energy source which is becoming more costly already today. Other methods of methane production include coal gasification and biogas (i.e. off-gassing collected from biowaste and garbage), but coal is more likely to be liquefied directly into gasoline, and current as well as projected biogas production is rather insignificant [Clark 2006]. The volumetric energy content of methanol is about half that of gasoline, and thus more liquid methanol has to be transported when compared to gasoline. Additionally, with regards to internal combustion engines, methanol can not start a cold motor and thus must be mixed with gasoline. Also, it burns with an invisible flame and is toxic, and is thus inferior to ethanol for large scale automotive applications. It should be mentioned here that the toxicity of alternative fuels is often contemplated while forgetting about or ignoring the inherent severe toxicity of established and widespread systems such as gasoline [Scala 1988].

There are many other valid alternative fuels that are being researched at the moment which are not mentioned here. This should by no means be interpreted as limiting the valid choices we have for our future energy to only the ones mentioned here.

2.5. Conclusion

Although hydrogen is one of the figureheads of fuel cell development, it is not certain how significant the energy carrier hydrogen will become for widespread applications. The energy losses associated with hydrogen conversion, storage, and transport and the low energy density per volume make large scale hydrogen production energetically inefficient, but also offer a great potential for technological advancement. In the unlikely case that there will be an excess of electricity production not consumed, this energy could theoretically be used to produce hydrogen by electrolysis, but modern battery technologies are also energetically favourable for energy storage. Due to their weight, and short-term storage capacity, even batteries require further technological developments. Any energy storage will always have to compete with its own production energy source as shown by Bossel et al., [Bossel 2003, Bossel 2005]. It can consequently be said that hydrogen will play a role in decentralised units operating on freely available (once capital costs are amortised) alternative energy supplies. These uses could include residential fuel cell systems or small refuelling systems for vehicles. The role hydrogen might play on a large scale is not entirely known yet. A stronger dependence on renewable sources producing ethanol as a liquid fuel is likely, but usage of arable land for fuel production as well as conversion of potentially edible food into inedible fuel have to be addressed, especially if ethanol is to replace a more significant part of the global gasoline supply as an automotive fuel. As the presented ideas show, there can be no one single easy answer to world energy and climate problems, and our future energy will be provided by a much larger variety of energy sources and carriers as used today. Among these will likely be all of the energy carriers presented in this chapter.

3. The Establishment of Global Warming as a Scientific Fact

3.1. Introduction

Controversy about climate is often claimed for the effect of populist media. However, peer reviewed literature generally agrees on the establishment as global warming as a scientific fact.

3.2. Human Impact on Global Warming

Taking the example of a study investigating planetary plant growth changes during the past 20 years by satellite imagery and modelling, the planet's growths seem to have greened in the northern hemisphere between 1982 and 1994 in spring and summer, and have been browning ever since from the equatorial region outwards [Sanders 2005]. Annual snapshots at the same date were taken and compared. While an increase in planetary temperature temporarily increased green growth and thus the overall intake of CO_2 by plants in 1982-1994, earth's ability to take up CO_2 and use it for growth now appears to be decreasing, possibly due to further increase in temperatures causing plants to dry out. The greenery is browning, due to effects such as forest clearing, soil erosion, fires, insect propagation, and droughts [Hebda 1995, Hebda 1997]. But while these models indicate that these changes might be due to the increase in greenhouse gas emissions, the authors also state that non-CO_2 cycles such as methane and sulphate cycles have not yet been implemented into the model. Furthermore, changes in plant distribution with rising temperatures, increase in fires and disease, or insect pests and resulting erosion dust effects in the oceans will have to be taken into account to arrive at a more precise conclusion about the causes of the observed effects. This study is but a small example of global climate changes and difficulties in evaluating these.

Additionally, one must take into consideration that the current controversy about human induced climate change mainly exists in the media, non-peer reviewed journals and fictitious literature [Lomborg 2001, Essex 2003,Crichton 2004]. As Levitt and Dubner 2005 state, all humans respond to incentives, and it turns out enjoying controversial stories is more interesting than repeatedly reading the same dire story of our future investigated by scientists and engineers from different angles.

Former US vice president Al Gore [Gore 2006], in fact, has gathered data on over 800 peer-reviewed journals and a similar number of media reports and other sources. He found that 100% of the authors in peer-reviewed scientific articles agree that there is scientific evidence that we are seeing the beginnings of significant climate changes, all of them mostly introduced by human activities such as greenhouse gas emissions. In contrast, only 50% of the media reports state human induced changes as a fact, whereas the rest report on conflict and disaccord in the scientific community. In order to make these reports sound more interesting, the media may intentionally try to create a controversy that does not exist within the scientific community.

There are many models that focus on the impact of fossil fuel consumption on global warming [Turner 1999]. Depending on the model, average temperature increases between 2°C and 6°C have been predicted [Matsuoka 1995, Plambeck 1996, Valdalbero 2004]. In contrast, there are non-peer reviewed reports that claim that most climate models as well as actual temperature and CO_2-level measurements are wrong, and that even a significant increase in human-made CO_2 emissions will not have any effect on the climate, postulating CO_2 as a beneficial aerial fertilizer [e.g. Idso 2000]. This conclusion coincides with that of the abovementioned study on global greenery coloration. However, the effect of increased CO_2 is reversed when increased average temperatures create droughts killing more plant volume than the addition of CO_2 in the air potentially increases [Sanders 2005]. Also, most studies focus on the changes in CO_2 levels and temperature over the past millennia. A copious study of the impact of humans on the environment has been compiled by the Intergovernmental Panel on Climate Change [IPCC 2001]. If the worst predictions of this report come true, we will increase our CO_2 level from 300 ppm in 1900 to 370 ppm today to over 1000 ppm in only one more century hence, and the global average temperature will increase 6 °C over the next 300 years. Compared to geological levels, these increases are rather small [Ruddiman 2000, Hieb 2006]. We do live in a cold time, with average global temperatures up to 10°C lower than historical, geological levels [e.g. Smith 2000, Mann 1995, von Storch 2006, Berner 2001, Scotese 2006]. It might consequently be a natural occurrence if the planetary temperature increases. However, although the comparison with prehistoric data suggests that the expected climate changes are not unprecedented on earth, it should be noted that airborne carbon levels are rising more steeply than ever recorded, and that current CO_2 levels have been reported to be at the highest level in the last 650,000 years [Black 2005].

Only one such notable increase in CO_2 levels exists in the geological records of the Permian era. In that era, the atmospheric CO_2 levels increased to 2000 ppm (CO_2 levels 700% higher than today) after a long time at low levels similar to those of today. This rise was accompanied by a rise in average surface temperature from 18°C to about 24°C. The timescale for that rise, however, is still a few million years, and not the projected few centuries it will take in our age. Other than that occurrence, the relationship between airborne CO_2 and global temperature does not seem to be one of directly coupled behaviour on a geological scale. Figure 1 compares atmospheric CO_2 levels relative to today's level (100%) with changes in surface temperature. The data for CO_2 levels are derived from ice core drillings, and the temperature and climate zones were determined by mapping the past positions of the continents and plotting on these maps the distribution of certain rock indigenous to specific climatic zones. Certain types of rocks, for example coals, need abundant rainfall to form under certain climatic conditions. As the presence of ocean-floor clathrates frozen out of the atmosphere at some prehistoric time show, many other types of GHG such as this methane can have an impact on the climate.

Even though both temperature and CO_2 changes look steep in Figure 1, the time scale of these changes is several million years, and thus very gradual as compared to the changes we observe today [Bradley 2000, Jones 2001]. The IPCC reports, e.g. [IPCC 2005] and more dramatically more recent versions of the report show the trend of surface temperature and airborne CO_2 level changes on a more recent timescale. The predictions of different levels over the next century derive from several scenarios created by the IPCC, varying between immediate, global, significant reduction of CO_2 exhaust to continuation of today's trend over the next century. These graphics are not without controversy, with regard to the used resources and methods [McIntyre 2005]. But even corrections for increased noise levels and higher historic CO_2 levels creating a more u-shaped curve as opposed to "hockey stick" curves can not eliminate the evidence that the global mean temperature is rising. This does not mean that every single point on earth is getting warmer, only that the average temperature does, in fact, increase.

The increased CO_2 levels predicted even by [IPCC 2001, IPCC 2005] will reach a maximum of 300% relative to today's level, as opposed to even higher values shown in Figure 1. However, this change may happen in only one century, an unprecedentedly rapid change, even on the prehistoric scale from Figure 1. And not only is the rate of change very rapid, but also the repercussions of this heating will take much longer times to manifest than might be expected. Like any chemical system, the atmosphere is bound to reach energy equilibrium. Depending on the effect, this process can take many millennia, a timescale no government plan will ever include. Even if we stopped producing any CO_2 today, the climate would change significantly in the near (~centuries) future as a result of atmospheric changes already occurring.

The IPCC furthermore showed the times to equilibrium for several effects influenced by climate change, including the expansion of water volume due to both thermal expansion and melting of ice reservoirs. These effects, although already observed today, will most likely continue to effect us over several hundred years [Combes 2003]. Today's observed glacier recessions are severe in their dimensions and impacts on the land that depends on their constant water flow year-round [Barnett 2005]. Large land areas will have severe drought problems if their main source of water will dry out during summer [Moore 2005, Barnett 2005b]. Saskatchewan, British Columbia, and Washington are just some of the western Canadian and American provinces and states directly afflicted by the slowly vanishing Columbia ice fields in Canada, but the effect is observed globally and many more areas are affected. The effects of melting glaciers, however, are negligible when compared to the expected impacts of the polar ice melting on increasing sea levels and disturbing delicate fresh/saltwater balances in the oceans.

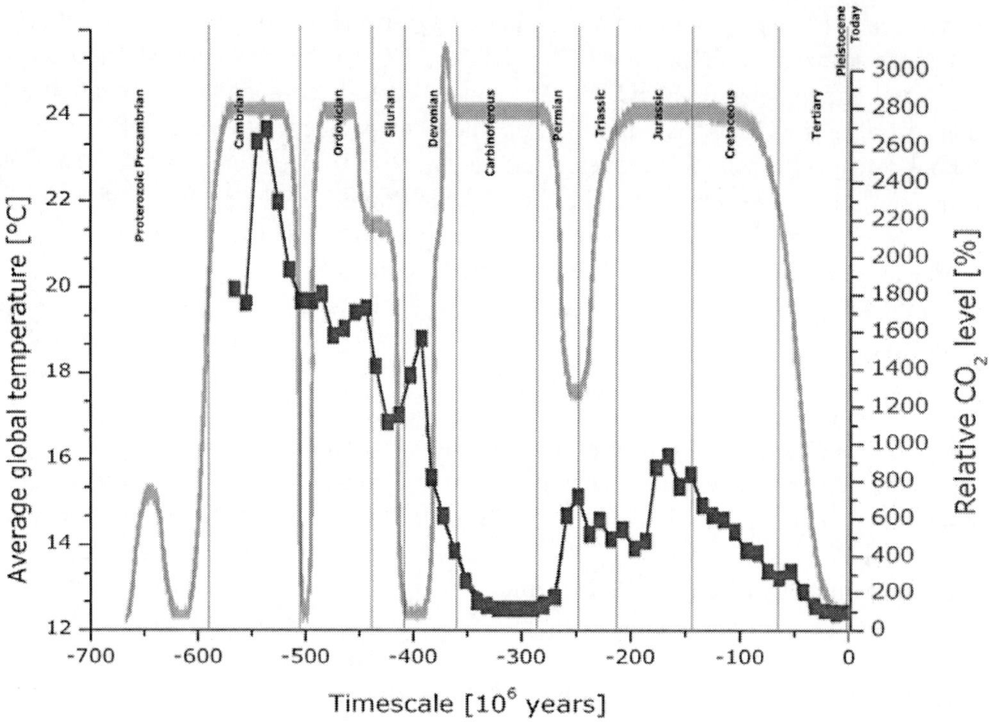

Figure 1. Changes in atmospheric CO_2 levels (rectangles) and average global temperature (continuous line) on a geological scale. Figure created by the author using data with kind permission from the American Journal of Science and the original authors: Atmospheric CO_2 levels from [Scotese 2006], average global temperature from [Mann 1995]. Relative CO_2 levels = 100 * CO_2 levels (time) / CO_2 levels (20th century average: ~295ppm).

3.3. Conclusion

Aerial carbon dioxide levels correlated with measured increases in average temperatures over prehistoric geological timescales (Figure 1) might lead to a conclusion that global temperature is not necessarily influenced by increased levels of aerial CO_2. Indeed, there are many factors that can have an influence on the global temperature besides greenhouse gases, such as aerial dust particles, average distance from the sun, volcanic activity, or surface reflectivity difference between trees, ice, and bare earth. On a shorter timescale, however, the relation between CO_2 levels and atmospheric temperature is strongly evidenced. Many of the effects attributed to global warming today are still relatively small compared to the potential impacts of CO_2 and temperature levels that might continue to increase with time [IPCC 2001, Fleming 2004]. But they give us an early indication of effects that we might have to deal with in the near future [Flannery 2006]. Although the suggested link between increase in CO_2 emissions and global warming might not be considered an established scientific fact in the media, the scientific data accumulated over the past decades *strongly* evidence such a relation, e.g. [von Storch 1999, Turner 2006, Schellnhuber 2004, IPCC 2001 & 2005].

Or, as von Storch et al. 2005 recently put it; "calling apocalyptic climate theories into question is becoming more and more difficult for sceptical scientists, but meanwhile, the public is getting increasingly tired of being fed a diet of fear".

4. CO_2 Emissions Calculations due to Electricity Generation Based on Various Sources

4.1. Introduction

The World Resources Institute (WRI) has created a comprehensive database of greenhouse gases and other climate-relevant indicators in their Climate Analysis Indicators Tool [CAIT 2006]. The following subsection will correlate the data from this source to other sources to try to validate them.

4.2. Electricity Generation

Focussing on CO_2 emissions only, values for CO_2 contributions for specific sectors can be extracted from [CAIT 2006]. Similar figures from the WRI exist for all years from 1985-2002. Looking at the most recent data, approximately 43 percent of the total CO_2 emissions are due to electricity production and heating.

The total CO_2 emissions due to electricity production approximated from [CAIT 2006] is 42.7%. The total contribution of electricity production to global green house gas emissions is thus 42.7% minus green house gas emissions from heating. The current values for global annual energy production for electricity are 16.7 Peta Watt Hours (PWh) and for heat 10.2 Exa Joule (EJ) [IEA 2003].

In the following calculations an attempt will be made to derive the CAIT data from different independent sources. Based on 1999 data from the US Department of Energy (DOE), annual CO_2 production by electricity generation in the US reached 2.3 billion metric tons resulting from a total of 3,700 GkWh (=TWh) of electricity, at approximately 0.62 kg CO_2 per kWh [DOE 2000, Statemaster 2006, NP 2001]. According to the Japanese Research Institute of Economy, Trade and Industry, the US created more than 24% of the annual worldwide CO_2 production in 2000, about 2 billion metric tons [Yoshihide 2003]. The US contribution had increased to 25.3% in 2003 [NM 2006, WRI 2006].

The UK Department of Environment, Food and Rural Affairs and the National Energy Foundation have developed a list of conversion factors that approximate the amount of CO_2 produced by different fuels [DEFRA 2001]. For the UK fuel mix in 2001, they established a conversion factor of 0.43 kg CO_2 per kWh [NEF 2001]. This factor, however, can vary for different fuel mixes from 0.37 to 0.69. Consequently, the actual fuel mix is of great importance [Otaki 2006, DOE 2006c].

The International Energy Agency, Institute for Energy Statistics has estimated the total global annual electricity production in 2003 to be 16,742 GkWh. This represents a less than

33% difference from the figures for the US from the US National Priorities Project Database [NP 2001] in 2001 (3,150 GkWh) and the CIA Factbook 2003 [WFB 2003] for the global figures of the remaining world (9,230 GkWh), which add up to a combined global annual electricity production of 12,380 GkWh.

Global CO_2 emissions were also estimated by the Intergovernmental Panel on Climate Change [IPCC 2001, IPCC 2005], with significant room for various development scenarios. When comparing such data, it has to be taken into account that one ton of carbon is the equivalent of approximately 3.67 tons of CO_2.

The values presented above are used in Table 3 to determine the fraction of total CO_2 emissions created by electricity generation.

The calculation in Table 3 suggests that the total contribution to global CO_2 production due to electricity production is in the range of 18.2% for lowest CO_2 fuel mix and global electricity production in 2001, to 46.2% for highest CO_2 fuel mix and global electricity production in 2003. This correlates well with the WRI data and aforementioned considerations outlining that the total contribution of electricity is 47.2% minus CO_2 emissions due to heat production, giving us similar numbers to the ones presented in Table 3.

Table 3. Calculations of the CO_2 emissions due to electricity production

Unit CO_2 per unit electricity (kg/kWh)	0.37-0.69 (UK fuel mix: 0.43) [NEF 2001, Otaki 2006]
Annual global electricity production in 2001 (GkWh)	12,380 [Statemaster 2006, WFB 2003]
Annual global electricity production in 2003 (GkWh)	16,742 [IEA 2003]
CO_2 emissions from electricity production (Gt)	4.58-11.6 (calculated from the values above)
Total global CO_2 emissions (Gt CO_2)	25.1 [WRI 2006, IPCC 2001]
Fraction of total emission due to electricity generation (%) (Lowest/highest total CO_2 emissions compared to lowest/highest emissions from electricity)	*18.2-46.2%*

4.3. Conclusion

Basic calculations to visualise the contribution of electricity generation and transportation were shown. In these calculations, electricity generation, including all sources and consumers in industry, the public sector, and the residential sector, is the top contributor to CO_2 production, and was thus analyzed in detail. Since fuel consumption for heating is often combined with electricity usage in some sectors, these values are difficult to extricate. The numbers presented are estimates, which vary depending on the source of the data, making precise projections very difficult. In one aspect, however, all predictions coincide. The CO_2 emissions are increasing, and there are palpable effects on the environment that might detrimentally influence us for more than just the next century.

5. Summary and Overall Conclusion

The present chapter investigated four separate issues that all tie into one theme: Resource diversification and global climate change.

Amongst novel discoveries of fossil fuel reserves it seems unlikely that we will face shortages any time soon. Global politics seldom look more ahead than the next legislative period, which is simply not far ahead enough to prepare society for potentially severe shortages of the types of cheap fossil fuels available today. This does not even include political instabilities in the source countries for each of those fuels.

Unfortunately, there is no ready-made alternative at hand to date. Hydrogen based energy storage has huge potential, but requires significant further research and development. Electricity distribution in overland lines remains an efficient method to deliver electric energy, but precludes mobile applications. Energy storage in batteries is also very efficient, but requires heavy systems, and modern batteries do not store energy efficient, especially at elevated temperatures, indicating another area of significant benefit from further research and development. Initial attempts to market electric automobiles in the late nineties in the US failed, but modern post 2008 financial crisis vehicle models start to offer a large variety of hybrid and non-fossil fuel engine technology alternatives.

Media reports sometimes try to diffuse the united scientific opinion that climate change is happening and that it is happening due to the impact of our own actions. Although opinions vary vastly of how to deal with the problem, the fact remains that we will be witnessing climate change unprecedented in human history within this century. The news of climate change is far from novel, although they have become more urgent during the past years. . Governments are capable of introducing incentives for reduced GHG emissions, but it remains with each individual to actually effect a lifestyle change. As shown clearly by Levitt and Dubner [Levitt 2005], humans only react to incentives. These have to be set by governments or other organisations which have the power and instruments necessary to set examples and show the right signals to catalyse a change in lifestyle in all population.

However, recent global financial problems seem to have put a damper on necessary sustainable energy research and development. And with the current steep growth rates in China and India, accompanied by massive expansion on coal power plants, GHG emissions are more likely to increase over the next decades rather than to decline.

References

Allen, M., (2006). How far can you drive on a bushel of corn?, *Popular Mechanics,* May, 74-81.

API, American Petroleum Institute, (2006). *Information on artificially low oil price increases compared to everyday items such as food,* www.conocophillips.com/newsroom/other_resources/energyanswers/oil_profits.htm (last accessed October (2006))

Asendorpf, D., (2004). *Die Zeit,* 42, 07 October.

Badiali, M., (2006a). The Next American Oil Boom, *Daily Wealth,* 14 April (2006).

Badiali, M., (2006b). The U.S. Govt's Secret Colorado Oil Discovery, *The Oil Report,* Stansberry & Associates, September (2006).

Barnett, T.P., Adam, J.C., Lettenmaier, D.P., (2005). *Some Impacts of Global Warming on Regional Water Availability*, University of Washington, Department of Land Surface Hydrology, Surface Water Group, 16 June.

Barnett, T.P., Adam, J.C., Lettenmaier, D.P., (2005b). Potential impacts of a warming climate on water availability in snow-dominated regions, *Nature,* November, 438[17], 303-309.

Berner, R.A., Kothavala, Z., (2001). Geocarb III: A Revised Model Of Atmospheric CO_2 Over Phanerozoic Time, *Amer. J. Science,* 301, February (2001), 182–204.

Black, R., (2005). CO_2 highest for 650,000 years, 24 November (2005), BBC News, UK.

Bossel, U., (2005). On the Way to a Sustainable Energy Future, *European Fuel Cell Forum,* Oberrohrdorf, Switzerland, October.

Bossel, U., Eliasson, B., Taylor, G., (2003). The Future Of The Hydrogen Economy: Bright Or Bleak?, *Cogeneration and Distributed Generation Journal,* 18[3], 29-70.

BP, British Petroleum, (2005). *Putting energy in the spotlight:* BP Statistical Review of World Energy, UK.

Bradley, R., (2000). Paleoclimate: 1000 Years of Climate Change, *Science,* 26 May, 288[5470], 1353-1355.

Cait, (2006). *Climate Analysis Indicators Tool,* Version 3, World Resources Institute, Washington, USA, (2006).

Campbell, C.J., (2000). *Peak Oil,* Presentation at the Technical University of Clausthal, December.

Clark, W.W., Rifkin, J., (2006). A green hydrogen economy, *Energy Policy,* 34[17], 2630-2639.

Coimbra C., Esteves, P.S, (2004). *OPEC Review,* 28[2], 87-106.

Combes, S., Prentice, M.L., Hansen, L., Rosentrater, L., (2003). *Going, Going, Gone, Climate change & global glacier decline, Global Warming is melting glaciers in every region of the world, putting millions of people at risk from floods, droughts and lack of drinking water,* WWF publications, IPMA Worldwide Press, 26 November.

Crichton, M., (2004). State of Fear, Harper Collins, December.

DC, Daimler Chrysler, (2004). 360 Grad – Umweltbericht (2004), *Daimler Chrysler AG,* Stuttgart, Germany.

DEFRA, (2001). Department of Environment, Food and Rural Affairs, London, UK, www.defra.gov.uk, data revised March (2001), last accessed October (2006).

DOE, (2000). Department of Energy, Environmental Protection Agency: *Carbon Dioxide Emissions from the Generation of Electric Power in the United States in (2000),* Washington, USA.

DOE, (2006). US Department of Energy, *Naval petroleum and oil shale reserves department,* www.fossil.energy.gov/programs/reserves, last accessed September (2006).

DOE, (2006b. U*S Department of Energy: Energy Efficiency and Renewable Energy,* http://www1).eere.energy.gov/biomass, 25 August (2006).

DOE, (2006c). *Energy Information Administration: Electric Power Industry Net Generation in (2004),* www.eia.doe.gov/cneaf/electricity/epa/figes2.html, last accessed September (2006).

Econ, (2003). *Methane and climate change - Swamp thing or monster of the deep,* The Economist, 17 April.

Essex, C., McKitrick, R., (2003). Taken by Storm, The Troubled Science, Policy and Politics of Global Warming, Key Porter Books, May.

Ewe, T., (2004). *Bild d. Wissensch.,* 3, 86-93.

Flannery, T., (2006). *The Weather Makers, The History & Future Impact of Climate Change,* Text Publishing, Melbourne, Australia.

Fleming, P., (2004). New Report Benchmarking Air Pollution from Top 100 Electric Companies Shows Carbon Dioxide Pollution Increasing, *Ceres Press Release,* April 14.

GeoCan, Geological Survey of Canada, (2006). http://gsc.nrcan.gc.ca, last accessed October (2006).

Gore, A., (2006). *An Inconvenient Truth: The Planetary Emergency of Global Warming and What We Can Do About It,* Atlantic Monthly Press.

GP, (2006). *Information on shale products and developments from Greenpeace:* http://greenpeace.org.au/climate/causes/criminals/shaleoil/overview.html, last accessed September (2006).

Grotelueschen, F., (2004). *Bild d. Wissensch.*, 3, 94-97.

Hansen, S., (2003). Pipeline Politics: The Struggle for control of the Eurasian Energy Resources, *Clingendael International Energy Programme,* The Hague, The Netherlands.

Haq B.U., (1998). Natural gas hydrates: searching for the long-term climatic and slope-stability records, *Geol. Soc. London, Spec. Publ.,* 137, 303-318.

Hebda, R.J., (1995). British Columbia Vegetation and Climate History with Focus on 6 ka BP, In: Paleogeography and Paleoecology of 6000 yr BP in Canada, ed. H. Jetté. *Géographie physique et Quaternaire,* 49, 55-79.

Hebda, R.J., (1997). Impact of climate change on biogeoclimatic zones in British Columbia and the Yukon, in: eds. E. Taylor, B. Taylor: *Responding to global climate change in British Columbia and the Yukon,* Vol. 1 of Canada Country Study: Climate impacts and adaption.

Hieb, M., (2006). *Climate and the Carboniferous Period,* www.geocraft.com/WVFossils/Carboniferous_climate.html, last accessed June (2006).

Hoel, M., Kvemdokk, S., (1996), *Resour. Energ. Econ.,* 18, 115-136.

Idso, C.D., Idso, K.E., (2000). *There Has Been No Net Global Warming for the Past 70 Years,* CO_2 Science, 3 [13], 1 July.

IEA, (2003). *International Energy Agency,* Institute for Energy Statistics: Electricity in World in (2003), www.iea.org, last accessed October (2006).

IEA, (2005). *Earth Trends: Data Tables, Energy, Consumption by Sector and International Energy Association: Energy Consumption by Sector, (2005),* www.earthtrends.org, last accessed October (2006).

Inui, Y., Urata, A., Ito, N., Nakajima, T., Tanaka, T., (2006). *Energy Conversion and Management*, 47[13-14], 1738-1747.

IPCC, Intergovernmental Panel on Climate Change, (2001). *Climate Change (2001): Synthesis Report,* Geneva, Switzerland.

IPCC, Intergovernmental Panel on Climate Change, (2005). *Carbon Dioxide Capture and Storage,* (2005), Geneva, Switzerland.

Jones, P.D., Osborn, T.J., Briffa, K.R., (2001). *The Evolution of Climate Over the Last Millennium, Science,* 292, 662-667.

Kammen, D.M., (2006). *Scientific American, September,* 84-93.

Krajick K, (1997). *The crystal fuel - crystallized gas hydrates discovered under the sea floor,* Natural History, May.

Kvenvolden, K.A., (1988). Methane hydrates and global climate, *Global Biochemical Cycles,* 3, 221-229.

Kvenvolden, K.A., (1993). *Rev. Geophys.,* 31, 173-187.

Kvenvolden, K.A., (1998). *Geol. Soc. London, Spec. Publ.*, 137, 9-30.

Levitt, S.D., Dubner, S.J., (2005). Freakonomics, *A Rogue Economist Explores the Hidden Side of Everything,* William Morrow, April.

Lomborg, B., (2001). *The Sceptical Environmentalist, Measuring the Real State of the World,* Cambridge University Press, August.

Mann, M.E., Park, J., Bradley, R.S., (1995). Global interdecadal and century-scale climate oscillations during the past five centuries, *Nature,* 378, 266-270.

Markussen, P., Henriksen, N., Nissen, F. (eds.), (2004). *Energy Together: three possible scenarios for energy development until 2025: free market, crisis and grassroots,* Elsam A/S, Fredericia, Denmark.

Matsuoka, Y., Kainuma M., Morita T., (1995). *Energ. Policy,* 23 [4-5], 357–371.

McIntyre, S., Mckitrick, R., (2005). *The M&M Critique of the MBH98 Northern Hemisphere Climate Index, Update and Implications, Energy & Environment,* 16[1], 69-100.

Moore, R.D., Stahl, K., (2005). Glacier Recession: Implications for Late-Summer Streamflow and Water Temperature, in: "Hydrology for the Orographically Challenged", *A Workshop on Predictions in Ungauged Basins in Mountainous Regions,* 28-30 November, Manning Park, BC, Canada.

Natmast, Nationmaster Statistics, (2006). www.nationmaster.com, last accessed October (2006).

NEF, (2001). *National Energy Foundation* (2001), Milton Keynes, UK, www.nef.org.uk/energyadvice/co2calculator.htm, last accessed October (2006).

NM, *Nationmaster Statistics,* (2006). www.nationmaster.com, last accessed October (2006).

NP, (2001). *National Priorities Project Database* (2001) http://database.nationalpriorities.org/cgi-bin/WebObjects/nppdatabase.woa, last accessed October (2006).

NREL, National Renewable Energy Laboratory, (2006). National Center for Photovoltaics, www.nrel.gov/highperformancepv, *Information of new photovoltaic systems last accessed August* (2006).

O'Neill, G. et al., (2003). *The Foundation for the Economics of Stability, Ireland: Before the wells run dry, Will oil from tar sands enable the global supply to be maintained?,* Douthwaite, R., (Ed.), Green Books.

Ogele, C., McVay, D.A., Lee, W.J., (2006). Quantifying Uncertainty in Original-Oil-In-Place Estimates From Volumetric and Material Balance Methods, in: Proc. (2006) *American Association of Petroleum Geologists Annual Convention, Risk Analysis and Portfolio Management,* April 9-12.

OGNews, (2003). Drilling outlook is for decline in US, steady gain in Canada, *Oil and Gas J.,* 101[4], January 27.

Otaki, T., (2006). *Energy providers at odds over emissions standards,* Asahi News, www.asahi.com, last accessed January (2006).

Plambeck, E.L., Hope, C., (1996). *Energ. Policy,* 24 [9], 783-793.

Pool, R., (2006). *Grow and go – the sweetest solution, Institution of Engineering and Technology,* Engineering Communities Transport.

RFA, Renewable Fuel Association, (2006). *Ethanol Facts,* www.ethanolrfa.org, last accessed July (2006).

Rose, L., (2006a). Steam engine technology of the future: fuel cells running on alcoholic cocktails, *Science Creative Quarterly,* March 22, (2006), http://www.scq.ubc.ca/?p=198, last accessed October (2008).

Rose, L., (2007a). Surface technologies (2006) — *Alternative energies and policy options Energy Policy,* 35[12], 6106-6111.

Rose, L., (2007b). Thermec (2006). *Energy Policy,* 35[6], 3541-3545.

Rose, L., (2008). Review of Fueling Our Future — An Introduction to Sustainable Energy, *Energy Policy* 36[3] 1243-1244.

Ruddiman, W. F., (2000). *Earth's Climate Past and Future,* W.H. Freeman Publishing, New York, USA.

Sanders, R., (2005). *Faster carbon dioxide emissions will overwhelm capacity of land and ocean to absorb carbon,* UC Berkeley News, 2nd August.

Scala, R.A., (1988). *Motor Gasoline Toxicity, Fundam. Appl. Toxicol.,* 10[4], 552-562.

SCC, Sierra Club of Canada, (2006). *A planetary citizen's guide to the global climate negotiations,* Climate Action Network Canada.

Schellnhuber, H.J., Crutzen, P.J., Clark, W.C., Hunt, J., Claussen, M., Held, H. (Eds.), (2004*). Earth System analysis for sustainability*, MIT press, November.

Schultz, M.G., Diehl, T., Brasseur, G.P., Zittel, W., (2003). Air Pollution and Climate-Forcing Impacts of a Global Hydrogen Economy, *Science,* 302, 624-627.

Scotese, C.R., (2006). Paleomap Project, www.scotese.com, last accessed August (2006)

Shell International, (2001). Exploring The Future: Energy Needs, Choices and Possibilities, *Scenarios to 2050, Global Business Environment.*

Shiga, H., Shinda, K., Hagiwara, K., Tsutsumi A., Sakurai M., Yoshida, K., Bilgen, E.: Int. J., (1998). *Hydrogen Energ.,* 23 [8], 631-640.

Smith, D.J., Desloges, J.R., (2000). Little Ice Age History Of Tzeetsaytsul Glacier, Tweedsmuir Provincial Park, British Columbia, Canada, *Géographie physique et Quaternaire*, 54[2], 135-141.

Statemaster, (2006). *Information on electricity consumption in the US* (2001): www.statemaster.com/graph/ene_tot_ele_con-energy-total-electricity-consumption, last accessed October (2006).

Stern, N., (2006). *Stern Review on the economics of climate change,* HM treasury, Government of the UK, Independent Reviews.

Toronto, (1988). Consensus introduction to: Our Changing Atmosphere: Implications for Global Security, *The Changing Atmosphere Conference*, Toronto, 1988.

Tromp, T.K., Shia, R.L., Allen, M., Eiler, J.M., Yung, Y.L., (2003). Potential Environmental Impact of a Hydrogen Economy on the Stratosphere, *Science,* 300, 1740-1742.

Tügel, H., (2005). *GEO Magazin* 7, 84-91.

Turner, R.J.W., Clague, J.J., (1999). Temperature Rising, Climate change in Southwestern British Columbia, Addendum to Vol.1, Canada Country Study, *Environment Canada,* Miscellaneous Report 67.

Turner, R.J.W., Clague, J.J., (2006). Temperature Rising: Climate Change in Southwestern British Columbia, Poster of the GVRD, www.gvrd.bc.ca/climate.

Valdalbero, D.R.D. (ed.), (2004). European Commission, Community research: World Energy, *Technology and Climate Policy Outlook,* 1-139, Office for Official Publications of the European Communities, Luxembourg.

Van Herle, J., Membrez, Y., Bucheli, O., (2004). *Journal of Power Sources*, 127[1-2], 300-312.

von Storch, H., Stehr, N., (2005). *How Global Warming Research is Creating a Climate of Fear, Spiegel international,* 4, January 24.

von Storch, H., Stehr, N., (2006). *Anthropogenic Climate Change, A Reason For Concern Since The 18th Century And Earlier, Geogr. Ann.,* 88A[2], 107–113.

von Storch, H., Zwiers, F.W., (1999). *Statistical Analysis in Climate Research,* Cambridge University Press.

Vuuren, D.V., Yun, L., Vries, B.D., Kejun, J., Graveland, C., Fengxi, Z., (2003). Energy and emission scenarios for China in the 21st century - Exploration of baseline development and mitigation options, *National Institute of Public Health and the Environment,* Bilthoven, The Netherlands.

WFB, (2003). CIA World Factbook, www.cia.gov/cia/publications/factbook/index.html, last accessed September (2006).

Witthuhn, B., (2004). *Bild d. Wissensch.,* 3, 98-101.

WRI, *World Resources Institute,* (2006). Washington DC, USA, www.wri.org, last accessed October (2006).

Yoshihide, S., (2003). *Kyoto Protocol as a Diplomatic Issue,* (2003), 11, Japanese Research Institute of Economy, Trade and Industry.

Zerbe, J.I., (1991): *Liquid Fuels from Wood – Ethanol, Methanol, Diesel, World Resource Review,* 3[4], 406-414.

In: Energy Policy: Economic Effects, Security Aspects... ISBN 978-1-60692-894-3
Editor: Noah B. Jacobs

Chapter 5

Renewable Energy Legislation in Latin America to Ease Dependency on Fossil Fuels

Elizabeth Lokey
The University of Colorado at Boulder, Boulder, CO, USA

Abstract

Latin American countries have increasingly turned to fossil fuel-generated electricity to fill demand growth in recent years as the hydro applications that comprise 73% of the region's energy generation portfolio have proven unreliable in years of drought. These countries, which experienced a wave of restructuring in the 1990s, cannot attract private investors that are willing to construct hydro applications because these sources of new generation necessitate huge capital expenditures. Recent fossil fuel prices have caused some countries to consider how to steer new capacity additions away from fuel oil generation with high operating costs. Many of these countries have implemented new or strengthened renewable energy legislation to promote generation with stable operating costs. Countries have pursued diverse renewable energy policy incentives like production tax credits in Argentina, feed-in tariffs in Ecuador, and portfolio mandates in Chile. The impact this legislation will have on future capacity additions will depend on how it is incorporated in the existing energy sectors and the longevity of the support. Countries like Brazil that have strong renewable energy laws may neglect this new legislation as domestic future fuel prices drop in response to new fossil fuel discoveries like the large off-shore oil deposit found in late 2007. The decision made by these countries on how to fulfill future demand growth is critical for climate change mitigation. However, an international Mechanism to involve developing countries in the effort to mitigate climate change ironically may be rendered useless if a Latin American country has taken domestic steps to promote renewable energy applications.

Introduction

Latin America's electrical grid is dominated by hydro power, which comprises 73% of the region's generation portfolio [1]. This hydro development was mainly implemented by the state-run utilities that were able to procure loans from the government to erect these capital-intensive structures with long payback times. Hydro applications made sense for the region, which, as a whole and in general, experiences plentiful rainfall and has varied topography.

This model of state-run, capital-intensive energy applications with long payback times no longer applies because of privatization that has swept the region. Now, these countries cannot attract private investors that are willing to construct large new hydro applications because of the huge capital expenditures required [2]. Recent fossil fuel prices have caused some countries to consider how to steer new capacity additions away from fuel oil generation with high operating costs. Also, in some countries like Colombia, Peru, and Chile severe droughts have made hydro electricity unreliable [3], [4], and [5].

Some countries are considering renationalizing the energy sector to provide the requisite funding for new hydro applications. Also, most of the Latin American countries have begun implementing new or strengthened renewable energy legislation to promote generation with stable operating costs. Countries have pursued diverse renewable energy policy incentives like production tax credits in Argentina, feed-in tariffs in Ecuador, and portfolio mandates in Chile. Each of these renewable energy policies has associated pros and cons and some are better suited than others for particular political and economic environments.

The impact this legislation will have on future capacity additions will depend on how it is incorporated in the existing energy sectors and the longevity of the regulatory support. Countries like Brazil that have strong renewable energy laws may fail to provide sustained, long-term support for renewables as domestic future fuel prices drop in response to new fossil fuel discoveries like the large off-shore oil deposit found in late 2007 [6].

The choice these Latin American countries make will chart their course into the future for the lifetime of the new capacity additions, which can be over 100 years for hydro applications. This long-term decision can have profound global warming impacts since this region is undergoing tremendous growth in the electrical sector. However, an international Mechanism called the Clean Development Mechanism (CDM) to involve developing countries in the effort to mitigate climate change ironically may be rendered useless if a Latin American country has taken domestic steps to promote renewable energy applications. The CDM can also complicate sectoral laws to promote mitigation activities and usually cannot be used by state-run utilities.

Table 1. Non-Technical Electrical Losses before and after Privatization [8]

	Year Privatized	Pre-Privatization	Post-Privatization
El Salvador	1998	11.3%	9.3%
Guatemala	1998	13.5%	15.5%
Nicaragua	2000	30.1%	28.78%
Panama	1998	17.88%	10.42%

PRIVATIZATION

During the 1990s, Latin American hydro development halted as the region experienced a wave of privatization. Many sectors from electrical to telecommunications were transferred to the private sector. The impetus behind this shift was twofold. Firstly, in most countries, the government could no longer make costly upgrades and capacity additions due to poor money management and high levels of debt. Secondly, countries hoped that through privatization, customers could receive better service more cheaply [7]. By transferring this responsibility to the private sector, the hope was that better service with fewer technical and non-technical distribution losses would result. The results of privatization have been mixed, with some countries receiving worse service but most countries enjoying more reliable electricity and less expensive rates as non-technical losses from electricity theft has been decreased.

The governments of each Latin American country, with the exception of Mexico, Costa Rica, and other countries to varying degrees which still have primarily state-controlled electrical sectors, created energy policy to promote new market entrants, often debundled the vertically-integrated state-run companies, established market managers, policy-making groups, and a wholesale market. The process of creating policy has been one of trial and error. Some countries, like Uruguay, have privatized the market, but not created enough of an incentive to have new market entrants because the country's small amount generation needs perhaps lends itself to a natural monopoly [9]. Honduras, one of the first countries to become privatized in 1994, has also failed to attract sufficient investors to fulfill the capacity demands [2].

The way in which each of these countries went about privatizing the energy sector differs dramatically. In Costa Rica, independent power producers (IPPs) have a limited role in the market and can only comprise 30% of the market [10]. Those that do enter the market must be renewable generators only and owned at least 35% by national entities [11]. Likewise, Mexico offers IPPs 85% of the state-run utility's avoided cost of generation. Or, IPPs can structure a power purchase agreement (PPA) with a shareholder, but must pay between 15-30% of the energy payment they receive in a transmission tariff to the state-run utility (Comisión Federal de Electricidad- CFE) [12].

Chart 2. Privatization Schemes in Select Countries[1]

Country	Privatization Scheme
Argentina	Price-based energy auction
Brazil	Distributors aggregate generation needed
Chile	Cost-based energy auction
Costa Rica	30% IPP penetration; IPPs must be renewable generators with 35% Costa Rican ownership
Honduras	Limited PPAs with large consumers; ENEE offers fixed price unless generation solicited
Mexico	IPPs enter market under self-supply scheme or get paid 85% of CFE's avoided costs
Panama	2-phased energy transition with single purchaser model before wholesale market

More open markets like Honduras offer a hybrid between a state-run and privatized market by allowing IPPs to create PPAs only when they have been designated as "authorized entities," of which there is only one. In this country IPPs can negotiate PPAs with the state-run company (Empresa Nacional de Energía Eléctrica- ENEE) if there has been a bid for a new capacity addition or accept fixed energy payments from ENEE if there has not [2].

[1] See endnotes in the paragraph above for references.

Panama introduced a phased transition to an open market that has proven successful. The first stage lasted from February 1997 to February 2002 and consisted of a Single Power Purchaser model whereby generators had to sell 100% of their capacity to the state-owned transmission company who then resold the generation to distribution companies without making a profit. Then Panama moved into a wholesale market with an energy auction for least-cost generation after 2002 [13].

Brazil has a power pool whereby distributors aggregate the new generation needed and pay fixed, long-term prices to generators [14]. Chile has a very open market with the option for generations to engage in PPAs with large customers, accept node prices from the grid, or enter the wholesale power market [15]. Argentina has a wholesale power market, but unlike most of these markets in the region that are least-cost based, Argentina's is price-based. Argentina's auction has price caps that prevent generators from offering a cost that is too high [5]

Often these newly-privatized countries are not able to attract investment because the current price of electricity, which is usually based on hydro resources that operate very cheaply, is low. New generators must compete against these low-cost resources with their new generation. Other countries, like Nicaragua, have had less new market entrants than the systems needs because of the country's political and economic risks [16].

This situation has led to a lack of capacity additions, blackouts, and power shortages throughout the region that are reminiscent of the 1990s when the state-run utility could not provide adequate supply [17]. In order to reduce these supply shortages, the government has begun contracting emergency generation from fossil-fuel burning sources. This trend is especially pronounced in Central America where many sites of "rented" generation are springing up. These sites typically burn bunker fuel oil #6, which is highly polluting and expensive. The government is entering into PPAs with these entities for high rates of between 23-30¢ USD/kWh. The government contracts these generators because they can provide generation that is quickly installed, and they do not have to enter into long-term agreements with the generators since the facility components like the gas or diesel turbine can be easily moved to another location [2].

Hydro and other types of renewable generators do not receive this type of preferential treatment. They are discouraged by these PPAs as they are still forced to compete with the existing low-cost hydro generation on the grid in energy auctions. In most cases, non-hydro or biomass renewable energy like wind must compete against fossil fuel generation that earns capacity as well as generation payments [18].

Private generators have also been discouraged from building new hydro facilities in some countries because of droughts that have plagued areas of South America. Colombia, Peru, and Chile have experienced severe El Niño and La Niña-related droughts since the late 1990s that caused outages [3], [4], and [5]. Some countries like Peru even implemented legislation to discourage hydro development in order to diversify the country's generation sources, reduce these outages, and support the natural gas industry. During 1998-2000, President Fujimori of Peru banned all new hydro development and put in place strong incentives for natural gas plants [19], [20], [21], [22], and [23].

Recent high fossil fuel prices at $119 per barrel of crude oil (as of 4/30/2008), however, have made this type of new fossil fuel-based generation in most countries undesirable [24]. Chile, which is currently facing a natural gas supply shortage from Argentina, has had electricity prices jump from $40 USD/MWh in the April of 2004 to $250 USD/MWh in April of 2008 as the country has had to convert most of its natural gas facilities to accept petroleum-based products [25].

Renewable Energy Legislation

In response to these high fossil fuel prices, most Latin American countries have begun to promote renewable energy with incentives and mandates. This energy, which usually has fixed fuel costs for biomass and no fuel costs for hydro, wind, and other types of generation, is now desirable. To avoid the problems with drought, countries like Chile and Brazil are only allowing hydro below 20 MW in Chile and 30 MW in Brazil to qualify for the renewable energy incentives and mandates [26] and [27]. By having smaller hydro applications, the effects of drought could be minimized as installations would be more dispersed.

Like privatization, each country has taken its own path towards promoting renewables. These various schemes reflect the diversity in each country's philosophies and investment climate. Chile, which was one of the first countries in the region to become privatized and has a strong history of a neo-liberal, competitive marketplace, has chosen to create a mandate that requires generators to source 5% of their residential sales form renewable energy sources by 2010 and 10% by 2024. Generators who fail to comply with this legislation will pay a penalty of $27/MWh or $40/MWh for three consecutive years of non-compliance [28] and [29]. Transmission tariffs are reduced for renewable energy applications under 20 MW [26]. The non-compliance fees should help to command a higher price for renewable energy, but the country shied away from providing a direct incentive for this generation since a subsidy would be inconsistent with Chile's neoliberal practice of allowing the marketplace to operate with minimal governmental intervention.

Argentina has chosen to support renewables through a production tax credit (PTC) [30]. This credit is a per kilowatt hour payment given to renewable generators by the federal government that is in addition to payments they get on the spot market or in PPAs. Any renewable installations built while the PTC is active earn the PTC for the number of years that it is guaranteed, which is usually 10 years. Some Argentine provinces like Buenos Aires and Chubut add an additional PTC for generation that can be summed with the federal subsidy [31].

Ecuador has chosen to use a feed-in tariff to stimulate renewable energy development. This tariff, as shown in the chart below, is set at different levels for different types of technologies. Already, this feed-in tariff has prompted much growth, especially in the area of biomass from sugarmills, which burn bagasse. These sugarmills are now not only self-supplying themselves, but also making their processes more efficient to be able to sell excess generation to the grid and earn the feed-in tariff [32] and [33].

Beyond these major policy instruments to promote renewable energy, there have been some other incentives for these industries. Most Central American countries allow equipment for these applications to be exempt from import taxes and the installations usually do not have to pay income tax for the first ten years of operation. Occasionally, renewable energy will receive priority dispatch and streamlined permitting processes as well as transmission tariff breaks. See Chart 2 for a comprehensive description of the renewable energy legislation in each country.

Chart 2. Feed-in Tariff Prices [34]

Generation Source	Price (¢ USD/kWh) in continental territory	Price (¢ USD/kWh) in Galapagos Islands
Wind	9.39	12.21
Photovoltaic	52.04	57.24
Biomass and Biogas	9.67	10.64
Geothermal	9.28	10.21
Hydro up to 5 MW	5.8	6.38
Hydro between 5 and 10 MW	5	5.5

Beyond these major policy instruments to promote renewable energy, there have been some other incentives for these industries. Most Central American countries allow equipment for these applications to be exempt from import taxes and the installations usually do not have to pay income tax for the first ten years of operation. Occasionally, renewable energy will receive priority dispatch and streamlined permitting processes as well as transmission tariff breaks. See Chart 2 for a comprehensive description of the renewable energy legislation in each country.

Chart 3. Summary of Renewable Energy Legislation [35]

Country	Legislation
Argentina	Production tax credit of 1.5 Peso ¢/kWh for wind, hydro under 30 MW, biomass, and geothermal, and .9 Peso ¢/kWh for solar; 15-year exemption from income taxes [30]
Bolivia	Standards for PV installations [36]; several rural energy programs [1] and [37]
Brazil	Phase I: 1422 MW wind, 1191 MW small hydro, and 685 MW by 2008 with a 60% local component requirement Phase II: 10% of overall generation by 2022; 90% local component requirement suggested [27]
Chile	Short Law I: Guaranteed grid access and reduced transmission rates for projects under 20% [26]; Short Law II: 5% of country's generation for residential customers by 2010 [38]; Proposed Short Law III: 8% of residential, industrial, and commercial customers energy sourced from renewable sources by 2024 [29]
Colombia	No income or value-added tax on importations for the first 10 years of operation; other taxes on salaries, research equipment, and machinery exempted for "first-of-a-kind" projects [39] and [40]; 35% income tax waiver for projects that give 50% of CERs to community development [41]
Costa Rica	Former feed-in tariff for small renewable energy [1]; ICE will only buy renewable generation from IPPs, IPPs must have a 35% Costa Rican ownership structure, IPPs can make up only 30% of market, IPPs must sign contract with ICE for generation price, and relinquish plant operations to ICE after 18 years of operation in a Build, Operate, and Transfer agreement [42], [43] and [11]
Dominican Republic	Renewable energy fund from 5% of hydrocarbon sales [44]; subsidized financing during 10 years for a part of the capital required for installations not exceeding 50MW; 100% exemption from import tariffs on the equipment and tools needed for renewable energy generation 10-year exemption from income tax on revenues obtained in generation of renewable energy; tax incentive for self-production [45]; dispatch priority [46]

Country	Legislation
Ecuador	Feed-in tariff for hydro under 10 MW and wind, solar, geothermal, and biomass under 15 MW; applies until 2% penetration of renewables in grid [34]; renewables get guaranteed dispatch and exoneration from import and income taxes [47]
El Salvador	No taxes on renewable energy projects below 10 MW for 10 years; Projects between 10-20 MW are exempt from taxes for five years; projects do not have to pay taxes on CER revenues from projects [48]; projects under 5 MW have streamlined permitting [49]; fund for renewable energy and soft loans being considered [50]
Guatemala	No import tax for equipment, no income tax for first 10 years; retailers are obligated to buy electricity from private generators [43]; distributors have to buy generation for projects under 5 MW [51]
Honduras	For applications below 50 MW, no import tax for equipment and no income tax for first 10 years; state distribution company (ENEE) has to buy electricity from generators; ENEE pays its short-term marginal avoided cost + 10% for generation; transmission tariff set at .01 ¢ USD/kWh; systems under 3 MW do not need a generation license; generators can sell directly into the Central American grid or to large consumers; permits may take a maximum of 2 months to process [52]
Mexico	Renewable energy can be "banked" and counted towards fulfilling the goal of satisfying customer's demand for the month [53]; accelerated depreciation for renewable investors [54]; lower transmission tariffs for renewables based on capacity factor [55]; new LAFRE law debated that would allow first dispatch and a green fund to support renewables, accelerated depreciation on profits, and lower transmission capacity charges base on the average capacity of the renewable source [56]
Nicaragua	No import tax for equipment, no income tax for first 7 years [57]; a maximum of 5.5 to 6.5¢ USD/kWh paid for hydro generation [58]
Panama	Projects under 500 kW do not pay equipment taxes; projects under 10 MW do not pay transmission and distribution charges; up to 5% of project costs can be reimbursed if the project contributes to national infrastructure development and up to 25% can be reimbursed based on the carbon reductions the project represents; projects greater than 10 MW can only obtain half of the carbon reduction potential; currently projects cannot receive both CERs and reimbursement for project costs based on carbon reduction [59]
Peru	Renewable energy exempt from value-added tax [60]; combined heat and power systems get priority dispatch [61]; geothermal resources promoted [61]; bill being discussed that would add feed-in tariffs and accelerated depreciation being formulated [62]
Uruguay	Elicitation for 20 MW of wind, 20 MW of small hydro, and 20 MW of biomass and 2008 bid for 26.2 MW; least-cost bid process only among renewable generators [63] and [64]

CER: Certified Emission Reduction, ENEE: Empresa Nacional de Energía Eléctrica, ICE: Instituto Costarriquence de Electricidad, IPP: Independent Power Producer, LAFRE: Ley para el Aprovechamiento de Fuentes Renovables de Energía.

PROS AND CONS OF RENEWABLE ENERGY LEGISLATION CHOICES

As evidenced from Chart 3, there is no regional integration of renewable energy policies. The variety of mandates and financial incentives for generation have various positive and negative implications with regards to policy efficiency and effectiveness. The three main policies used to promote renewable energy penetration in Latin American markets are described and discussed below.

Renewable Energy Mandate

A renewable energy mandate requires existing and new generators to source a certain percentage of their generation from renewable sources by a given target year. Binding mandates typically involve non-compliance fees that can serve to create a premium for renewable energy. If the mandate hopes to promote a certain type of technology, it can create a set-aside for it, requiring that a percentage of the overall renewable energy requirement be sourced from a given technology [65]. Twenty-five of the U.S. states now have voluntary or binding renewable energy mandates [66].

These mandates can also be made in terms of MW capacity additions. Uruguay, Brazil, and the U.S. state of Texas have mandates of this type. These mandates achieve a set amount of installations, but fail to take into account electrical growth rates.

Renewable energy mandates continue to allow market pressures to lower the cost of generation to exist. However, a mandate without set asides for different types of renewable energy does not help promote a diversity of technology types. Also, some administrative costs are imposed as regulated entities must be checked for compliance [65].

Production Tax Credit

PTCs can be a useful way to help renewables compete with other types of generation technologies by providing the difference in generation costs. But, a PTCs can also be a detriment to an industry if they are not consistently supported by the federal government. In the U.S. the PTC for wind generation has expired and been renewed several times, causing a boom and bust in the wind industry [67]. If a developer begins the process of measuring wind speeds on a site, buying or renting the land, earning the requisite land, generation, and environmental permits, and sources the wind turbines, but does not begin generation on the site before the PTC has expired, then he is not eligible for it. The time required to procure permits and turbines (especially during this global turbine shortage period) is so uncertain, that often wind and other renewable developers cannot rely on the PTC.

Also, countries with an unstable currency are at risk of having the PTC be meaningless if it is fixed in federal legislation that cannot be easily adjusted to reflect devaluation. This situation occurred in Argentina in 2002 when the Argentine Peso was devalued 30% to the US Dollar. The PTC was reset years later in 2006 to 1.5 Peso ¢/kWh for wind, hydro under 30 MW, biomass, and geothermal, and .9 Peso ¢/kWh for solar [30]. But, the PTC now fails to provide complete investor confidence as it could again become meaningless if the currency is devalued.

Feed-in Tariff

Feed-in laws require distributors to buy renewable energy at a fixed rate that is higher than the average wholesale market price and usually close to the retail price of electricity. Some feed-in laws are structured to pay different rates for different types of renewables, which are based on the cost of generation [65].

The benefit of feed-in laws is that they are an effective way of promoting renewables since developers can get loans for development more easily because the bank is assured that generation will be paid for a set price. However, critics of feed-in tariffs say they are not an efficient way to allow renewables to penetrate the market. Giving generators a set price for electricity provides little incentive for innovation that would lower generation costs since generators are not competing against each other in an energy auction [65]. Also, phasing feed-in tariffs out can be problematic as generators rely on the preferred tariff for their existence and financial viability. Phasing out the tariffs may involve waiting for the lifetime of the plants built under the tariff structure to expire. Germany, Denmark, and the U.S. with its Public Utilities Regulatory Policies Act (PURPA) have implemented feed-in tariffs [65].

Comparison

It is difficult to judge which of these renewable energy stimuli is best since the answer to this question must consider the environment where it is implemented. Chile is able to successfully implement a mandate because it has a stable economic climate and is ripe for foreign investors. Therefore, it is likely that the mandate will be filled by competitive bids from renewable generators. Using a direct subsidy in Chile may have incurred unnecessary "deadweight loss" or revenues that were used to pay generators that were in excess of the amount that generators had to pay to source the required amounts of generation.

Trying Chile's scheme to promote renewable energy development in Nicaragua, however, may prove to be a failure. In Nicaragua, new generators may have been hesitant to enter the marketplace until the final moment before non-compliance fees for regulated, existing generators came into effect. Existing generators probably would have to pay a premium for this generation that would be close to the non-compliance fee in order to incite generators to invest in such an economically and politically-risky country. These new generators would expect a higher rate of return on their investment than in a country like Chile because of these additional risks. In this example, using a mandate over a direct incentive may not result in less "deadweight loss" and could lead to a lag in developer interest.

Ecuador may have recognized this phenomenon when it chose to implement a feed-in tariff to promote renewables. While perhaps not the most elegant and efficient policy instrument, the feed-in tariff is appropriate for this country given its economic crisis of 1998 and rapid succession of presidential changes since 1995 [68].

Therefore, the best policy choice for each country is case and site specific. The country's current electrical sector structure, portfolio mix, and investment climate should be considered when making this decision.

Future Developments

The inability of many Latin American countries to structure policies that promote sufficient capacity additions draws into question the sustainability of power sector reform in the region. In addition to creating incentives for renewable energy development, some countries, like Ecuador, have begun to move towards renationalization of the electrical sector. President David Correa of Ecuador created a Ministry of Electrification and Renewable Energy in 2007 in order to allow the government to develop several large new hydro applications [69] and [70]. Costa Rica still continues to rely heavily on its state-run utility, ICE, for 875 MW of hydro capacity additions from 2006-2016 [57].

Just as the region's move to privatization is not fixed and could involve iterations of restructuring policies, several countries changed their renewable energy legislation multiple times in an attempt to stimulate the desired amount of renewable energy. Mexico passed regulations between 2001 and 2006 that promoted renewables through allowing renewable energy in contracts with CFE to be counted over a period of time of a month which is more favorable for renewables [53], accelerated depreciation on profits [54], and lower transmission capacity charges based on the average capacity of the renewable source [55]. However, because these laws have not stimulated adequate growth in the renewables market, the Law for Utilization of Renewable Energy (LAFRE), which would set renewable energy goals, give renewables priority dispatch, create a $55 million USD fund for renewable energy, and streamline permitting for small projects, is being debated in the Mexican Congress[56].

Likewise, Chile passed Short Law I with reduced transmission tariffs for renewables [26]. Then, when that law did not promote the desired development, it passed, Short Law II to require generators to source 5% of their energy from renewable sources by 2010 [38]. Then, just two years after Short Law II, Short Law III was passed to require generators to source 10% of their energy from renewable sources by 2024 [29]. Panama also had to revise renewable energy provisions in the 1997 Law 6 with a new Law 45 in 2007 to promote development [59]. The future will most likely bring more renewable energy legislation modifications. This period of trial and error creates a moving target for investors and developers that is hard to follow and does not provide an environment of regulatory certainty.

The longevity of the current renewable energy reforms will depend on the way each law is structured and amount of long-term support each receives. Chile's renewable mandate sets a medium-term goal of 2024 for 10% renewable generation that provides for some certainty in the renewable market. Uruguay's periodic tenders for renewable energy of 60 MW in 2006 and 26.2 MW in 2008 tend to promote only short-term development that cannot sustain an industry that hopes to open operations in the country [63] and [64]. Brazil has a combination of the short and long-term strategies with its two elicitations for discreet amounts of renewable generation in its Phase I of the PROINFA legislation. Then, Phase II calls for a 10% of the country's generation to be sourced from renewables by 2022. Brazil's legislation also takes steps to make renewable generation a permanent part of the country's industry by stipulating a local component requirement of 60% for Phase I and a predicted 90% for Phase II [27].

The recent high costs of fossil fuels have prompted this renewable energy legislation in the region; whether or not the policies are renewed and strengthened or abandoned will be determined largely by what happens to the prices of these commodities in the future. Brazil made a huge discovery of petroleum off its coast in December of 2007 [6]. This discovery has not yet impacted the renewable energy legislation and may not ever given the opportunity cost of using the petroleum domestically instead of exporting it. Other countries, with high political and economic risks like Honduras, may abandon their recent efforts to promote renewable energy when and if prices stabilize.

Climate Change Implications and the CDM

The implications that these actions have on lessening or enhancing climate change will be important. Although the Latin American energy sector only contributes to 4.9% of global greenhouse emissions [71], the rapid rate of sustained annual electrical growth in these countries that is as high as 7% in some countries like Peru [72], compared to an annual electrical demand growth rate of 1.7% in the U.S. from 1996-2006, means that this contribution to global emissions will continue to increase [73].

Additionality Complications with Renewable Energy Legislation

Renewable energy projects have the potential to earn additional revenues through the United Nations Framework Convention on Climate Change (UNFCCC)'s Clean Development Mechanism (CDM). The CDM allows projects that produce emission reductions that are additional to what would have occurred in a business-as-usual case (known as "additional") to count towards the reduction targets of Annex I countries of the Kyoto Protocol [74]. In this way, the CDM presents an enormous opportunity for renewable energy projects from technologies that are underdeveloped or regions that have financial or other types of barriers for project implementation to earn additional financing.

Unfortunately, the CDM does not complement the incentives and mandates of each country, but instead competes with them. If a project were to occur because of a national mandate or because domestic incentives make it financially viable, then it cannot qualify for the CDM because the UNFCCC logic is that the project would have occurred even in the absence of the CDM. The emission reductions sold from CDM projects are meant to be emissions reductions that would not have otherwise occurred and therefore can be used to fulfill binding reduction goals [74].

The UNFCCC CDM Executive Board (EB) has not made a clear ruling that allows individual projects promoted by a host country to be eligible for the CDM. The EB tried to address this issue by making a decree in 2005 that said "the baseline scenario need not take these policies into account if the policy was implemented since the adoption of the CDM Modalities and Procedures in November, 2001 [75]." This decision, however, only impacts the number of emission reductions that renewable energy projects that can prove they are additional can earn. It is silent on whether or not these projects can prove their "additionality." Therefore, the CDM creates a perverse incentive for countries to not implement policies that promote renewable energy.

Additionality Complications in the Programme of Activities

In July of 2007, the EB once again tried to address this pitfall by creating a Programme of Activities (PoA), which promotes activities in specific industrial sectors, allowing multiple projects implemented at different times that comply with a governmental regulation or private sector initiative to be registered together. This PoA, therefore, provides an incentive for developing countries to devise policies that promote greenhouse gas reductions [76]. A host country policy itself cannot qualify for CDM, but programs implemented under or as a result of the policy can. However, the rules of the programmatic CDM are not clear with regards to whether or not the CDM project would be able to demonstrate regulatory additionality [77].

Currently, a project that fulfills a national mandate like a renewable portfolio standard and attempts to register as a PoA may not be able to prove its additionality. Countries with financial incentives for projects that mitigate greenhouse gases and attempt to register as a PoA may also face hurdles trying to show that the project needed to rely on CDM revenues for existence or financial additionality. These complexities are probably part of the reason why no PoA project has yet been registered. Furthermore, this programmatic CDM ruling still does not change the rules for individual renewable energy projects that try to prove additionality [78].

The CDM's exclusion of host country-supported projects has already caused some countries like Panama to take the CDM into account when devising renewable energy legislation. Panama's laws to support renewable energy provide aggressive incentives that cover up to 25% of the initial project costs for renewable energy generation. However, this law prevents CDM projects from earning both this domestic financial incentive and Certified Emission Reductions earned under the CDM [59]. Other countries are hesitant to make renewable energy legislation too strong since they are being advised by carbon brokers of the deterrent it will create for utilization of the CDM [79].

For greenhouse gas emissions to be reduced through and apart from the CDM, a clear decision that allows renewable energy legislation to exist apart from and not compete with the CDM needs to be made. Devising this decision will be difficult as the integrity and purpose of the CDM would be compromised if the reductions sold from projects were not additional. Perhaps this recent wave of renewable energy legislation represents the transition from the CDM to developing country baseline targets that would eliminate or drastically change the nature of the CDM.

Additionality Complications for State-run Electric Power Companies

The additionality criterion can not only complicate development in countries with renewable energy legislation, but it can also create headaches for state-run electrical entities that try to apply for CDM registration. These state entities are bound, by law, to provide the least-cost source of electrical generation for the country's citizens. Many CDM projects prove their additionality through a financial argument that shows that without CDM revenues, the project would not have been economically viable. Since state-run entities are not allowed to consider these revenues in the least-cost planning process since they are not certain, state entities cannot prove that any renewable energy project that they construct is financially additional [80].

Also, state entities are usually bound to implement only new projects that are in the country's capacity expansion plan. Only small-scale projects that do not significantly impact a country's grid that are not in this plan can be pursued by the state entity [53]. However, the description of the project in the country's expansion plan makes it seem as though it is business-as-usual and would have been pursued with or without CDM revenues, which complicates the process of proving additionality.

Therefore, private renewable developers are at an advantage in that they are able to better access CDM funds. Countries like Costa Rica and Mexico with primarily state-run electrical sectors are at a disadvantage to earning CDM revenues for energy applications [53, 80].

CONCLUSION

Restructuring of the electrical sector has had profound impacts on the Latin American electrical generation mix. The large hydro applications of the past, which comprise the majority of the region's generation, are no longer being constructed to fulfill new demand requirements. Now fossil fuel generation is being contracted in short-term PPAs for high prices to fulfill emergency demand requirements as the newly-privatized electrical sectors have not, in general, been able to elicit the requisite capacity additions. Some countries have considered renationalizing the electrical sector in order to provide the financial backing for new hydro applications. Other countries are pursuing renewable energy legislation to provide incentives that will allow the private sector to take on these large, capital intensive new capacity additions.

The impact this renewable energy legislation has and will continue to have will be as varied as the actual policies used to promote this type of generation. The future prices of fossil fuels will determine the political will that makes these programs sustainable or temporary. Renewable energy policy interaction with international greenhouse gas mitigation programs like the Clean Development Mechanism is complex. Unless new clear guidance is provided for future project developers, CDM development could be deterred in countries with renewable energy legislation because of additionality requirements. Likewise, countries with sectoral energy policies for mitigation and state-run electrical sectors could face challenges in registering energy-related, CDM projects.

REFERENCES

[1] Energy Sector Management Assistance Program, Latin America and the Caribbean Region (LCR): *Energy Sector – Retrospective Review and Challenges* June 15, 2007

[2] Salgado, G., CDM Consultant, former Designated National Authority of Honduras. September 11, 2007: Tegucigalpa, Honduras.

[3] Pombo, C., *Regulatory Reform in Colombia's Electric Utilities.* Revista de Economia de Universidad del Rosario, May 2001.

[4] Joval, J.R., *The impact of the 1997-1998 El Niño on the Andean Community of Nations.* El Centro Regional de Información sobre Desastres.

[5] Arango, S., I. Dyner, and E. Larsen, *Lessons from deregulation: Understanding electricity markets in South America. Utilities Policy,* September 2006. 14(3): p. 196-207.

[6] CBS News, *Brazil Oil Find May Be World's 3rd Largest.* April 14, 2008.

[7] Millan, J. and N.-H.M.v.d. Fehr, *Keeping the Lights On: Power Sector Reform in Latin America.* 2003, Washington, D.C.: Inter-American Development Bank.

[8] World Bank, *Benchmarking Data of the Electricity Distribution Sector in Latin America and the Caribbean 1995-2005.* 2005.

[9] Tasende, D., *Director of Renewables,* UTE. November 27, 2007: Montevideo, Uruguay.

[10] Villa, G., *Director of Energy within Ministerio de Ambiente y Energy,* Costa Rica. September 27, 2007: San Jose, Costa Rica.

[11] Broide, A., *Development Manager for Mesoamerica Energy.* September 26, 2007: San Jose, Costa Rica.

[12] Ley del Servicio Público de Energía Eléctrica, *Ley del Servicio Público de Energía Eléctrica,* in Articulo 3°. December 23, 1992.

[13] Millán, J., The Power Sector in: *Panama, in Profiles of Power Sector Reform in Selected Latin American and Caribbean Countries.* September 1999, Inter-American Development Bank.

[14] Development, O.f.E.C.a., *OECD Economic Survey of Brazil 2005: Regulation of the Electricity Sector in OECD Economic Survey of Brazil.* 2005.

[15] Millan, J., *The Power Sector in Chile, in Profiles of Power Sector Reform in Selected Latin American and Caribbean Countries*, I.-A.D. Bank, Editor. 1999.

[16] Renewable Energy and Energy Efficiency Partnership, *LAC Policy Descriptions: Nicaragua.* Sustainable Energy Policy Initiative.

[17] Rudnick, H. and J. Zolezzi, Electric sector deregulation and restructuring in Latin America: lessons to be learnt and possible ways forward. *IEE Proceedings - Generation,* Transmission, and Distribution, March 2001. 148(2).

[18] Cardona, E.Z., *Gerente General de Colegio de Ingenicios Mecanicos Electricistas y Quimicas* (former director of AHPPER). September 10, 2007: Tegucigalpa, Honduras.

[19] Ministerio de Energia y Minas (Direccion de Electricidad), *Ley que Modifica Diversos Artículos de la Ley de Concesiones Eléctricas.* December 14, 1999.

[20] Ministerio de Energia y Minas (Direccion de Electricidad), *Ley que Modifica Diversos Artículos y Definición Anexa de la Ley de Concesiones Eléctricas.* September 24, 1998.

[21] Ministerio de Energia y Minas (Direccion de Electricidad), *Ley de Promoción del Desarrollo de la Industria del Gas Natural* June 3, 1999.

[22] Netherlands CDM Facility, *Poechos I Project Design Document.* 2005.

[23] Center for Energy Economics in the Bureau of Economic Geology at the University of Texas in Austin, *Case Study: Gas and Power in Peru.* March 27, 2006.

[24] Energy Information Administration, *This Week in Petroleum.* April 30, 2008.

[25] Frias, C.A., *Especialista Area Ingenieria.* November 18, 2007: Santiago, Chile.

[26] Ministerio de Economia Fomento y Reconstruccion, Ley Corto I: Regla Sistemas de Transporte de Energia Electrica, Establece un Nuevo Regimen de Tarifas para Sistemas Electricos Medianos, *y Introduce Adecuaciones que Indica a la Ley General de Servicios Electricos.* March 13, 2004: Diario Oficial de la Republica de Chile.

[27] Valle, C.d. *Renewable Energy Policy Brazil. in Centro Clima: Center for Integrated Studies on Climate Change and the Environmen*t. Available through: Renewable Energy Policy Network for the 21st Century.

[28] Iglesias, R. *Proyecto de Ley para el Desarrollo de las Energias Renovables no Convencionales. in CORFO Chile Invest.* November 16, 2007. Santiago, Chile: Comision Nacional de Energia.

[29] Ministerio de Economía Fomento y Reconstrucción de Chile, Ley N° 20.257, *Comision Nacional de Energia de Chile,* Editor. April 1, 2008: Diario Oficial de la Republica de Chile

[30] El Senado y Cámara de Diputados de la Nación Argentina, *Regimen de Fomento Nacional para el Uso de Fuentes Renovables de Energia Destinada a la Produccion de Energia Electrica:* Ley 26,190, in Boletin Oficial July 5, 2000.

[31] Garcia, A., *Project Developer for ABO Wind.* November 22, 2207: Buenos Aires, Argentina.

[32] Carrión, R., *CONELEC Administrator in Planning.* October 26, 2007: Quito, Ecuador.

[33] Coviello, M.F., Renewable Energy Sources in Latin America and the Caribbean: Two years after the Bonn Conference. *ECLAC, GTZ, and One World,* April 2007

[34] Neira, D., Van Den Berg, Berend y De la Torre et al., *El Mecanismo de Desarrollo Limpio en Ecuador: Un diagnostico rapido de los retos y oportunidades en el Mercado de Carbono.* Banco Interamericano de Desarrollo and Ministerio del Ambiente and Corporacion Interamericana de Inversiones, 2006.

[35] Lokey, E., Identifying and Overcoming Barriers to Renewable Energy Clean Development Mechanism Projects in Latin America. Ph.D. *Dissertation at the University of Colorado at Boulder,* May 2, 2008.

[36] Programa Electricidad de Ministerio de Hidrocarburos y Energia. *Normativa. April 2008* [cited; Available from: http://www.hidrocarburos.gov.bo/07_NORMATIVA/normativa.php.

[37] Programa Electricidad de Ministerio de Hidrocarburos y Energia. *Electricidad para vivir con dignidad.* April 2008 [cited; Available from: http://www.hidrocarburos.gov.bo/07_PLAN/plan.php.

[38] Ministerio de Economia Fomento y Reconstruccion, *Ley Corto II: Modifica el Marco Regulatorio del Sector Electrico*. May 19, 2005.

[39] *Unidad de Pleanación Minerio Energética,* Decreto No. 3683. December 19, 2003.

[40] Cardonas, A., *Administrator Ministerio de Energia y Minas October 10,* 2007: Bogota, Colombia.

[41] Ministerio del Interior y Justicia, *Decreto 2755.* September 30, 2003: Diario Oficial 45,326.

[42] Millán, J., *The Power Sector in: Costa Rica, in Profiles of Power Sector Reform in Selected Latin American and Caribbean Countries.* September 1999, Inter-American Development Bank.

[43] CEPAL and GTZ, *Fuentes Renovables de Energia en America Latina y el Caribe: Situacion y Propuestas de Politica.* May 19, 2004.

[44] Congreso Nacional de La Republica Dominicana, *Ley de Hidrocarburos* 112-00. 2000.

[45] Congreso Nacional, Ley No. 5707 *sobre Incentivo al Desarrollo de Fuentes Renovables de Energía y de sus Regímenes Especiales.* May 2007.

[46] Congreso Nacional de La Republica Dominicana, *Ley General de Electricidad Ley* No. 125-01. 2001.

[47] Ley de Beneficios Tributarios para Nuevas Inversiones Productivas, Generacion de Empleo, *y Prestacion de Servicios.* November 18, 2005, Registro Oficial.

[48] Coviello, M.F., Renewable Energy Sources in Latin America and the Carribbean: Two Years After the Bonn Conference. *United Nations Economic Commission for Latin America and the Caribbean*, April 2007.

[49] Sanchez, I.A., *Estudio Sobre la Apllicacion del Mecanismo para un Desarrollo Limpio en El Salvador*. Ministerio de Medio Ambiente y Recursos Naturales, La Cooperación Internacional de Japón, and Universidad Centroamericana, June 2006.

[50] Red de Oficinas Económicas y Comerciales de España en el Exterior. El Salvador aprueba una Ley con incentivos para inversiones en Energia Renovable. November 10, 2007 [cited; Available from: http://www.oficinascomerciales.es/icex/cda/controller/pageOfecomes/0,5310,5280449_5282927_5284940_4030184_SV,00.html

[51] Ruiz, O., *Head of the Center of Information and Promotion of Renewable Energy, Ministerio de Energia y Minas*. September 7, 2007: Guatemala City, Guatemala.

[52] Decreto 70-2007, in *La Gaceta: Diario Oficial de La Republica de Honduras Comision Nacional de Energia,* Editor. October 2, 2007: Tegucigalpa, Honduras.

[53] Barnes de Castro, F., *Commissioner of Comision Regulatoria de Energía.* August 30, 2007.

[54] World Bank, *Project Information Document: Appraisal Stage for La Venta III.* April 26, 2006.

[55] Comisión Regulatoria de Energía, *Resolución Num.* RES/140/2001. 2001.

[56] Klimek Alcaraz, O., *Ley para el Aprovechamiento de las Fuentes Renovables de Energía,* Mexican Parliament, Editor. February 2006.

[57] Synergy de la Comunidad Europea, *Metodologías para la implementación de los mecanismos flexibles de Kioto – Mecanismo de Desarrollo Limpio (MDL)*- Guía Latinoamericana del MDL.

[58] Tellez, R.N. and M.A.A. Zeas, *Ley Para la Promocion de Energia Electrica de Fuentes Renovables, Asemblea Nacional de Nicaragua*, Editor. April 14, 2005: Normas Jurídicas de Nicaragua.

[59] Ministerio de Economia y Financas de Comision de Politica Energetica de Panama, *Legislative Assembly,* Law 45. August 4, 2004.

[60] Ministerio de Energia y Minas (Direccion de Electricidad), *Decreto Supremo: Aprueban Reglamento de la Ley que amplía los alcances del Régimen de Recuperación Anticipada del Impuesto General a las Ventas a las Empresas de Generación Hidroeléctrica* March 29, 2007.

[61] Ayon, H., Gerente de Finanzas de Paramonga. November 7, 2007: Lima, Peru.

[62] Gutiérrez, D.C., *Organizacion Supervisor de la Inversion de Energia administrator.* November 6, 2007: Lima, Peru.

[63] Kasprzyk, M., *Designated National Authority in the Ministerio de Vivienda, Ordenamiento, Territorial, y Medio Ambiente, Division de Cambio Climatico.* November 27, 2007: Montevideo, Uruguay.

[64] Parte Uno - Pliego de Condiciones Particulares para Pliego de Condiciones y Especificaciones para la realización de CONTRATOS DE COMPRAVENTA DE ENERGIA ELÉCTRICA por parte de UTE a proveedores instalados en el territorio nacional que produzcan dicha energía utilizando como fuente primaria energía eólica, de biomasa, o de pequeñas centrales hidráulicas, in P37637, *Administracion Nacional de Usinas y Trasmisiones Electricas,* Editor. March 6, 2008.

[65] Komor, P., *Renewable Energy Policy.* 2004, New York: Diebold Institute for Public Policy.

[66] Bird, L., L. Dagher, and B. Swezey, *Green Power Marketing in the United States: A Status Report.* December 2007.

[67] U.S. Energy Information Administration, *Production Tax Credit for Renewable Electricity Generation Issues in Focus,* 2005(AEO2005).

[68] The Heritage Foundation and The Wall Street Journal. *Index of Economic Freedom 2008* [cited; Available from: http://www.heritage.org/index/country.cfm?id=Ecuador.

[69] Oliva, P. and A.R. Moreno, *CONELEC Administrators in the Unidad de Gestion Ambiental.* October 26, 2007: Quito, Ecuador.

[70] Bustamente, M.M., *Administrator for CENACE.* October 26, 2007: Quito, Ecuador.

[71] The World Bank. WDI (2008) *World Development Indicators*. [cited; Available from: www.worldbank.org.

[72] Public Broadcasting Systems. *Peru: Commanding Heights.* [cited; Available from: http://www.pbs.org/wgbh/commandingheights/lo/countries/pe/pe_economic.html.

[73] Energy Information Administration, *World Total Net Electricity Generation, Most Recent Annual Estimates,* 1980-2006. September 13, 2007.

[74] Fenhann, J., et al., *CDM Information and Guidebook 2nd Edition. UNEP/Capacity Development for CDM,* June 2004.

[75] Figueres, C. and E. Haites, Policies and Programs in the CDM. *International Institute for Sustainable Development (IISD) in the context of the Development Dividend Project,* July 2006.

[76] Figueres, C. and K. Newcombe, *Evolution of the CDM: Toward 2012 and Beyond.* World Bank, July 2007.

[77] Sales, R. and B.K. Sabbag, *Legal Compliance with Environmental Requirements Impacting Assessment and Demonstration of Additionality in Clean Development Mechanisms: A Legal Review under the UNFCCC, the Kyoto Protocol and the Brazilian Legal Framework on Climate Change.* Baker & McKenzie's Brazilian Environmental and Climate Change Practice Group, January 2006.

[78] CDM UNFCCC. *Project Search.* May 1, 2008 [cited; Available from: http://cdm.unfccc.int/Projects/projsearch.html.

[79] Estrada, M., CDM *Consultant for Terracarbon.* August 18, 2007: Mexico City, Mexico.

[80] Cordero, F. and G. Mayorga, *Instituto Costarricense de Electricidad.* 2007.

In: Energy Policy: Economic Effects, Security Aspects... ISBN 978-1-60692-894-3
Editor: Noah B. Jacobs

Chapter 6

BIOMASS UTILIZATION AND PRICE SETTING STRATEGIES UNDER THE CLEAN DEVELOPMENT MECHANISM

Yoshihiro Yamamoto
EcoTopia Science Institute, Nagoya University, Japan

ABSTRACT

This chapter offers an assessment of projects to reduce greenhouse gas emissions, in particular projects involving biomass utilization, under the Clean Development Mechanism (CDM) of the Kyoto Protocol, as well as an analysis of price setting strategies for investing firms and host country governments involved in CDM projects. Drawing on a published database of registered CDM projects and project design documents, the assessment presented here shows that biomass utilization projects are particularly prevalent—and are of great significance—in Asia. To analyze price setting strategies, we employ a simple microeconomic model in which a foreign firm sets the price at which it purchases certified emission reduction credits obtained in a CDM project, and a host country government owning or regulating a utility or electricity distribution company sets the price at which it purchases the electricity the project produces. The model shows that the valuation of the social benefit of projects by host country governments plays a critical role in price setting strategies.

1. INTRODUCTION

Article 12 of the Kyoto Protocol establishes a Clean Development Mechanism "to assist Parties not included in Annex I in achieving sustainable development and in contributing to the ultimate objective of the Convention, and to assist Parties included in Annex I in achieving compliance with their quantified emission limitation and reduction commitments." (Note that "Parties" here refers to signatories to the United Nations Framework Convention on Climate Change [UNFCCC] treaty). The concrete procedures of the CDM are primarily stipulated in the Marrakesh Accords produced at the seventh session of the Conference of the Parties (COP 7) in 2001.

In general, the CDM is designed to enable entities in Annex I Parties (primarily developed nations) to invest in projects aimed at reducing greenhouse gas emissions in non-Annex I Parties (primarily developing nations), and in exchange receive tradable emission permits, or certified emission reductions (CERs). CERs are issued after the completion of a specified process that includes the submission of a project design document (PDD) as well as approval, validation, registration, monitoring, verification, and certification of the project. (For detailed information about the CDM, see the UNFCCC website.)

The first registration of a project under the CDM occurred in November 2004, and since then projects have been registered at an increasing rate, resulting in a recent total of 1162 projects. CDM projects vary widely in type (defined by factors such as inputs and outputs), scale, and location. The first section of this chapter offers an assessment of the current situation of CDM projects, in particular those utilizing biomass, which account for almost one fifth of all registered projects. The data for this assessment are drawn from the database of the Institute for Global Environmental Strategies (IGES), which is one of two organizations that receive and disseminate CDM data under an agreement with the UNFCCC secretariat (the other is the UNEP Risoe Center). Our assessment shows that, as indicated above, one fifth of all registered CDM projects are biomass utilization projects, and that moreover three-fourths of these are located in Asia, and more than one half are small-scale.

Under the CDM, CERs issued for a given project may be sold to firms in Annex I Parties that are not necessarily the same entities that invested in the project, and purchasing entities can apply these CERs to meeting their qualified emission limitation and reduction commitments. If an entity in an Annex I Party is interested in purchasing CERs from a project in which it has invested, however, it faces a dilemma with regard to pricing: as the price of CERs decreases, the volume of CERs issued, project outputs such as electricity, and hence, distributed profits decrease as well.

The governments of CDM project host countries often face a dilemma with regard to the pricing of project outputs as well. The electricity produced by projects such as hydroelectric or biomass projects (designed to substitute for conventional, fossil-fuel electrical generation) is often purchased by utilities or electricity distribution companies that are owned or regulated by host country governments. As the price at which such a utility (frequently a government agency), purchases this electricity decreases, the benefits resulting from the project -- such as improvements in the local environment, greenhouse gas emissions reductions, increased labor opportunities, and regional economic development -- decreases as well.

Thus, the CDM involves two important price settings. These price settings interact, since the sum of the resulting prices affects the volume of CERs issued and energy produced.

The second section of this chapter presents an analysis of price setting strategies for entities such as foreign firms that both invest in CDM projects and purchase CERs, and for host country governments that own or regulate utility or electricity distribution companies. To analyze these strategies, we employ a microeconomic model to investigate two scenarios. In the first, a firm in an Annex I Party that is an investor in a CDM project first commits to purchasing CERs produced by the project at a given price, and after this the government sets the price at which it will purchase electricity produced by the project. In the second scenario, the order of these two price settings is reversed. The model that we employ is a simple one; it is a direct application of a sequential-moves game similar to the Stackelberg model, which appears in standard microeconomic and industrial organization textbooks such as Varian (1992) and Shy (1995). One of the main findings that our model yields is that the valuation by the host country government of the social benefit of a project plays a critical role in price setting.

The model presented here may relate to that of Al-Saadon et al. (1996). In their model, a foreign firm and a host firm set up a joint venture. The foreign firm is the sole producer of the input, for which it sets the price. The government of the host country, however, imposes a production tax on the joint venture in order to maximize social welfare. The foreign and host firms negotiate the distribution of equity shares. Al-Saadon et al. investigate and compare negotiation outcomes for two cases, one in which the foreign firm sets the input price, and the other in which the government sets the tax rate. For each case, Al-Saadon et al. compare the outcomes when the negotiation between the foreign and host firms is completed before the price or tax rate setting, and when it is completed after the setting.

Like that of AL-Saadon et al., our model posits a project involving a foreign firm and a host country government in which the foreign firm sets a CER price or the government sets a wholesale electricity price according to their respective interests. However, our study focuses on the interaction between the foreign firm and the government, rather than the negotiation between the foreign and host firms over the distribution of equity shares.

2. Biomass Utilization and Small-Scale Projects under the CDM

2.1. Biomass Utilization under the CDM

This section presents an assessment of CDM projects registered with the CDM executive board. This assessment is based upon both the IGES database and project design documents (PDDs) of registered projects, which are available on the website of the UNFCCC.

The number of CDM projects has increased steadily since the first project was registered in November 2004 (Figure 1). A total of 1162 projects had been registered as of September 18, 2008.

The IGES database assigns each project to one of 16 categories, based on the description of project activity in the PDD of the project and the methodology employed in drawing up the PDD. The categories comprise the following: hydroelectric power generation, biomass utilization, biogas production, wind-power generation, methane recovery and utilization, waste heat/gas utilization, energy efficiency improvement, fuel switch, N_2O decomposition, activities related to cement manufacturing, HFC reduction, methane production avoidance, other renewable energies use, activities related to transportation, afforestation and reforestation, and PFC reduction. The category of methane recovery and utilization mainly comprises landfill gas recovery projects, with or without power generation, whereas the methane production avoidance category generally comprises composting projects. The category of other renewable energies use principally includes solar power generation and geothermal generation projects. Note that cattle excreta or livestock waste are usually considered biomass, but that the IGES database categorizes projects employing these inputs not as biomass utilization, but rather as biogas production.

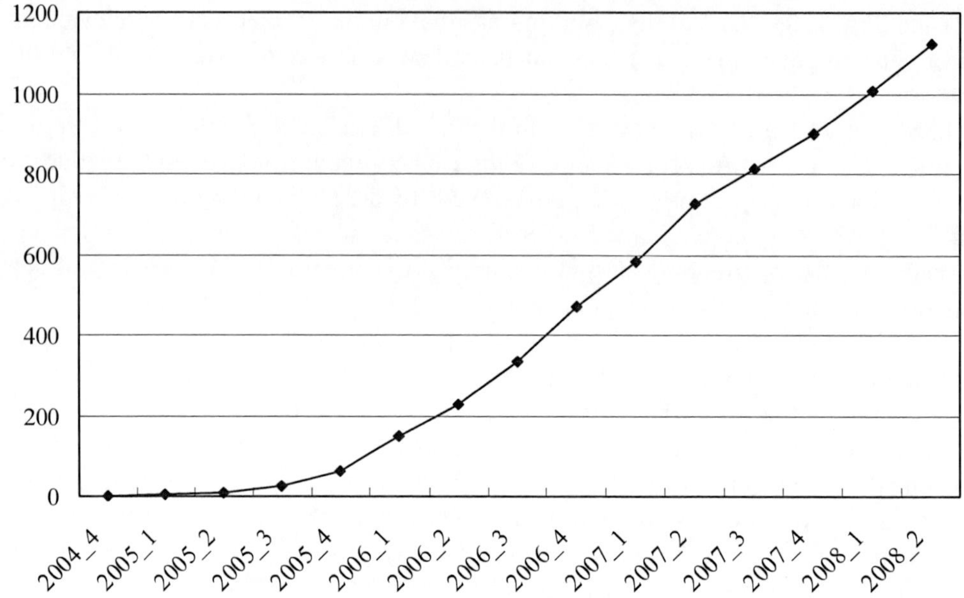

Figure 1. Cumulative quarterly totals of registered CDM projects since inception of CDM.

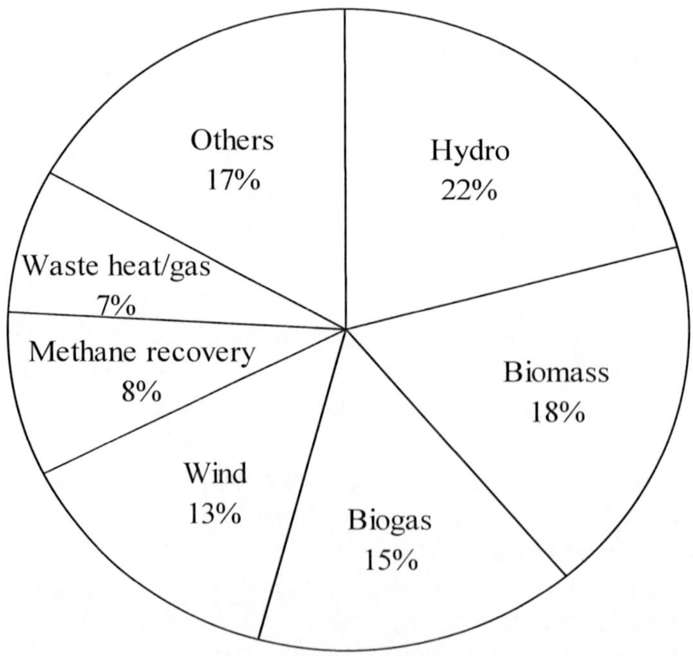

Figure 2. CDM projects by category.

Figure 3. CDM projects by host country.

Hydroelectric power generation projects constitute the largest share of registered projects with 22 percent of the total (Figure 2). Biomass utilization projects constitute the second largest share with 18 percent. Biogas production projects follow with 15 percent, wind-power generation projects with 13 percent, methane recovery and utilization projects with 8 percent, and waste heat/gas utilization projects with 7 percent of the total.

Among countries hosting CDM projects, India has the largest share of registered projects with 31 percent (Figure 3). China hosts the second largest share with 23 percent. Brazil follows with 12 percent, and Mexico with 9 percent. With regard to continental regions, Asia has the largest share of registered projects with 63 percent of the total, followed by South and Central America with a combined 32 percent.

Of the total of registered biomass utilization projects, India has the largest share with 57 percent (Figure 4), followed by Brazil with 17 percent, Malaysia with 7 percent, and China with 5 percent. Among continental regions, Asia has the largest share of biomass utilization projects with 75 percent of the total, followed by South and Central America with a combined 24 percent. Comparing the Asian share of all registered projects with that of biomass utilization projects (63% to 75%), one observes a concentration of biomass utilization under the CDM in Asia, which suggests an abundance of biomass resources in the region.

Biomass utilization accounts for 210 of the total of 1162 CDM projects registered, which amounts to an 18 percent share, as mentioned above. Examination of the PDDs of the biomass utilization projects reveals that some projects use a single type of biomass which is clearly identified, but that others use multiple types which are not specifically identified in their PDDs. Rather, the multiple types of biomass used in such projects are glossed as "other" in the IGES database, in which information from the PDDs is compiled.

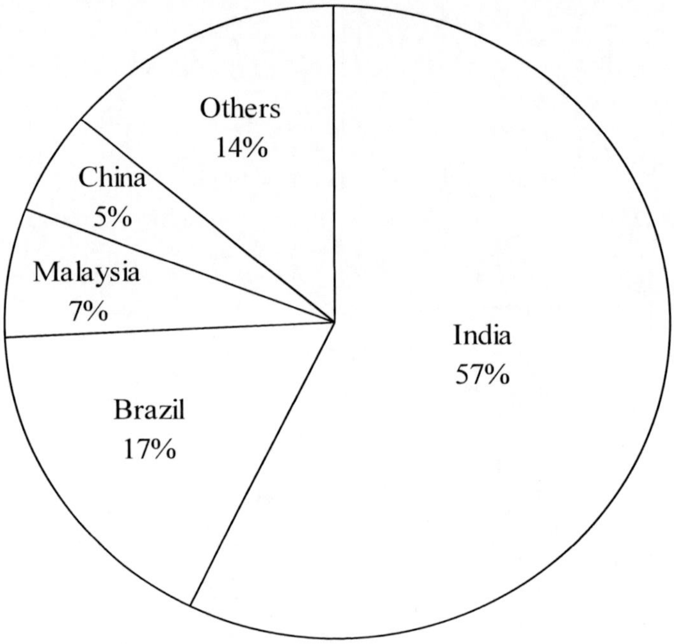

Figure 4. CDM biomass utilization projects by host country.

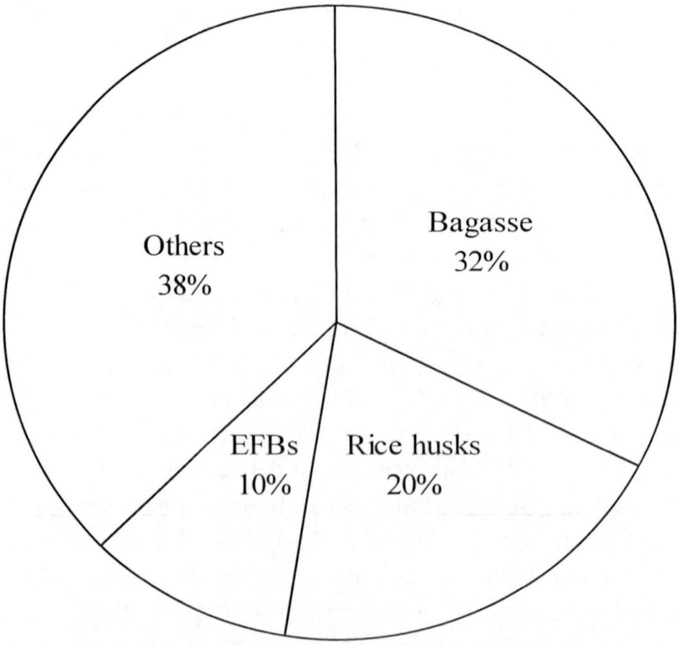

Figure 5. Biomass utilization project inputs.

The biomass most widely employed by projects using a single type of biomass is bagasse, a by-product of a sugar production process, which is used in 32 percent of the 210 biomass utilization projects (Figure 5). The next most widely used biomass is rice husks, which are used in 20 percent of projects. Empty fruit bunches (EFBs), a solid waste generated during a palm oil extraction process, are used in 10 percent of projects, the next largest share. Asia produces large amounts of bagasse, rice husks, and EFBs, and this is reflected in the large share of Asian biomass utilization projects using these inputs. Other biomass projects use waste wood, saw dust, corn residue, black liquor from a pulping process, briquettes, cotton stalks, soya bean stalks, wheat straw, coconut residue, coffee husks, prosopis juliflora, and mustard crop residue, among other products.

More than 90 percent of biomass utilization projects produce electricity, and approximately three-quarters of these sell the electricity to electricity distribution companies or utilities. Thus, approximately 70 percent of biomass utilization projects generate electricity that is grid-connected. Electricity produced by such projects that is not sold is used in-house. The PDDs do not show the tariffs on electricity sold by projects. It is often the case in developing countries that wholesale power markets are monopsonies, and that electricity distribution companies and utilities are state-owned or regulated.

2.2. Small-Scale CDM Projects

Some types of projects are best suited to a small scale. For example, a project that uses agricultural residues such as rice husks may be most efficient as a small-scale operation since the increased cost of transporting rice husks to a central facility outweighs the benefits of operating such a project at a large scale. Nonetheless, small-scale projects would be difficult to develop under the CDM if the transaction costs for registration were the same irrespective of project scale, since profit margins are smaller for small-scale projects.

To promote small scale projects, the CDM therefore defines small-scale project types for which it provides reduced transaction costs. Three types are defined by output constraints, as follows (FCCC/KP/CMP/2006/10/Add.1):

a) Type I project activities are such that renewable energy project activities shall have a maximum output capacity of 15 MW (or an appropriate equivalent).
b) Type II project activities or those relating to improvements in energy efficiency which reduce energy consumption, on the supply and/or demand side, shall be limited to those with a maximum output of 60 GWh per year (or an appropriate equivalent).
c) Type III project activities shall be limited to those that result in emission reductions of less than or equal to 60 kt CO_2 equivalent annually. Small-scale afforestation and reforestation projects are also defined under the CDM as follows (FCCC/KP/CMP/2005/8/Add.1):
d) Small-scale afforestation and reforestation project activities are those that are expected to result in net anthropogenic greenhouse gas removals by sinks of less than 8 kilotonnes of CO_2 per year and are developed or implemented by low-income communities and individuals as determined by the host Party.

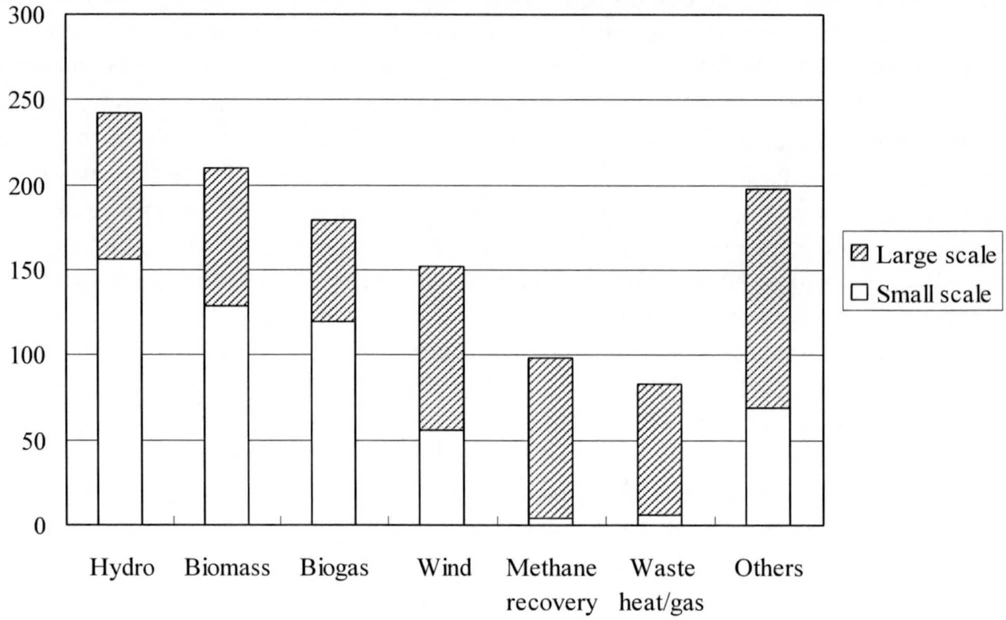

Figure 6. Scale of CDM projects by project type.

The modalities and procedures for registering small-scale CDM projects are simplified as follows in order to reduce transaction costs (FCCC/KP/CMP/2005/8/Add.1): projects or portfolios may be bundled at the stage of PDD drafting, validation, registration, monitoring, and verification and certification; the requirements for the PDD are reduced; baseline methodologies are simplified; monitoring requirements and plans are simplified; and the same designated operational entity may undertake validation, and verification and certification.

In addition, small-scale CDM projects must be registered within four weeks—as opposed to eight weeks for large-scale CDM projects—after the date of receipt by the EB of a request for registration, unless a Party involved in the project or at least three members of the EB request a review of the project proposal (FCCC/KP/CMP/2005/8/Add.1).

Moreover, it is easier for small-scale CDM projects to verify their additionality than for large-scale projects to do so. Participants in small-scale CDM projects have only to show that the project would not have occurred outside the CDM due to at least one of the following: an investment barrier, a technological barrier, prevailing practice, or other barriers (Appendix B of the simplified modalities and procedures for small-scale CDM project activities).

Analysis of the data in the IGES database shows that approximately 46 percent of total registered CDM projects are small-scale. As discussed above, hydroelectric power generation, biomass utilization, and biogas production are the three most common types of registered project. The majority of projects of these types are small-scale; 64 percent of hydro power generation, 61 percent of biomass utilization, and 67 percent of biogas production are small-scale projects (Figure 6). The difference of these shares implies that the ratio of small-scale to large-scale CDM projects may vary according to project type.

The CDM establishes incentives to promote small-scale projects, as discussed above. On the other hand, some countries have adopted measures of their own to promote small-scale power generation. According to the World Bank (2008), several developing counties have Small Power Producer (SPP) programs to streamline the development process for small renewable-energy power projects. These SPP programs accord with the CDM provisions and institutions that promote small-scale projects, though they have been developed independently. Such programs "will often include the development and implementation of standardized power purchase agreements, tariff setting procedures, interconnection regulations, and technical standards" (World Bank 2008). The SPP programs of five Asian nations, Thailand, Indonesia, India, Sri Lanka, and Vietnam, share key similarities; notably, these nations either deliberately or de facto employ a standardized Power Purchase Agreement (PPA), as well as an avoided cost concept to establish the SPP tariff structure (World Bank 2008). As Ferry (2006) argues, tariff design includes renewable premiums, tariff floors and subsidies, price bidding, renewable exemptions, tariff incentives, capacity tariffs, and a standardized PPA.

As discussed above, the tariffs on electricity generated in biomass utilization projects and sold to utilities and distribution companies and the prices of CERs obtained from CDM projects are not shown in PDDs. Nonetheless, the tariff on grid-connected electricity and the price of CERs have a great effect on the viability of CDM projects.

3. Microeconomic Analysis

This section presents an investigation of price-setting strategies for these tariffs and prices. Suppose that a firm in an Annex I Party, firm F, and a firm in a non-Annex I Party, firm H, establish a joint venture, firm J, that undertakes a CDM project in firm H's country. Firm J is a special-purpose company established for this project which generates electricity using biomass such as bagasse, rice husks, or EFBs. The electricity produced by firm J is sold to a grid company at a price, r. This grid company is owned or regulated by the government of the host country, so it is plausible that the government determines r. The CERs that are obtained as by-products are purchased exclusively by firm F at a price, p, that firm F determines. Let us assume that firm F values one unit of CER at p_0. It is furthermore plausible, arguably, that firm F restricts this CER price to being less than or equal to its valuation, i.e., $p \le p_0$, because the department of the firm that deals with foreign direct investment, and the department that deals with environmental concerns such as the purchase of CERs, have independent accountings. The marginal cost for firm J of biomass power generation is assumed to be cx when the output is x, and its fixed costs are f, with c and f both being positive constants. The reason that the marginal cost increases with output levels is that the greater the output, the more costly it becomes to gather biomass, and the less efficient the boiler used for power generation. The profits of firm J are divided between firms F and H, according to a sharing rule determined outside the model; let us assume that they distribute profits equally between themselves.

Under these assumption, the profits of firm J, π_J, are

$$\pi_J = (p + r)x - \left(\frac{cx^2}{2} + f \right) \tag{1}$$

By the first order condition, the optimal output levels are $(p+r)/c$, and hence the profits of firm J are $(p+r)^2/(2c)-f$. In order for the project to be implemented, $p+r \geq \sqrt{2cf}$ must hold.

Assume that $p_0 < \sqrt{2cf}$, since otherwise a CDM project could be undertaken even if the government did not set the wholesale electricity price from a political viewpoint, that is, to make the project feasible.

The government determines price r so as to maximize the sum of firm H's profits, the utility or distribution company's payoffs, and social benefit. The payoffs for the distribution company are represented by $(r_0 - r)x$, where r_0 is either the cost of conventional generation if the company generates electricity by itself, or the wholesale price of conventional electricity if it purchases electricity from another power generation company. Let us normalize r_0 to zero for analytical simplicity. Social benefit includes improvement of the local environment, reduction of greenhouse gas emissions, and regional economic development with increased labor opportunities. Under these assumptions, the objective function of the government to maximize with respect to r is

$$\frac{1}{2}\left[(p+r)x-\left(\frac{cx^2}{2}+f\right)\right]-rx+bx \tag{2}$$

where $x=(p+r)/c$ from the optimization by firm J as characterized in equation (1), and b is a positive parameter that represents the government valuation of a unit of social benefit produced by this project.

The following subsections show the calculation of the optimal price p for firm F and r for the government in two circumstances: one in which firm F determines p before the government determines r, that is, in which firm F is the leader and the government the follower; and the other in which the reverse holds, that is, the government is the leader and the firm is the follower. Note that it is the follower that must set its strategic variable so as to satisfy $p+r \geq \sqrt{2cf}$.

3.1. CER Price Set First

Suppose firm F determines p before the government determines r. In this case, we first calculate the optimal r as a function of p. The maximization problem of the government is

$$\underset{r}{Max}\ \frac{1}{2}\left[(p+r)x-\left(\frac{cx^2}{2}+f\right)\right]-rx+bx\text{, with } x=\frac{p+r}{c} \tag{3}$$

$$\text{s.t. } p+r \geq \sqrt{2cf}$$

Using Lagrange multipliers, the solution may be obtained as follows (Appendix A.1):

(a1) $r = \dfrac{2b-p}{3}$ when $\dfrac{3\sqrt{2cf}}{2} - b \le p < 2b$

(a2) $r = \sqrt{2cf} - p$ when $p < \dfrac{3\sqrt{2cf}}{2} - b$ and $p < \sqrt{2cf}$

(a3) $r = 0$ when $p \ge 2b$ and $p \ge \sqrt{2cf}$

Next, let us calculate price p that firm F determines. Since $p \le p_0$ is assumed, case (a3) is ruled out, and cases (a1) and (a2) are transformed into (a1′) and (a2′) respectively.

(a1′) $r = \dfrac{2b-p}{3}$ when $\dfrac{3\sqrt{2cf}}{2} - b \le p < p_0$

(a2′) $r = \sqrt{2cf} - p$ when $p < \dfrac{3\sqrt{2cf}}{2} - b$ and $p < p_0$

Case (a1′) may be analyzed as follows. The maximization problem of firm F is

$$\underset{p}{Max}\ \frac{1}{2}\left[(p+r)x - \left(\frac{cx^2}{2} + f\right)\right] + (p_0 - p)x, \text{ with } x = \frac{p+r}{c} \text{ and } r = \frac{2b-p}{3} \tag{4}$$

s.t. $p \le p_0$

$$p \ge \frac{3}{2}\sqrt{2cf} - b$$

This can be solved as $p = 0$ and $r = 2b/3$ when $b \ge 3\sqrt{2cf}/2$, and $p = 3\sqrt{2cf}/2 - b$ and $r = b - \sqrt{2cf}/2$ when $3\sqrt{2cf}/2 - p_0 \le b < 3\sqrt{2cf}/2$ (Appendix A.2).

The same analysis applies to case (a2′), yielding the solution $p = 0$ and $r = \sqrt{2cf}$ only when $b \le 3\sqrt{2cf}/2$ (Appendix A.2).

Comparing the results of cases (a1′) and (a2′), yields

$$p = 0 \text{ and } r = \frac{2b}{3} \text{ when } b > \frac{3\sqrt{2cf}}{2}$$

$$p = 0 \text{ and } r = \sqrt{2cf} \text{ when } b \le \frac{3\sqrt{2cf}}{2}$$

The output levels of the electricity are $\sqrt{2cf}/c$ when $b \le 3\sqrt{2cf}/2$ and $2b/(3c)$ when $b > 3\sqrt{2cf}/2$. Note that $2b/(3c) > \sqrt{2cf}/c$ when $b > 3\sqrt{cf}/2$. In other words, the output levels are greater when the government places a higher value on social benefit.

In summary, when firm F sets the CER price before the government sets the wholesale electricity price, firm F sets the price at the lowest level, i.e., zero, irrespective of the valuation of the social benefit of the project by the government. In contrast, the government sets the electricity price at a higher level if it values the social benefit highly, and the higher the valuation, the higher the price, so as to increase output levels. If the government places a relatively low value on social benefit, it sets the price low to increase the viability of the project.

3.2. Wholesale Electricity Price Set First

Now suppose that the government determines r before firm F determines p. In this case, we first calculate the optimal p as a function of r.

The maximization problem of firm F is

$$\underset{p}{Max}\, \frac{1}{2}\left[(p+r)x - (\frac{cx^2}{2} + f)\right] + (p_0 - p)x \text{, with } x = \frac{p+r}{c} \tag{5}$$

s.t. $p + r \ge \sqrt{2cf}$

$p \le p0$

We solve for p only in the case in which $p_0 < \sqrt{2cf}/2$, namely, the case of a relatively low p_0. The case $p_0 > \sqrt{2cf}/2$, is complicated to analyze, though not difficult.

Using Lagrange multipliers yields the following (Appendix A.3).

(b1) $p = 0$ when $r \ge \sqrt{2cf}$

(b2) $p = \sqrt{2cf} - r$ when $\sqrt{2cf} - p_0 \le r < \sqrt{2cf}$

Next, we calculate the price r determined by the government. Case (b1) may be analyzed as follows. The maximization problem of the government is

$$\underset{r}{Max}\, \frac{1}{2}\left[(p+r)x - \left(\frac{cx^2}{2} + f\right)\right] - rx + bx \text{, with } x = \frac{p+r}{c} \text{ and } p = 0 \tag{6}$$

s.t. $r \ge \sqrt{2cf}$

Using Lagrange multipliers yields the solution (Appendix A.4) (c1) $r = \frac{2b}{3}$ and $p = 0$ when $b \geq \frac{3\sqrt{2cf}}{2}$

(c2) $r = \sqrt{2cf}$ and $p = 0$ when $b < \frac{3\sqrt{2cf}}{2}$

It follows that the government payoffs, π_G, are $\pi_G = b^2/(3c) - f/2$ in case (c1) and $\pi_G = -2f + b\sqrt{2cf}/c$ in case (c2).

The same analysis applies to case (b2), yielding the solution (Appendix A.4)

(c3) $r = \sqrt{2cf} - p_0$ and $p = p_0$

Hence, the government payoffs are $\pi_G = -2f + \sqrt{2cf}(b + p_0)/c$ in case (c3).

Next let us compare the government payoffs (c1)–(c2) and (c3). Showing that the payoffs in (c3) are greater than those in (c2) is straightforward. To compare the payoffs in (c1) with those in (c3), we subtract π_G in (c3) from that in (c1) to obtain

$$\begin{aligned} &\left(\frac{b^2}{3c} - \frac{f}{2}\right) - \left[-2f + \frac{\sqrt{2cf}(b + p_0)}{c}\right] \\ &= \frac{1}{12c}\left[\left(2b - 3\sqrt{2cf}\right) + 2\sqrt{3\sqrt{2cf}\,p_0}\right]\left[2b - \left(3\sqrt{2cf} + 2\sqrt{3\sqrt{2cf}\,p_0}\right)\right] \end{aligned} \tag{7}$$

Since $b \geq 3\sqrt{2cf}/2$ in case (c1), the first term is always positive. Let b_0 be

$$b_0 = \frac{3}{2}\sqrt{2cf} + \sqrt{3\sqrt{2cf}\,p_0}$$

It follows that if $b \geq b_0$, equation (7) is greater than or equal to zero, that is, π_G in (c1) is greater than or equal to π_G in (c3). If $3\sqrt{2cf}/2 \leq b < b_0$, equation (7) is negative, that is, π_G in (c3) is greater than π_G in (c1). Note that b_0 becomes smaller as the cost parameters, c and f, or firm F's valuation of a unit of CER, p_0, becomes smaller.

This yields the following result:

$r = \frac{2b}{3}$ and $p = 0$ when $b \geq b_0$

$r = \sqrt{2cf} - p_0$ and $p = p_0$ when $b < b_0$

The output levels of electricity are $2b/(3c)$ when $b \geq b_0$ and $\sqrt{2cf}/c$ when $b < b_0$. Note that when $b \geq b_0$, $2b/(3c) > \sqrt{2cf}/c$. In other words, as in the case analyzed in the previous subsection, output levels are greater when the government values the social benefit of a project more highly than when it does not.

In summary, when the government sets the wholesale electricity price before firm *F* sets the CER price, the government sets the price at the lowest level so as to increase project viability if it places a relatively low value on the social benefit of the project; in contrast, the government sets the wholesale electricity price at a higher level if it values the social benefit more highly. On the other hand, firm *F* sets the CER price at the lowest level, i.e., zero, if the government values the social benefit of the project highly; if the government places a low value on the social benefit of the project, firm *F* sets the CER price at the highest level, i.e., p_0.

3.3. Results

When the valuation of social benefit by the government is relatively low, both firm *F* and the government will set prices— the CER price for firm *F* and the wholesale electricity price for the government—as low as possible, irrespective of which is the leader and which the follower. They will do so because firm *F* gains more by setting a lower price for the CERs it purchases, and because the government is passive in supporting the project. Accordingly, the leader, firm *F* in subsection 3.1, and the government in subsection 3.2, sets its price as low as possible, and the follower sets the lowest price possible as well, to increase the likelihood that the project will be implemented.

In contrast, when the government puts a high value on the social benefit of the project, the government sets a higher wholesale electricity price, while firm *F* sets the CER price as low as possible, irrespective of which is the leader and which the follower. They do so because in this case, the government is eager for the project to produce a higher output, yielding greater social benefit, while firm *F* increases its earnings by setting the price low. In this case, firms *F* and *H* can earn more than in the previous case. Consequently, firm *F* will be more willing to invest in a CDM project in countries where the host government is more interested in the CDM.

There exists a threshold in the valuation of social benefit that the host country government places on projects whether the government is the leader or follower, though the threshold value is different when the government is the leader from when it is the follower. In either case, the government sets the price of electricity in proportion to its valuation of social benefit if the social benefit exceeds the threshold, and sets a constant electricity price if the social benefit is below the threshold.

4. Conclusion

As the assessment of current CDM projects above has shown, hydroelectric power generation, biomass utilization, and biogas production account for more than half of all registered projects. Given that projects using cattle excreta and livestock waste are not included as biomass projects in the IGES database, though in the usual sense they are considered such, biomass projects may be understood to account for a major share of activity under the CDM.

Favorable sites and opportunities for CDM projects will become scarce as more projects are implemented. The types of CDM projects that will be implemented in the future will depend on the further development of technologies to reduce greenhouse gas emissions. Accordingly, research and development in such technologies, and technology transfers to developing nations, will play key roles in climate change mitigation through the CDM.

The analytical model that we present here shows that the volume of emissions reduction through the CDM is dependent on the strategies of foreign investor firms and host country governments. These strategies are shaped by the design of the CDM. For instance, small-scale CDM projects, which are promoted by newly established provisions and institutions of the CDM, are prevalent in major project categories such as hydroelectric power generation, biomass utilization, and biogas production. Studies concerning the function of the CDM and its institutions are therefore of great importance.

APPENDIX

A.1. Solution for Maximization Problem (3)

The Lagrangian is

$$\frac{1}{2}\left[(p+r)x-\left(\frac{cx^2}{2}+f\right)\right]-rx+bx+\alpha\left(p+r-\sqrt{2cf}\right),\text{ with } x=\frac{p+r}{c} \tag{A1}$$

where $\alpha \geq 0$. Differentiating equation (A1) with respect to r and α, we obtain the following conditions.

$$\frac{2b-p-3r}{2c}+\alpha \leq 0,\ r\left(\frac{2b-p-3r}{2c}+\alpha\right)=0,\text{ and } r \geq 0 \tag{A2}$$

$$-\sqrt{2cf}+p+r \geq 0,\ \alpha\left(-\sqrt{2cf}+p+r\right)=0,\text{ and } \alpha \geq 0 \tag{A3}$$

From (A2), if $\alpha = 0$, then $r = 0$ when $p \geq 2b$ and $r = (2b-p)/3$ when $p < 2b$. When $p \geq 2b$, since $r = 0$, $-\sqrt{2cf}+p+r \geq 0$ only if $p \geq \sqrt{2cf}$. When $p < 2b$, since $r = (2b-p)/3$, $-\sqrt{2cf}+p+r \geq 0$ only if $p \geq 3\sqrt{2cf}/2-b$.

On the other hand, if $\alpha > 0$, then $r = \sqrt{2cf}-p$ from (A3). Thus, $\alpha = \left(3\sqrt{2cf}/2-b-p\right)/(2c)$ from (A2), which is positive only if $p < 3\sqrt{2cf}/2-b$.

A.2. Solution for Maximization Problem (4)

From equation (4), the Lagrangian is

$$\frac{1}{2}\left[(p+r)x-\left(\frac{cx^2}{2}+f\right)\right]+(p0-p)x+\beta(p0-p)+\gamma\left(p+b-\frac{3}{2}\sqrt{2cf}\right),$$

with $x=\dfrac{p+r}{c}$ and $r=\dfrac{2b-p}{3}$ (A4)

where $\beta\geq 0$ and $\gamma\geq 0$. Differentiating equation (A4) with respect to *p*, *β* and *γ*, we obtain the following conditions.

$$-\frac{2(2b+5p-3p_0)}{9c}-\beta+\gamma\leq 0,\ p\left[-\frac{2(2b+5p-3p_0)}{9c}-\beta+\gamma\right]=0,\text{ and } p\geq 0 \quad \text{(A5)}$$

$$-p+p_0\geq 0,\ \beta(-p+p_0)=0,\text{ and } \beta\geq 0 \quad \text{(A6)}$$

$$p+b-\frac{3}{2}\sqrt{2cf}\geq 0,\ \gamma\left(p+b-\frac{3}{2}\sqrt{2cf}\right)\geq 0,\text{ and } \gamma\geq 0 \quad \text{(A7)}$$

If $\beta=0$ and $\gamma>0$, then $p=3\sqrt{2cf}/2-b$ from (A7). Thus, $-p+p_0\geq 0$ only when $3\sqrt{2cf}/2-p_0\leq b<3\sqrt{2cf}/2$.

If $\beta>0$ and $\gamma>0$, $p=p_0$ from (A6) and $p=3\sqrt{2cf}/2-b$ from (A7), which is valid only when $b=3\sqrt{2cf}/2-p_0$.

If $\beta=0$ and $\gamma=0$. If $2b-3p_0>0$, then $p=0$ from (A5). Thus, $p+b-3\sqrt{2cf}/2\geq 0$ only when $b\geq 3\sqrt{2cf}/2$. If $2b-3p_0\leq 0$, then $p=(-2b+3p_0)/5$ from (A5). It follows that $p+b-3\sqrt{2cf}/2<0$ due to $p_0<\sqrt{2cf}$, which contradicts (A7).

If $\beta>0$ and $\gamma=0$, then $p=p_0$ from (A6), which contradicts (A5).

As for case (a2′), the Lagrangian is

$$\frac{1}{2}\left[(p+r)x-\left(\frac{cx^2}{2}+f\right)\right]+(p_0-p)x+\lambda(p_0-p)+\mu\left(\frac{3}{2}\sqrt{2cf}-b-p\right),$$

with $x=\dfrac{p+r}{c}$ and $r=\sqrt{2cf}-p$ (A8)

where $\lambda \geq 0$ and $\mu \geq 0$. The derivative of (A9) with respect to p is $-\sqrt{2cf}/c - \lambda - \mu$, which is always negative. Hence, $p = 0$. It follows that $3\sqrt{2cf}/2 - p - b \geq 0$ only when $b \leq 3\sqrt{2cf}/2$.

A.3. Solution for Maximization Problem (5)

The Lagrangian is

$$\frac{1}{2}\left[(p+r)x - \left(\frac{cx^2}{2} + f\right)\right] + (p0 - p)x + \delta(p0 - p) + \eta\left(p + r - \sqrt{2cf}\right), \text{ with } x = \frac{p+r}{c} \quad \text{(A9)}$$

where $\delta \geq 0$ and $\eta \geq 0$. By differentiating equation (A9) with respect to p, δ and η, we obtain the following conditions.

$$-\frac{3p - 2p0 + r}{2c} + \delta - \eta \leq 0,\ p\left(-\frac{3p - 2p0 + r}{2c} + \delta - \eta\right) = 0, \text{ and } p \geq 0 \quad \text{(A10)}$$

$$p + r - \sqrt{2cf} \geq 0,\ \delta\left(p + r - \sqrt{2cf}\right) = 0, \text{ and } \delta \geq 0 \quad \text{(A11)}$$

$$p_0 - p \geq 0,\ \eta(p_0 - p) = 0, \text{ and } \eta \geq 0 \quad \text{(A12)}$$

If $\delta > 0$ and $\eta = 0$, then $p = \sqrt{2cf} - r$ from (A11). Thus from (A12), it must hold that $\sqrt{2cf} - p_0 \leq r \leq \sqrt{2cf}$. Then, $\delta = \left(3\sqrt{2cf} - 2p_0 - 2r\right)/(2c)$ from (A10), which must be greater than zero. Thus, it must hold that $r < -p_0 + 3\sqrt{2cf}/2$.

If $\delta > 0$ and $\eta > 0$, $p = \sqrt{2cf} - r$ from (A11) and $p = p_0$ from (A12). This is the case only when $r = \sqrt{2cf} - p_0$.

If $\delta = \eta = 0$, then $p = 0$ when $2p_0 - r < 0$ and $p = (2p_0 - r)/3$ when $2p_0 - r \geq 0$ from (A10). If $p = 0$, it must hold $r \geq \sqrt{2cf}$ from (A11), and if $p = (2p_0 - r)/3$, it must hold $r \geq -p_0 + 3\sqrt{2cf}/2$ from (A11). Thus in the case where $\delta = \eta = 0$, $p = 0$ when $r > 2p_0$ and $r > \sqrt{2cf}$, and $p = (2p_0 - r)/3$ when $r \leq 2p_0$ and $r \geq -p_0 + 3\sqrt{2cf}/2$.

If $\delta = 0$ and $\eta > 0$, $p = p_0$ from (A12). Then $\eta = -(p_0 + r)/(2c)$ from (A10), which is negative and a contradiction.

In summary, we obtain

$p = 0$ when $r > 2p_0$ and $r > \sqrt{2cf}$

$$p = \frac{2p_0 - r}{3} \text{ when } r < 2p_0 \text{ and } r \geq \frac{3\sqrt{2cf}}{2} - p_0$$

$$p = \sqrt{2cf} - r \text{ when } \sqrt{2cf} - p_0 \leq r < \sqrt{2cf} \text{ and } r < \frac{3\sqrt{2cf}}{2} - p_0$$

There are two possibilities to be considered; one is $2p_0 < \sqrt{2cf}$ and the other $2p_0 > \sqrt{2cf}$. It is assumed in the text that $p_0 < \sqrt{2cf}/2$, and on this basis (b1) and (b2) are obtained.

A.4. Solution for Maximization Problem (6)

From equation (6), the Lagrangian is

$$\frac{1}{2}\left[(p+r)x - \left(\frac{cx^2}{2} + f\right)\right] - rx + bx + \varphi\left(r - \sqrt{2cf}\right), \text{ with } x = \frac{p+r}{c}, \text{ and } p = 0 \quad \text{(A13)}$$

where $\varphi \geq 0$. Differentiating (A13) with respect to r and φ, we obtain the following conditions.

$$\frac{2b-3r}{2c} + \varphi \leq 0,\ r\left(\frac{2b-3r}{2c} + \varphi\right) = 0, \text{ and } r \geq 0 \quad \text{(A14)}$$

$$r - \sqrt{2cf} \geq 0,\ \varphi\left(r - \sqrt{2cf}\right) = 0, \text{ and } \varphi \geq 0 \quad \text{(A15)}$$

If $\varphi = 0$, $r = 2b/3$ from (A14). It must then hold that $b \geq 3\sqrt{2cf}/2$ since $r - \sqrt{2cf} \geq 0$ from (A15).

If $\varphi > 0$, $r = \sqrt{2cf}$ from (A15). From (A14), we then obtain $\varphi = \left(3\sqrt{2cf} - 2b\right)/(2c)$, which must be positive. Thus it must hold that $b < 3\sqrt{2cf}/2$.

As for case (b2), the Lagrangian is

$$\frac{1}{2}\left[(p+r)x - \left(\frac{cx^2}{2} + f\right)\right] - rx + bx + \theta\left(\sqrt{2cf} - r\right) + \omega\left(r - \sqrt{2cf} - p_0\right),$$

$$\text{with } x = \frac{p+r}{c}, \text{ and } p = \sqrt{2cf} - r \quad \text{(A16)}$$

where $\theta \geq 0$ and $\omega \geq 0$. Differentiating (A16) with respect to r, θ and ω, we obtain the following conditions.

$$-\frac{\sqrt{2cf}}{c} - \theta + \omega \geq 0,\ r\left(\frac{\sqrt{2cf}}{c} - \theta + \omega\right) = 0, \text{ and } r \geq 0 \tag{A17}$$

$$\sqrt{2cf} - r \geq 0,\ \theta\left(\sqrt{2cf} - r\right) = 0, \text{ and } \theta \geq 0 \tag{A18}$$

$$r - \sqrt{2cf} + p_0 \geq 0,\ \omega\left(r - \sqrt{2cf} + p_0\right) = 0, \text{ and } \omega \geq 0 \tag{A19}$$

If $\omega = 0$, $-\sqrt{2cf}/c - \theta + \omega$ is always negative, contradicts A17).

If $\theta = 0$ and $\omega > 0$, $r = \sqrt{2cf} - p_0$ from (A18). Then $\omega = \sqrt{2cf}/c$ from (A17), which is positive. Additionally, $\sqrt{2cf} - r = p_0$, which is positive.

If $\theta > 0$ and $\omega > 0$, $r = \sqrt{2cf}$ from (A18) and $r = \sqrt{2cf} - p_0$ from (A19), which is a contradiction.

References

Al-Saadon, Y and Das, S. (1996). Host-country policy, transfer pricing and ownership distribution in international joint ventures: A theoretical analysis. *International Journal of Industrial Organization*, *14*, 345-364.

IGES (2008). IGES CDM Project Database. http://www.iges.or.jp/en/cdm/index.html

Ferry S. with Cabraal A. (2006). *Renewable Power in Developing Countries: Winning the War on Global Warming*. Tulsa, PennWell.

Shy, O. (1995). *Industrial Organization*. Cambridge: MIT Press.

Varian, H. (1992). *Microeconomic Analysis, Third Edition*. New York: W.W. Norton.

World Bank (2008). REToolkit: A Resource for Renewable Energy Development. http://www.worldbank.org/retoolkit

In: Energy Policy: Economic Effects, Security Aspects... ISBN 978-1-60692-894-3
Editor: Noah B. Jacobs

Chapter 7

INSTITUTIONALLY EMBEDDED: THE DILEMMA IN THE FORMULATION OF SUSTAINABLE ELECTRICITY POLICY IN HONG KONG

Alex Y. H. Lo
Research School of Social Sciences, The Australian National University, Australia

ABSTRACT

The development of electricity policy in Hong Kong has illustrated the arduous task of incorporating environmental principles into a traditionally growth-led model. Attempts to reform the obsolete policy framework prove ineffective due to some path-dependent institutional constraints which are inherited from its development model and intensified by the recent political contests. These have precluded a timely transformation of the relations between the power supply and demand. The state has failed to tackle these structural impediments and tends to focus more on technical solutions than truly collaborative initiatives. The minimal progress in local democratic development and the planned extension of the monopolist status of the existing utilities have also created public distrust against the state and the utilities. Environmental initiatives then become technocratic or rhetoric. The analysis reveals that the ecological rationalization of the electricity policy has been confined by the inherent institutional dilemma. Capacities for developing a sustainable electricity policy have diminished as some existing institutions resist changing, despite preliminary forms of responsive institutions are established.

1. INTRODUCTION

Hong Kong, formally known as Hong Kong Special Administrative Region (HKSAR), has been criticized as lagging behind among developed economies in the pursuit of sustainable development (Mottershead, 2004). The economy suffers from certain deeply embedded institutional constraints precluding effective policy learning required of a sustainability discourse (Gouldson et al., 2008; Hills, 2004). In many aspects, the conditions for change are immature despite the widespread recognition of the declining environmental quality. Policy responses tend to be technocratic or rhetoric:

> Hong Kong has mostly seen technical forms of policy learning, with some limited forms of conceptual or social learning being pursued not to transform or to supplement existing institutions and approaches but instead merely to reinforce and legitimize them. (Gouldson et al., 2008, p. 328)

There is an emphasis of technical or economic measures by the policymakers while participatory initiatives are treated as a publicity tool or a means to legitimize their predefined policy goals or regulatory approaches. Innovative institutions for sustainability become vulnerable to the existing ones wherever there are significant conflicts between them. For Gouldson et al. (2008, p. 328), the scope of policy learning in Hong Kong is limited because it has encountered 'embedded, reactive institutions that do not change until the pressure for change becomes irresistible'. It is suggested that there are limited improvements from promoting reflexive and responsive institutions as the embedded structural resistance forces remain operative.

Capacities for environmental policy reform diminish when the institutions are locked-in to existing approaches. There are always resistance forces arising from the actors who have vested interests from the existing institutions, leading governments to adopt reactive rather proactive policies (Jänicke, 1990). Yet, the reluctance to change may also come from those who are not expected to directly benefit from the existing institutions. The legitimacy (or illegitimacy) of the potential institutional change may constitute another set of resistance force based on the social/political positions of the involved actors. This is likely to happen particularly when the environmental policy reform becomes an entirely political issue which seems to be extended from another more rigorous social or political debate. As a result, the environmental contributions associated with the proposed change are no longer the key issue despite reflexive and responsive institutions – a prerequisite of sustainability policy – are present. Capacities for change are bounded to the interests and expectations of individual actors whose aggregate benefits do not necessarily contribute to the common good. The assumption that policy learning would take place in reflexive societies with rational, responsive institutions may then fail when the preconditions for change are not satisfied (Gouldson et al., 2008).

This paper opens an examination of the electricity policy of Hong Kong through a discourse analysis. It begins with a review of the institutional contexts of Hong Kong, followed by a critical analysis of the government's treatment of the issue. The inherited institutional constraints are identified and the capacities of the existing institutions to enhance sustainability are discussed.

2. PAST AND PRESENT: THE CONTEXT OF HONG KONG

2.1. The Environment

Located at the southeast of China, the territory of Hong Kong extends over 1100km^2 and has a population approaching 7 million leading to a population density of 6,330/km^2 in 2007. The economy grew rapidly since 1960s but collapsed when the Asian financial crisis took place in 1997. During the last decade, it had experienced mixed economic fortunes as the GDP growth rate dropped by 6% in 1998 but peaked at 9.5% in 2000. To date it has rebounded from the economic difficulties reaching a per capita GDP of HK$233,358 (approx. US$30,000) as at 2007.

Hong Kong is characterized by a unique institutional context in developing sustainability strategies. Declining environmental quality as an externality of its economic miracle has been extensively documented (Barron, 1996; Barron & Steinbrecher, 1999; Hung, 1997). In particular, the territory-wide air pollution problems arising from the gaseous emissions associated with the power industry have been frequently criticized. This is coupled with the growing vehicular activities and the trapping of air pollutants between the numerous high-rise buildings. The recent concerns about the 'wall effect' as a result of this have illustrated the public intolerance over the uncontrolled vertical expansion at the expense of ambient air quality. The acute marine water pollution and waste management problems are also unresolved awaiting more public funding to proceed.

Hong Kong was under British colonial rule prior to the handover of sovereignty to People's Republic of China in 1997 and is currently bound to a 'One country, Two systems' constitutional framework. Unlike many other countries, international forces play little role in the environmental-political economy of Hong Kong as it is never a sovereign state. Despite being economically mature, Hong Kong is not subject to Kyoto Protocol because it is recognized as a part of China - a non-Annex I country. It is likely that internal processes in relation to the ongoing power contests have stronger influences on the transformation to a sustainable state. Its institutional features, development mode and land constraints constitute a complex set of confines to the endeavour. Although there are some positive signs of change, the local sustainability discourse has been jeered as 'a means to an end rather than an end itself' (Hills, 2004, p. 30). Pursuit of sustainability is taken as a way to assure continual ('sustainable') economic development. This is partly related to the historical development of the economy and the needs of restoring economic stability and the public's confidence as a matter of political reality.

2.2. The Growth-oriented Model

The ecological unsustainability of Hong Kong is related to the capitalist development strategies that make its economic success possible. The British colonial government had intentionally developed the city as a regional commercial center since it exercised its rule in mid-19th century. Hong Kong people, on the other hand, tended to consider themselves as living on someone else's lands and, as a tradition, manifested as sort of single-minded economic animals with a primary focus on economic well-beings. Thanks to their common expectations, the economy enjoyed a fascinating period during the later half of the last century as the ideal of growth-first society was widely accepted. Civil society is yet acculturated in this 'city of strangers' (Ng, 2002, 2004).

The growth-oriented model had been vividly printed in the economic policy framework and was adopted in many areas. The governance principle of 'positive non-interventionism' dictated a convention of minimal governmental involvement in many economic aspects, including electricity issues. There is comparatively little direct intervention over the monopolist electricity industry which is subject to a rate-of-return regulation that covers a set of standards, limits or caps to enable self-adjustment by the utilities. It is believed to be effective to save bureaucratic resources in monitoring and assessing their performance and therefore no integrated energy bureau or department ever exists in the territory.

Nevertheless, the interventionist role had expanded in some aspects throughout the same period. Hills (1985) observed that the free market environmentalism did not take any major role in the environmental discourse arisen as early as the 1980s. Like many other jurisdictions, Hong Kong government began with a 'command-and-control' approach in handling environmental problems associated with rapid industrialization (Barron, 1996; Hills & Barron, 1997). However, the powerful industrial lobby held strong oppositions against the increasing environmental controls on sewage, air and waste matters because it would run counter to the *laissez-faire* promise. Although the tension was relaxed owing to the subsequent industrial relocation to Pearl River Delta Region since the mid-1980s, transboundary air and water pollution problems remain as an unresolved issue. Uncontrolled economic expansion collided with the newly arrived environmentalism for the first time.

There are long-lasting implications of such a market-led model. Heavy reliance of free market system suggests that attempts to challenge its shortcomings may manifest itself, or conspiratorially be seen, as a distortion of free market operation and an attack to economic development. Implementation of environmental policies is vulnerable to the oppositions from powerful interested parties. Governmental treatments to protect private sector from stringent environmental regulations, on the other hand, are often immediately equaled to a disobedience to the tenets of sustainable development. Compromising vested private interests for environmental reasons is costly to the non-interventionist government.

Hong Kong adopts a low-tax system and the government has imposed loose regulations on business enterprises to stimulate market-led economic activities. Public expenditure is highly dependent of the revenue from selling government lands as a compensation for the limited tax revenue from the other side. Social welfares and other government services are effectively derived largely from lands revenue. Possible shrinkage in government services may result if the land-based approaches cease. Alternatively, an upward adjustment in tax rates may help but it is conceivably more difficult in view of the potential political reactions. Also, there is an urgent need to increase public resources used in aiding the lower end of the society in recent years. There is a struggle between re-routing the environmentally damaging, growth-led development path and maintaining the growth model that breeds the elites and voters.

2.3. The Political Economy

Formulation of environmental policy in Hong Kong has been restricted by the unbalanced power distribution, both past and present. During the British rule, there was no universally elected legislature until 1990 and the government leaders were appointed by London; environmental decision-making was therefore highly centralized. Presently, democracy is still fairly limited because the senior officials are designated, or quasi-appointed, by Beijing. In spite of the growing public demand and frequent protests, universal suffrage as a key step towards local democracy is disallowed by Beijing. These institutional patterns draw a couple of implications. The remnant colonial governance traditions are to some extent at odds with the notion of public participation underlying sustainable development. And, the present political conflict is masked by a claim to speed up economic growth which may potentially intensify some environmental threats.

Hong Kong experienced a harsh political climate after the return of sovereignty. It reached a climax when the unprecedented mass protest by nearly half a million people took place on July 1, 2003 (Cheung, 2005; Ma, 2005). Struggles for democracy rose to a new height as the public could no longer tolerate the poor performance of the administration and the deprivation of local autonomy affirmed by Beijing's unwelcome interventions in several events (Holliday et al., 2002). The government has failed to rescue the collapsing public support through economic means due to the economic and fiscal difficulties. The resulting public's discontent and distrust against the administration was followed by frequent and blistering criticisms. It is likely that these have been extended to other public affairs as well. The fact that the authority of the government was weakened and kept being challenged by the political actors and the community did limit its capacity of policy bargaining. As a remedy to its failure to implement political reform in accordance with the expectation of the public, the government turned the public attention to economic affairs by advertising the need to boost up the economy after the economic downturn. This has been made explicit in the Policy Address documents issued in the last two years (Tsang, 2006, 2007). An idea of absorbing political defies into economic preoccupation is obvious throughout the past few years.

The absence of full democracy in Hong Kong is but one of the barriers to a 'bottom-up', participatory sustainability strategy (Mottershead, 2004; Mottershead & La Grange, 2000). Following its traditions, the government is inclined to employ technocratic measures for managing the environment as an alternative to open, discursive and citizen-based approach. It is likely that there is a retreat of the rhetoric of administrative rationalism in the government agenda given its reluctance to transform itself from a controller to a facilitator regime (Hills, 2004; Hills & Welford, 2002). The government tends to manifest itself as the 'manager' for the environment and makes use of a series of 'command-and-control measures to tackle pollution problems on a 'top-down' basis. Environmental sustainability is understood as a deployment of technical tools to address the operational dimensions of industrial processes (Barron, 1996; Hills & Barron, 1997). Hills (2004) contended that this deeply entrenched administrative rationality is inconsistent to the very nature of sustainable development which calls into question many of the fundamental tenets of the former.

As discussed in the following sections, the early development of the electricity policy framework had been strongly affected by the pragmatic needs of fulfilling economic needs at that time. Since the 1990s, however, the policy structure inherited from the growth-led regime has become a source of critiques from the public sphere, prompted by the rising discontent over the unwelcome administration.

3. The Electricity Policy and the Policy Community

Mitigation of the environmental problems associated with electricity generation has reached the government agenda in the late 1980s. It was not until the 1990s that the expansionary electricity policy justified by the pursuit of rapid economic growth was strongly criticized as a source of both economic and environmental inefficiencies. Protests against the failure of the government in tackling the associated environmental problems called for a more aggressive policy approach. Institutionalization of environmental initiatives then became a hot topic in the debates of the post-2008 electricity market reform during the past few years. Certainly this is related to the growing awareness of environmental protection and civil society. The movement challenges the rationale behind the existing regulatory regime, creating demands for a thorough reform of the outdated electricity policy framework. An imbalance in the expectations between the supply and demand sides has emerged. Yet, finally, substantial and effective changes were not observed. The government decided to retain the monopolistic status of the power companies for at least ten years while relying on technical measures to cope with the problems.

The remainder of the paper is devoted to an investigation of the ways that different actors interpret the issues. The regulatory arrangement is first introduced followed by a discussion of the positions and responses of different actors. The conflicts between the existing institutions and actors are analyzed and the path-dependent institutional barriers are stressed.

3.1. Regulatory Arrangement

The Economic Development and Labour Bureau (EDLB)[1] was responsible for formulating energy policy to ensure that the energy needs of the community can be met. It collaborated with the Environment, Transport and Works Bureau (ETWB)[2] / Environmental Protection Department (EPD) that are involved in electricity-related environmental issues. The electricity policy remained apolitical, until recently. Traditionally it did not play a major role in the political agenda of the administration as the supply was highly reliable at reasonable prices and the government did not want to intervene in the ownership and management of the private utilities (Hills, 1991b). Since the 1990s, nevertheless, the situations became complicated. Criticisms against the rationale of the current control scheme grew as a result of the rising public concerns on tariff issues and the transparency of the policy system which was intensified by the democratic movement over the past two decades (Lam, 1997a).

Electricity in Hong Kong is provided by two privately owned utilities: CLP Power Company Limited (CLP) (formerly known as China Light and Power Company Limited) and Hongkong Electric Company Limited (HEC). Their electricity businesses are individually

[1] Its role as the regulator had been transferred to the newly established Environment Bureau.
[2] The ETWB has been reorganized and renamed as Environment Bureau in July 2007.

subject to a rate-of-return regulation, namely the Scheme of Control Agreements (SCAs). Introduced as early as the 1960s, the SCAs stipulate the financial arrangements of the utilities while they are entitled to provide reliable and adequate electricity services to the economy at reasonable prices. Financial affairs of their electricity business are subject to monitoring by the EDLB including periodic financial and annual tariff reviews. Despite bound to the SCAs, the two electric utilities are guaranteed a rate of return of 13.5% on their average debt-financed net fixed assets and 15% on their equity-financed assets. It is essentially a cost-plus regulation that any cost incurred can be recovered through annual tariff adjustment upon approval by the government. No modification to the SCAs can be made without mutual agreement of the signed parties.

There was a belief that minimal governmental intervention could maximize economic efficiency of power supply. In order not to directly intervene their operations, the administration developed a series of standards, limits or caps, such as the permitted rate of return, to allow self-adjustment by the utilities. It is intended to reduce bureaucratic resources for monitoring and assessing industrial performance by employing clear, predefined benchmarks for the authorities to determine compliance. Until very recently, no enforceable environmental requirement was specified in the agreements. There is no provision to obligate the two power companies to generate electricity using renewable energy sources but they are subject to the emission reduction targets stipulated by the government.

In 2005, the EDLB formally proposed a reform of the regulatory framework for the electricity market in view of the current SCAs coming to expire in 2008. A two-stage public consultation exercise was taken place to collect the public's views about various issues and possible options in relation to the future development of the electricity market (EDLB, 2005a, 2005b). The EDLB recommended a relaxation of the regulatory constraints about introducing new power supplies and imposition of more stringent control over the existing utilities. The consultation has stimulated a debate on the future prospect of the electricity market, mainly around market liberalization issues.

3.2. The Policy Community

The government traditionally tended to minimize its involvement in the local electricity industry (Hills, 1991a, 1991b). The electricity industry of Hong Kong has been dominated by private entities and the policy makers once believed that a reliable and adequate electricity supply was associated with minimal intervention. The government officials were sceptical to make big changes from the status quo and inclined to maintain a co-operative relationship with the power companies (Chan, 2006). This strategy had probably worked well prior to 1990s in terms of the contributions to rapid economic growth. Nevertheless, there is a series of criticisms over its failure of tackling the pressing environmental problems, particularly after 1997 when the local political environment became much more complicated. This prompted the government to look for some changes as a reaction, eventually putting the relationship with the power industry into a new shape.

There does not have a single independent agency within the government machinery to oversee all aspects of the electricity matters. Responsibilities are divided among a few loosely allied departments and quasi-official organizations. Each of them attends to a specific set of the issues and is expected to demonstrate some degree of intra-governmental co-operation.

The role of regulator of the electricity policy is principally represented by the EDLB which was designated to ensure the energy needs of the community to be met. Another key player is the ETWB / EPD which played a key role in actively managing electricity-related environmental problems. The Electrical and Mechanical Services Department (EMSD), on the other hand, serves to provide technical support for the manipulation of energy policy, including safety issues and energy efficiency programme. Apart from these, there are two advisory bodies involving in the energy policy community, namely, the Energy Advisory Committee for providing advice on energy policy and the CSD for promoting sustainability strategy in Hong Kong, including the promotion of renewable energy which has been identified as an area of concern under the First Sustainable Development Strategy for Hong Kong (CSD, 2005).

Electricity suppliers in Hong Kong are represented by the CLP and HEC. They were posited as an economic actor and assumed a key role when the economy grew rapidly since the 1960s, because of their importance in supporting industrial and commercial activities. The CLP and HEC entered into the SCAs since 1964 and 1979 respectively. They are granted no franchise to supply electricity in Hong Kong so that the market is open to new comers. However, new suppliers have never emerged and currently there is no direct competition between the two existing utilities. The utilities own and operate their respective generation, transmission and distribution assets in their service areas, suggesting that electricity provision in Hong Kong is highly centralized.

As regards environmental commitment, the two utilities are not completely irresponsive. They have shown positive attitudes to general environmental principles but often remain defensive when confronting trade-offs. They claimed that they must in some circumstances make 'difficult decisions' between social, environmental and economic interests (CLP, 2004b, p. 17). The 'difficult decisions', as they suggested, invariably involve impacts on supply reliability that the public are most concerned about (CLP, 2006; HEC, 2006). Their efforts on environmental improvement have made a certain degree of success in the mitigation of SO_2 and NO_x emissions. Promotion of renewable energy is an area of high priority in their environmental agenda, followed by some other supply-side technological measures for emission mitigation and educational programmes (CLP, 2004b; HEC, 2005). However, these attempts are either charismatic in the sense that they can easily draw the attention of the general public, or essentially supply-led involving little demand-side involvement which sounds more complicated and difficult to manage (their demand-side-management (DSM) activities are limited to large, industrial customers). The energy efficiency initiatives outlined in the CLP Environmental Manifesto appeared rhetoric and considered the inefficient use of energy as a matter of lack of awareness rather than lack of capacity to respond (CLP, 2004a).

The public of Hong Kong traditionally placed little attention to environmental issues (Barron, 1996; Hills, 1985). Yet environmental concerns on energy issues have been growing since the 1990s (Hills, 1991a). The public hold the view that the existing regulatory arrangement has failed to fulfill both economic and environmental objectives. Particularly after the economic downturn since 1997, they see no reason for the two large companies to keep earning an attractive, guaranteed return while remain irresponsive to the social concerns. The people nowadays not only expect reliable and adequate supply of electricity at reasonable prices but also better environmental performance. Nevertheless, when making trade-off is necessary, environmental goals are viewed not as important as expected. The electricity consumers usually rank environmental considerations far lower than economic aspects. In a survey, there were only 3.7% of the household respondents indicating environmental concerns as the most important aspect of electricity provision, compared to the 49.7% on supply reliability (Census and Statistics Department, 2004). A similar result was also obtained in another survey conducted for commercial and industrial consumers. It was also observed that there existed discrepancy between awareness and actions on the part of domestic consumers, but it looked more consistent for commercial and industrial consumers in another survey (EDLB, 2004). A total of 63.6% of the household respondents supported introduction of renewable energy by the government but only 31.9% would choose to use if available (Census and Statistics Department, 2004), compared to the 89.1% and 88.9% respectively for commercial and industrial consumers (EDLB, 2004). This reveals that the domestic consumers were more hesitant to commit, probably because they were more concerned about having to pay a higher tariff and the reliability issue associated with imported renewable energy from Mainland China (EDLB, 2005b). Given the predominant economic concerns, the EMSD (2002) admitted that it is too early to expect substantial public support for the use of sustainable energy sources. This dilemma is indicative of an inherent collective action problem associated with electricity provision that influences the behaviours of other actors, especially the elected officials.

The public's attitude has been amplified by political parties and pressure groups who are actively involved in the ongoing debates on behalf of public interests. Most of the political parties strongly support a liberalization of electricity market mainly because of the potential tariff reduction (and wider choice as well) (see e.g., Democratic Party, 2006). And this is advocated by some environmental NGOs to different extents (see e.g., Friends of the Earth, 2005; Greenpeace China, 2006b), in spite of the potential environmental drawbacks from lower tariffs. Given that the viability of these organizations depends on public's support, they have very incentive to accuse the utilities and the government of their failure to enhance economic and/or environmental efficiencies. In a Legislative Council (LegCo) meeting, it was commented that the utilities have *full* obligation to offer cheaper and cleaner electric power even without attractive returns as compensation (LegCo, 2006). These actors advocate a radical restructuring of the power industry and accept a supply-oriented perspective based on the principle of 'polluter pays' rather than 'user pays'. However, the arguments of the politicians are sometimes self-contradictory. For example, some LegCo members insisted on removing the SCAs which guarantee the return of the electric utilities; they, on the other hand, advocated a profit control scheme be imposed on the unregulated towngas company (Lam, 1997a). After all, this has significantly sharpened the provider-consumer relationship drawing a clear-cut boundary between the two sides.

4. Institutional Embeddedness and Sources of Conflicts

4.1. An Awkward Position

In the early years, the utilities had little conflict with the government and public because their business objectives were generally compatible with the social aspirations, i.e. satisfying people's basic needs. The industry was, in fact, subject to the pressure of increasing electricity supply to fuel the rapidly growing economy. A regulatory framework that could provide more incentives for capacity expansion was considered as acceptable.

However, the expansionary electricity policy has frequently gone challenged since the 1990s. The utilities were criticized for keep expanding the generating capacity, causing over-consumption of non-renewable energy resources and aggravating pollution problems. There is a public expectation that the power companies should demonstrate their responsibility to the society in the form of environment protection. Being private monopolists, the two utilities remain active but defensive in responding to the social pressure. Following the changes in political and economic conditions, the private and public interests become increasingly divided. Bound to the agreements the utilities believe that it is a zero-sum game to balance reliability, tariff reduction and environmental protection, which is however not shared by the public. This creates an oppositional relationship built upon distrust, ultimately discouraging the formation of consensual linkage.

The government officials manage to resolve the conflicts as the situation is one of inherent and inextricable, because if they:

> allow to set the price too high, they impose a burden on consumers. But if they set prices too low, they deprive the power companies of reasonable return of investment and capital to build new power plants. If they endorse the construction of power plants, they may threaten the public health and environment. But if they reject or defer proposals for new power plants, they may run the risk of insufficient capacity for future development. If they fail to encourage efficient use and conservation of energy, they aggravate the supply problems. But if they promote energy efficiency and conservation, it entails higher electricity tariff and hence exacerbates the charge burden of consumers (Lam, 1998, p. 58).

Clearly the government is still facing difficulties in searching for effective solutions to strike the balance. There is a tendency to view the utilities, and to a lesser extent the government itself, as both the causes and solutions of the problems. As a result, the government officials are prone to simply 'cap and control' the industry because it involves minimal intervention, lower regulatory costs and probably more tangible outcomes. Demand-side initiatives designated to bring about changes on the part of consumption and consultation exercises are uncommon especially before the 2000s.

4.2. SCA: the Environmentally Inconsistent Institution

As a bilateral contract formulated more than thirty years ago, the SCAs are often criticized as failing to follow the changes in the local environmental context. It has also created barriers to effective bargaining due to its long-term nature (15 years). In particular, it offers little incentive to the promotion of energy efficiency and wider supply sources.

The agreements were initially set up to encourage expansion of generating capacity in order to ensure stable and adequate electricity supply necessary for rapid economic growth in the early years. It links average net fixed assets to the utilities' permitted returns so that more investment in generation capacities means higher returns. The SCAs require each of the two utilities to set up a development fund to finance their acquisition of fixed assets and stabilize tariffs. Any shortfalls between the actual profit after taxation and the permitted return will be transferred to or from the fund. As long as there is unexpected growth in demand, the fund accumulates and a larger fund will then be available for acquiring fixed assets to meet the growing demand by which the utilities are entitled a higher permitted return. Although the arrangement has provided incentives for the utilities to improve operational efficiency, it has also encouraged over-expansion of generating capacity (Lam, 1997b). The utilities are happy to see a continual growth in demand in order to justify their expansion plan. They are therefore tempted to keep making over-optimistic demand forecast and uninterested in demand management initiatives (Hills, 1991a, 1991b).

Furthermore, there is little incentive to promote wider choice on power supply on the part of customers. As specified in Supply Rules of the two power companies, the customers must take all their required electricity from the utility and an internal generating unit that runs in parallel with the company's supply is prohibited; this effectively creates barriers to the introduction of renewable energy from other suppliers (Close, 1997). Since the incentive structure of the existing SCAs can only accommodate traditional energy sources, the power companies have little motivations to adopt renewable energy like landfill gas (EMSD, 2002). On-site generation using the environmentally friendly technologies and alternative energy sources are consequently under-developed. Lastly, the pricing mechanisms do not account for environmental costs associated with combustion of fossil fuel as it involves no carbon tax or like. Conventional methods of power generation are therefore made relatively cheaper and more attractive than those generated from renewable energy sources to which there was no direct or indirect subsidy (until recently the SCAs are renewed).

4.3. The Production-focused Strategy

There is a growing recognition that the regulatory settings are not conducive to the goal of improving air quality and the expansion of generating capacity has to be controlled by administrative means. The problem is viewed as one of demanding tougher environmental regulations imposed on the power industry so that the usual 'command-and-control' approach is self-justified. During the 1990s, a series of policy initiatives were developed with mixed results, as summarized by EPD:

- Requiring the power companies use the best practicable means to reduce emissions as required in the Air Pollution Control Ordinance (APCO) and at the same time enhance the operational efficiency of the power plants and the combustion and generation efficiencies;
- Requiring the power companies to use low sulphur coal for all the existing coal-fired generating unites;
- Banning the construction of new coal-fired generating units since 1997;
- Setting emission caps in any Specified Process Licenses (SPLs) issued or renewed to power companies under the APCO; and
- Encouraging power companies consider adopting the most effective economic tools (including emissions trading) to achieve the emission reduction targets. (EPD, 2006)

There is a focus on regulating the power industry by capping and controlling production aspects with the aid of technological capacity. The regulators believed that environmental objectives can be achieved by reconfiguring the industrial processes in compliance with their technical requirements. They followed an expert-led approach and held a technological optimistic perspective on the basis of a presumption that 'all consumption-related environmental problems can ultimately be solved through technical inventiveness' (Murphy & Cohen, 2001). Policy-makers tend to narrowly view environmental problems more as one of technical than an outcome of a complicated social process involving consumption as well (Murphy, 2001). The technologist's view often results in an emphasis of supply-led measures.

The environmental policy of Hong Kong has been described as being trapped in the framework of 'administrative rationalism' (Hills, 2004). The approach is one of administrative-led with an emphasis of the role of expert and technologic perspectives while demand-side participation is limited. It is unlikely that the administrative mind will diminish as the dominant role of 'command-and-control' approach had been reaffirmed in the Stage I Consultation Paper. For the government, the approach was seen as 'effective' and there seemed to have no immediate need to replace it given the low monitoring cost and intervention efforts required of the existing approach. In the Stage II Consultation Paper, the economic bureau made this explicit:

> Experience has shown that the mandatory licensing arrangements have been effective for improving the environmental performance of power companies in Hong Kong. We propose that environmental regulation continue to be applied in the future electricity market (EDLB, 2005b, p. 14).

While it is doubtful how 'effective' it has been and in what manner the regulatory approaches will 'continue' to lead in the environmental decision-making, this production rationality has been embodied in the First Strategy proposed by the CSD in which four specific measures are devised for achieving sustainable use of energy. In the action plan, there is a belief that the environmental practices taken by the producers will be analogously instrumental to encourage positive changes in consumers' behaviours. Educational initiatives are highly valued despite in most cases they are insufficient to bring about actions in practice due to value-action gap. Allowing third party access to electricity grid can also make renewable energy supply available but it does not mean profitable and hence viable, given the consumers' scepticism of shifting to renewable power as discussed above. The CSD attempted to build consensus among stakeholders through consultation exercises, but little was done to enable real demand-side commitment underpinning sustainable infrastructure that is essentially based on transformation of supply-demand dynamic.

In similar vein, the ETWB outlined two key environmental initiatives about energy in the Policy Agenda 06-07, i.e., continuing to monitor the power companies on the development of renewable energy and impose emission caps onto power plants (ETWB, 2006). This is echoed by the Action Blue Sky Campaign (Campaign) launched in 2006 by the EPD to demonstrate their commitment of improving air quality. As the Chief Executive (CE) of HKSAR had planned to go for re-election in the year after (2007), the Campaign, led by the CE, consequently manifested as a political act. One initiative of the Campaign was to emphasize the responsibility of the utilities in meeting emission targets. Dr. Sarah Liao, the former

Secretary for ETWB, thought that the first and foremost is to:

> set clear targets and milestones, taking into full consideration the need to strike a balance between protecting the environment and sustaining economic and social development, while harnessing technological advancement and economic tools to accelerate their attainment. (Liao, 2006)

The idea of technological optimism is clear enough. No matter for meeting political needs or truly for environmental reasons, these initiatives affirmed the belief that to 'command and control' should remain as the major responsibility of the government as it did twenty years ago. The unadjusted presumption has provided legitimacy to the use of tougher regulation which, they believe, was and will be capable of overcoming most of the electricity-related environmental problems.

On the other hand, consumption rationality is inadequately addressed. Policy initiatives with a focus on altering consumption patterns were given a place somewhere in the government's environmental agenda, they were, however, limited in numbers and effectiveness. Public involvement was absent in the electricity policy-making which was entirely determined by administrative decisions and the consumers were assumed to be captive recipients of electricity. The public was normally uninvolved, until recently (around the 2000s) a few consultation exercises were implemented.

During the 1990s, the government had shown some interests in promoting DSM which was once a policy objective stated in the Economic Services Bureau's (ESB) (the then EDLB) Policy Objective. However, the energy efficiency programmes are regarded as piecemeal as admitted later by the CSD (2006). In 2000, the EDLB entered into a short-term additional agreement with each of the utilities for implementing a DSM Programme during mid 2000-2003. The Programme targeted industrial customers and aimed to encourage efficient use of electricity so that installation of additional generating capacity by the utilities could be deferred or avoided, thereby reducing the environmental impacts arising from power generation. Both participating electricity users and the utilities were entitled to some financial incentives. The Programme consisted of two parts: the rebate programmes - providing rebate to non-residential customers to encourage installation of energy efficient lighting and air-conditioning equipment; and the non-rebate programmes – including time-of-use tariff programme, educational and promotional activities, and market survey to gauge the public's views on DSM. The program costs were recovered by a DSM charge collected from all consumers in the monthly electricity bills.

Comparing to the conventional supply-led measures, the Programme could be considered as a breakthrough attempt for its focus on consumption. However, it managed to get satisfactory performance. Although the energy savings had outperformed the expected levels, the actual capacity savings for both utilities had only slightly exceeded the targets approved by the authority prior to implementation, and there were some individual items performed far lower than the targets. Lee and Yik (2002) criticized that because the rebate rates were poorly designed there were diverse participation rates. Also, the Programme was regarded as 'largely cosmetic' and did not have significant effect on consumption (Bronstein, 2003). In fact, the annual energy savings were minimal as it accounted for only 0.21% of the electricity sales. More importantly, it failed to motivate the utilities to actively take part in demand management. In their respective accomplishment reports, both utilities expressed little interest in extending the programmes to residential customers as initially planned (CLP, 2003; HEC, 2003). No wonder that it is because the idea of DSM was at odd with the rationale behind the SCAs.

This, perhaps, has led the government to put more resources on general educational and informational measures instead. The Campaign, for example, has been taken merely as an educational and publicity programme. A series of awareness-raising measures were proposed by the EPD, including the establishment of a voluntary Energy Conservation Charter to encourage more efficient use of energy. It is pity that the proposal on energy efficiency and conservation had included every possible suggestion for individuals except practical incentives. It looked like an iteration of the environmental responsibility of every member of the public by using a checklist of 'dos' and 'don'ts', which is similar to the educational programmes the department implemented from time to time. There is a rhetorical focus on the role of consumers and insufficient considerations were placed on the practical barriers and the means to influence the embedded nature of consumption practices.

4.4. Restricted Reforms: Public Protests Against Private Interests

The elected politicians and the pressure groups have exerted lots of pressures on the government to limit the private interests of the utilities for the sake of public benefits (e.g., lower tariffs). As the SCAs are fairly protective, the utilities are often accused of not complying with environmental principles while making a huge sum of money. This is exemplified by the following two cases.

1. The DSM Programme

As mentioned, the utilities appeared uninterested in extending the DSM Programme due to the limited gains under the SCAs. In fact, at the beginning, a total incentive earning of HK$76 million (equivalent to 0.23% of the total sales revenue in 2001) was agreed between the EDLB and the utilities for implementing DSM activities for three consecutive years. It was proposed to permit the latter to get a compensatory return comparable to that they would have if they otherwise add generating capacity and transmission and distribution facilities to meet the demand (ESB, 1999). Yet it was soon abandoned due to the objection of some LegCo members who felt uncomfortable to legitimize such an extra profit and allow them to take an improvement of social and environmental well-beings as a business opportunity. In a LegCo meeting, one of them 'doubted the rationale of providing incentive earnings to the power companies which effectively guaranteed the power companies' profit level despite the implementation of DSM' (LegCo, 1999, para. 27). Eventually, the profit cut was accepted by the utilities on a condition that they would not be penalized in shortfall in target savings as previously agreed. Incentive to maximize programme performance was therefore reduced.

DSM involves program costs *plus* revenue losses and risks. An effective incentive mechanism should contain not only full cost recovery but also adjustment of lost revenues and an above-cost bonus to offset risks (Gellings & Chamberlin, 1993). Without reasonable compensations, the DSM Programme could hardly fall into the private interests of the utilities when they were still subject to the rate-of-return regulation. As a consequence, soon after the Programme was completed, both of them were inclined not to extend the Programme to residential sector as planned for the reason that public awareness on energy conservation has already been raised 'significantly'.

2. The Stage II Consultation

The second stage consultation about the proposed electricity market reform has been politicized with the social distrust over public issues being extended to the electricity affairs. In a LegCo panel meeting, it was suggested by a member of LegCo that:

> [i]nstead of discharging their corporate responsibility to improve emission performance, the power companies were using emission reduction facilities as tools for bargaining better terms for their SCAs, such as higher rate of return and longer duration, where were contrary to public aspirations. (LegCo, 2006, p. 11)

It was echoed by a green group:

> the power companies make profits by running this business. Pollution control is part of their production cost. It should be internalized, and the cost burden shouldn't be shifted to the consumers[3] (Greenpeace China, 2006a, p. 28, quoted a comment from the Secretary for ETWB).

According to this view, environmental improvement is by no means a 'sellable' product, but a non-compensable duty of electricity industry. It is the intention of these actors and the public as well that the utilities should be responsible for the largest part, if not all, of the costs of environmental initiatives. It was suggested by other LegCo members that it is impossible to work with the industry in the promotion of energy conservation without thorough reform of the policy and imposition of punitive measures (LegCo, 2005, p. 3187). There is general belief that the government should take a role of controller rather than facilitator or moderator in the electricity market (e.g., Greenpeace China, 2006b).

The confrontational positions of the politicians and pressure groups were taken as a justification of the technocratic regulatory approach. The government is motivated to manifest as a tough regulator as a response to the public's protests. The Secretary for ETWB said that if the utilities do not comply with environmental standards, the government would 'be tough, fine them, applying heavier fines to make them panic[4]' (Wong, 2006). This, however, almost led to a collapse of the government-utility relationship (Yung, 2006). It appears that the government tended to hide behind social consensus and acquiesce in confrontation rather than collaboration between the private and public sectors. The public consultations were more a strategic attempt to legitimize their technocratic approach rather than to incorporate the public's views (Gouldson et al., 2008). Intensification of legal and technical controls is thus favoured in such a political context, giving little incentive to the industry for investing in environmental innovations which require a freer market condition. As a result, collaborative measures were restricted to 'window-dressing' educational and promotional activities and demonstration of technological innovations to school children and media.

---

[3] Quoted from an interview aired on 26th February, 2006 with 'Hong Kong Connection', a documentary programme of Radio Television Hong Kong. The original comment was made in Chinese.

[4] My translation. The original comment was made in Chinese.

5. Conclusion

Hong Kong is characterized by its growth-led development path and the not-yet-democratic governance structure. For a long time, environmentally unfriendly industrial practices were legitimized by the exclusive economic objectives and dealt with in a piecemeal manner mechanically. The growth-led mentality inherited from colonial era is deeply entrenched and many existing institutions are still adhered to this philosophy. The executive-led administrative system also appears fundamentally at odds with a 'bottom-up' sustainability approach. Uncertainties in constitutional development have checked higher-level public participation. To respond the public's aspirations, short-term economic welfare measures, such as lower tariffs, are considered as a more desirable option than a sustainability-oriented reform that mainly benefits future generations. The development of electricity policy has remained growth-led. A higher priority for a stable and cheap power supply has registered in the government's agenda and this is generally consistent to the public's expectation. Technical forms of reform and minor remedial measures are in favour of fundamental transformation of the existing institutions. Electricity-related environmental strategies are manipulated on an *ex post* basis.

In Hong Kong, understanding environmental impacts as a technical problem seems to be a common practice of the regulatory agencies. New institutions are bounded to the existing ones or vested interests. The regulatory constraints (i.e. SCA) have profoundly impeded remedial and innovative measures conducive to environmental benefits. While big changes are needed, conservative measures are favoured as a matter of political reality. As the government has become much more susceptible to the reaction from the public sphere, it is tempted to employ 'hard' technologic policy tools that can more easily produce tangible results for assessment within a fixed period of time. Adoption (or merely demonstration) of renewable technology is more preferable than DSM initiatives because it is more 'eye-catching' and charismatic. The policy goal is more on strengthening supply capacity than that of demand. Practical incentive given to the electricity users and the means to shift consumption patterns are far from adequate. There is a strategic focus on the products or by-products (i.e. electricity and emissions) rather than the social process of power supply.

The institutional failure mainly stems from the regulatory constraints inherited from the development paths of Hong Kong. Environmental endeavour has never been given the same priority as economic growth. Particularly at the time of economic downturn, no one can afford to undermine growth for environmental reasons. Upward adjustment of tariff or applying energy tax can help energy saving but this is obviously not a politically desirable option. It is observed that the sustainability-oriented institutions prove inconsistent with its conventional regulatory structure and have been handled in a piecemeal manner. The regulatory constraints inevitably create critiques from the concerned parties accelerated by the growing public awareness and the not-yet-successful democratic movement. These then turns to an emphasis of the role of utilities as the contributors of the causes as well as the solutions and a view that environmental gains are a function of a unidirectional, 'one-to-one' relationship between the government and utilities. Moreover, the community has never been convinced that environmental protection should be taken as a business opportunity as long as the citizens' living is going to suffer. As a result, the government manifests as a reactive party and mechanistically deals with the problems. And the private interests of the utilities are posited as a grave conflict to the social objectives. An extension of the production-focused approach is thus adopted given the political agenda. Consequently, although there are some improvements in the power generation process, promotion of sustainable energy consumption remains rhetoric and ineffective. Communicative actions are taken as a way to gain public support and legitimize the conservative policy changes rather than a truly 'bottom-up' approach.

To conclude, the electricity policy of Hong Kong is institutionally embedded as the required changes for moving to a more sustainable state have been precluded by other existing institutional constraints or vested interests. The reflexive institutions appear ineffective since the institutional impediments remain unresolved. The supply-demand relations are ill-constructed in a way that everyone is fairly aware but nobody has the motivation of taking meaningful actions. For the public, cleaner energy is considered as too expensive; for the utilities, opening the power market is economically unattractive; while for the government, big change is politically too risky. The path-dependant institutional settings have restricted a timely transformation of the role of the actors necessary for achieving a more sustainable state.

REFERENCES

Barron, W. (1996) The environment and the political economy of Hong Kong, in: D. Mole (Ed.) *Managing the New Hong Kong Economy (*Hong Kong, Oxford University Press).

Barron, W., & Steinbrecher, N. (1999) *Heading Towards Sustainability?: Practical Indicators of Environmental Sustainability for Hong Kong* (Hong Kong, Centre of Urban Planning & Environmental Management, the University of Hong Kong).

Bronstein, S. (2003) *Competition in the Power Sector* (Hong Kong, Civic Exchange).

Census and Statistics Department (2004) *Thematic Household Survey Report No. 17* (Hong Kong, Social Surveys Section, Census and Statistics Department).

Chan, C. T. (2006) What are the alternatives available to Hong Kong in structuring the electricity supply industry?, *Energy Policy, 34*, pp. 2891-2904.

Cheung, A. B. L. (2005) Hong Kong's post-1997 institutional crisis: problems of governance and institutional incompatibility, *Journal of East Asian Studies, 5*, pp. 135–167.

Close, J. (1997) *The potential of building integrated photovoltaics (BIPV) in Hong Kong's architecture, its contribution to demandside management of the electricity supply and the legislation and incentives to facilitate its introduction.* Paper presented at the POLMET '97: Pollution in the Metropolitan and Urban Environment, 25-27 November 1997 (Hong Kong, The Hong Kong Institute of Engineers).

CLP Power Hong Kong Limited (2003) *DSM Programme Summary Accomplishment Report (July 2000 - June 2003)* (Hong Kong, CLP Power Hong Kong Limited).

CLP Power Hong Kong Limited (2004a) *CLP Environmental Manifesto* (Hong Kong, CLP Power Hong Kong Limited).

CLP Power Hong Kong Limited (2004b) *CLP Group Social and Environmental Report 2004* (Hong Kong, CLP Power Hong Kong Limited).

CLP Power Hong Kong Limited (2006) *CLP's Response to the Government's Consultation Paper on Future Development of the Electricity Market in Hong Kong - Stage II Consultation* (Hong Kong, CLP Power Hong Kong Limited).

Council for Sustainable Development (2005) *A First Sustainable Development Strategy for Hong Kong* (Hong Kong).

Council for Sustainable Development (2006) *Clean Air and Blue Skies - The Choice is Ours* (Hong Kong).

Democratic Party (2006) *The Democratic Party's Proposal to the Electricity Market: Response to the Stage II Consultation* (Hong Kong, Democratic Party) (in Chinese).

Economic Development and Labour Bureau (2004) *Electricity Supply in Hong Kong: Establishment Survey*, Paper submitted to the Energy Advisory Committee on September 2004 (Hong Kong).

Economic Development and Labour Bureau (2005a) *Consultation Paper on Future Development of the Electricity Market in Hong Kong: Stage I Consultation* (Hong Kong).

Economic Development and Labour Bureau (2005b) *Consultation Paper on Future Development of the Electricity Market in Hong Kong: Stage II Consultation* (Hong Kong).

Economic Services Bureau (1999) *Implementation of Demand Side Management*, Paper submitted to the Legislative Council Panel on Economic Services on December 1999 (Hong Kong).

Electrical and Mechanical Services Department (2002) *Study on the Potential Applications of Renewable Energy in Hong Kong: Stage 1 Study Report* (Hong Kong).

Environmental Protection Department (2006) *Future Development of the Electricity Market in Hong Kong - Environmental Regulation in the Future Electricity Market*, Paper submitted to the Legislative Council Panel on Economic Services on 23 September 2006, LC Paper No. CB(1)2229/05-06(01) (Hong Kong).

Environmental Transport and Works Bureau (2006) *2006-07 Policy Agenda: Policy Initiatives of the Environment, Transport and Works Bureau*, Paper submitted to the Legislative Council Panel on Environmental Affairs on 20 October 2006 (Hong Kong).

Friends of the Earth (2005) *Friends of the Earth's Response to the Future Development of the Electricity of Hong Kong (Stage I Consultation)* (Hong Kong, Friends of the Earth) (in Chinese).

Gellings, C., & Chamberlin, J. H (1993) *Demand-Side Management Planning* (Lilburn, GA, Fairmont Press).

Gouldson, A., Hills, P., & Welford, R. (2008) Ecological modernisation and policy learning in Hong Kong, *Geoforum, 39*, pp. 319-330.

Greenpeace China (2006a) *Hong Kong's Roadmap of Replacing Coal with Wind Energy By 2020: Greenpeace's Response to "Future Development of the Electricity Market in Hong Kong: Stage II Consultation"* (Hong Kong, Greenpeace China)

Greenpeace China (2006b) *Response to the Liberalization of the Electricity market of Hong Kong: Fully Liberalizing the Electricity Market in 2018*, Paper submitted to the Legislative Council Panel on Economic Services (Hong Kong, Greenpeace China) (in Chinese).

Hills, P. R. (1985) Environmental protection in a laissez-faire economy, *Built Environment, 11*(4), pp. 268-282.

Hills, P. R. (1991a) *The Hong Kong energy economy: private interest and public neglect,* Paper presented at the Energy Efficiency Symposium: An Economic and Environmental Imperative, 11-12 March 1991 (Hong Kong, Hong Kong Institution of Engineers).

Hills, P. R. (1991b) *The Household Energy Transition in Hong Kong* (Hong Kong, Centre of Urban Planning and Environmental Management, University of Hong Kong).

Hills, P. R. (2004) Administrative rationalism, sustainable development and the politics of environmental discourse in Hong Kong, in: T. Mottershead (Ed.) *Sustainable Development in Hong Kong* (Hong Kong, Hong Kong University Press).

Hills, P. R., & Barron, W. (1997) Hong Kong: the challenge of sustainability, *Land Use Policy, 14*(1), pp. 41-53.

Hills, P. R., & Welford, R. (2002) Ecological modernisation as a weak form of sustainable development in Hong Kong, *International Journal of Sustainable Development and World Ecology, 9*, pp. 315-331.

Holliday, I., Ma, N., & Yep, R. (2002) A high degree of autonomy? Hong Kong Special Administrative Region, 1997-2002, *The Political Quarterly, 73*(4), pp. 455-464.

Hongkong Electric Company Limited (2003) *DSM Activity Report* (Hong Kong, Hongkong Electric Company Limited).

Hongkong Electric Company Limited (2005) *Environment, Quality, Health and Safety Report 2005* (Hong Kong, Hongkong Electric Company Limited).

Hongkong Electric Company Limited (2006) *Submission on Stage II Consultation: Future Development of the Electricity Market in Hong Kong* (Hong Kong, Hongkong Electric Company Limited).

Hung, W. T. (1997) The politics of the environment in Hong Kong, in: P. Hills & C. Chan (Eds.) *Community Mobilization and the Environment in Hong Kong* (Hong Kong, Centre of Urban Planning and Environmental Management, University of Hong Kong).

Jänicke, M. (1990) *State Failure: The Impotence of Politics in Industrial Society* (Cambridge, Polity Press).

Lam, P. L. (1997a) *Competition in Energy* (Hong Kong, City University of Hong Kong Press).

Lam, P. L. (1997b) Incentive effects of rate-of-return regulation in Hong Kong, *Journal of Regulatory Economics, 12*, pp. 319-325.

Lam, S. F. R. (1998) *An Analysis of the Policy of Energy Efficiency and Conservation in Hong Kong in the 1990s,* Unpublished MPA thesis (Hong Kong, University of Hong Kong).

Lee, W. L., & Yik, F. W. H. (2002) Framework for formulating a performance-based incentive-rebate scale for the demand-side-energy management scheme for commercial buildings in Hong Kong, *Applied Energy, 73*(2), pp. 139-166.

Legislative Council (1999) *Minutes of Meeting of the Panel on Economic Services*, 24 May 1999 (LC Paper No. CB(1)2030/99-00) (Hong Kong).

Legislative Council (2005) *Official Record of Proceedings of the Council,* 14 December 2005 (Hong Kong).

Legislative Council (2006) *Minutes of Meeting of the Panel on Environmental Affairs*, 27 March 2006 (LC Paper No. CB(1)1298/05-06) (Hong Kong).

Liao, S. (2006) Comprehensive Strategy to Protect Our Environment (16 October 2006), *Action Blue Sky Blog*, Retrieved April 1 2007, Available from http://blog.epd.gov.hk/

actionbluesky/?blog=6&page=1&paged=2 (Hong Kong, Environmental Protection Department)

Ma, N. (2005) Civil society in self-defense: the struggle against national security legislation in Hong Kong, *Journal of Contemporary China, 14*(44), pp. 465-482.

Mottershead, T. (2004) Sustainable Development in Hong Kong: a road yet to be travelled?, in: T. Mottershead (Ed.) *Sustainable Development in Hong Kong* (Hong Kong, Hong Kong University Press).

Mottershead, T., & La Grange, A. (2000) Developing a strategic framework for sustainable development in Hong Kong: is an Earth Charter the answer?, *Public Administration and Policy, 9*(2), pp. 37-70.

Murphy, J. (2001) *Ecological Modernization: The Environment and the Transformation of Society*, OCEES Research Paper No.20 (Oxford, Oxford Centre for the Environment, Ethics & Society).

Murphy, J., & Cohen, M. J. (2001) Consumption, environment and public policy, in: M. J. Cohen & J. Murphy (Eds.) *Exploring Sustainable Consumption: Environmental Policy and the Social Sciences* (Amsterdam, Pergamon).

Ng, M. K. (2002) Sustainable urban development issues in Chinese transitional cities: Hong Kong and Shenzhen, *International Planning Studies, 7*(1), pp. 7-36.

Ng, M. K. (2004) Sustainable development and planning, in: T. Mottershead (Ed.), *Sustainable Development in Hong Kong* (Hong Kong, Hong Kong University Press).

Tsang, Y. K. (2006) Proactive, Pragmatic, Always People First, *Address by the Chief Executive at the Legislative Council Meeting on 11 October 2006* (Hong Kong, Printing Department).

Tsang, Y. K. (2007) A New Direction for Hong Kong, *Address by the Chief Executive at the Legislative Council Meeting on 10 October 2007* (Hong Kong, Printing Department).

Wong, K. M. (2006) Penalize the two power companies to make them panic if they do not take care of the environment, *Hong Kong Economic Times,* November 4, pp. A22 (Hong Kong) (in Chinese).

Yung, C. (2006) Cuts put all at risk, says power chief, *The Standard,* pp. A10, January 17 (Hong Kong).

In: Energy Policy: Economic Effects, Security Aspects... ISBN 978-1-60692-894-3
Editor: Noah B. Jacobs

Chapter 8

ELECTRICITY SECURITY AND CONTINGENCY PLANNING IN HIERARCHY AND NETWORK - EXAMPLES FROM SWEDEN

Jenny Palm*

Department of Technology and Social Change, Linköping University, S-581 83 Linköping, Sweden

ABSTRACT

This article discusses the emergency management strategies of municipal authorities for securing the electricity supply, according to a networked, or "governance", control and direction structure, and how this influences the relationship between electricity companies and Swedish municipalities. The article also discusses the role of the households during outages, and how they perceive their responsibility during an outage, but also how municipalities and grid companies understand the households' responsibility. Since electricity market deregulation, the responsibility for electricity supply security has become unclear; field studies of Swedish municipalities indicate that all actors still seem to be seeking to find their proper roles in the deregulated market. Municipalities still expect to exercise influence over private energy company decisions regarding prioritization of emergency power deliveries. Energy companies vacillate between emphasizing their need to regard economic factors and their sense of responsibility for providing a secure electricity supply to vital municipal functions (even though municipalities may lack contracts specifying this). The households' responsibility for being prepared for an outage was fuzzy for the actors. Both municipalities and grid companies expected the customers to have some preparedness. The household meant that they lacked any responsibility to be prepared for an outage even if they *need* to be prepared to survive.

---

* Fax +46 13 284461; Phone +46 13 285615. E-mail: jenpa@tema.liu.se

INTRODUCTION

The electricity market and the need for a secure energy supply have been debated in Sweden since Sweden experienced two powerful storms, "Gudrun" and "Per in" 2005 and 2007. These storms damaged the power grid and caused power outages in parts of southern Sweden. (SEA 2005; 2007a). Because of the experience of these storms, increased expectations have been expressed regarding the ability of both private energy companies and the government to organize responses and take the lead in emergencies. (Palm, 2008) Also, the unclear role of the households' responsibilities has been highlight. The role of the households in the emergency management system and their relationship and mutual dependency on municipalities and grid companies are not so thorough analyzed. In this article I will analyse not only the municipalities and grid companies' responsibility, but also how the households perceive their responsibility during an outage, and how municipalities and grid companies understand the households' responsibility.

Actors cooperation, coordination and responsibilities will be discussed in relation to the perspective that the society turn more and more into a network society. The thesis, "from government to governance," includes the idea that the power to govern the policy development in an area has been transferred from "the public" to autonomous networks where also private actors take part.

The article emanates from a case study in the county of Östergötland, which consists of 13 municipalities. The study focusing on how households that suffered outages during the storm Per, how the municipalities and the grid companies experienced the emergency management, coordination and cooperation and their own as well as other actors' responsibilities when it came to maintaining ordinary and everyday activities in organisations as well as in households. A small survey was conducted in Östergötland in which I asked households how they perceived the outage. The main purpose of this survey was to identify households that were willing to be interviewed about their experience of power outages. In total, I sent 200 surveys to households that had suffered outages during Per; of these, 62 households completed and returned the survey, and 15 of these were willing to be interviewed. The results of the survey are not statistically significant, and I will only cite them if they are relevant to our in-depth interviews. As well as the members of the households, I also interviewed 15 municipal administrators and four politicians working on emergency management and three executives of the three grid companies owning most of the distribution network in Östergötland. In total, 37 people were interviewed. I also undertook participatory observation of a collaboration meeting, in which politicians and officials in the municipality of Kinda could discuss outage-related issues with the grid company, E.on., in the region.

The article is organized as follows. First, I describe organization and deregulation of the Swedish electricity market. Then I consider earlier research on networking and households action space during crises. Thereafter comes the analyze of the results of the case studies of one county, Östergötland, in Sweden. The article concludes by discussing actors' networks and responsibilities in the electricity market during outages.

THE ORGANIZATION OF THE SWEDISH ELECTRICITY MARKET

The way Sweden organized its energy market through most of the twentieth century has it roots in the 1860s when its cities often owned the gas plants established in them. When the electricity system was established in the 1880s, Swedish municipalities continued to play a central role, by both producing and distributing electricity. As of 1920, all 110 Swedish cities had electricity plants; half of these plants were owned by the municipalities and half by private actors. (Kaijser 1986; Kaijser, Mogren and Steen 1988; Palm, 2006). This organization was maintained until the mid 1990s when the Swedish electricity market was deregulated.

The restructuring of the electricity market commenced in January 1992 when the production and sale of electricity were separated from network operations at a national level. Svenska Kraftnät was established and made responsible for the national grid. Svenska Kraftnät runs the national grid and has the role of transmission system operator (TSO). This entails ensuring that the power plants in the Swedish electricity system are all working together reliably and that production and imports correspond to consumption and exports.

In spring 1992, parliament adopted targets and strategies for electricity market reforms the objective of which was to increase competition on the electricity market. New legislation for electricity market reforms was proposed in 1993, and parliament passed it in 1994. After the general election the same year, however, the reforms were postponed. In autumn 1995, it was decided that the legislation would come into force on 1 January 1996 (Jacobsson 1999).

The reforms kept the production and sale of electricity separate from the transmission of electricity. Power trading and network operations could not be conducted by the same legal entity, entailing that trading and network operations would be kept separate in different organizations. The deregulation made it possible to produce and trade electricity in competition; network operations, however, were seen as a natural monopoly to be regulated and supervised by the authorities (Government Bill 2004/05:62). Today, the Energy Market Inspectorate of the Swedish Energy Agency is legally the regulator of electricity networks and the electricity market. It ensures that the network tariffs are reasonable and do not exceed the self-financing principle and that network operations do not subsidize other activities (Svenska Kraftnät et al. 2006).

Since deregulation of the electricity market in 1996, there has been consolidation of ownership and several municipalities have sold their energy companies to larger electricity utilities. In 1991 Sweden had 151 municipally owned electricity companies; as of 2002 these had become 101 grid companies and 52 sales companies (Bladh 2002).

In municipalities with municipally owned electricity companies, these companies are often key actors, because they are formally responsible for and own the infrastructure, i.e., the power plants and distribution networks. If the municipality lacks its own energy company, it acts mainly as a consumer of energy, buying related services from private companies. In this case, the municipality's main capacity to exert influence, besides as a consumer, lies in its planning monopoly and its ability to govern through and within actor networks (Wihlbord & Palm 2008).

The experience of the storm Gudrun led to new regulations and changes to the Electricity Act (Government Bill 2005/06:27). The new Act gives consumers better guarantees of a secure electricity supply and strengthens the electricity companies' obligation to secure the electricity supply and make contingency provisions. For example, the energy companies now must compensate consumers if an outage exceeds 12 hours (a matter previously unregulated by law), and starting in 2011, an outage exceeding 24 hours will be even more expensive for the energy companies.

Swedish emergency management in general, is based on responsibility for geographic areas, municipalities, county administrative boards, and the national government being responsible at the local, regional, and national levels, respectively. This means that these bodies are responsible for co-ordinating actors at their individual levels before, during, and after a crisis; they never encroach on the responsibilities, however, of another level of government. Swedish municipalities have a general social responsibility, for example to supply water, to their citizens and must ensure that they do not suffer prolonged power outages. Several municipalities have auxiliary generation capacity to use in case of emergency, most commonly in the form of diesel generating units. Such backup systems cannot handle large loads, so prioritization is often needed – a controversial issue in Sweden. Several studies have identified the need to prioritize electricity provision, supplying it to the most important users even in peacetime crises, and this need was further emphasized by experience with Gudrun and Per. Today, however, such prioritization has no legal basis. The Electricity Act is normally interpreted to state that no prioritizing can be done in peacetime, for example, when capacity shortage calls for load shedding. Another situation when prioritizing would be called for is during power restoration after a long outage. In practice, however, municipalities and energy companies prioritize certain customers today when absolutely necessary. Besides, according to Swedish law (SFS, 2006:544), municipalities are responsible for maintaining municipal services, even during crises. Some municipalities interpret this law as allowing municipalities to prioritize between their services and their citizens. Many municipalities have listed which actors and facilities should be given first priority, which second, third, etc. The Swedish Energy Agency also has the task of studying how legal, organizational, and technical barriers will be removed to make it possible to prioritize in peacetime crises. (Palm 2008).

The requirements households can have on the grid companies refer to both regulations and existing agreements between the household and the company. In government bill 2005/06:133 is it emphasised that citizens have a responsibility to take accurate measures to prevent and handle extra ordinary accidents. The Swedish Energy Agency (2007a) has interpreted this as the households have a responsibility to handle an outage, even if it is unplanned.

Electricity Supply and Contingency Planning in Network Structures

As discussed above, the Swedish electricity grid is still a monopoly, while the trade in electricity is deregulated. Grid operations are regarded as a natural monopoly and is still supervised by public authorities in Sweden. In this way, the authorities can still serve as a final court of appeal for fundamental issues. On the electricity market this mean that authorities still can influence the grid and especially the level of investments. Even though this monopoly situation still exist concerning the grid, the ideas following from a deregulation of the electricity trade can be expected to spill over also to the grid companies, which is what partly have happened, which I will come back to later in the analysis. (Palm, 2008)

The deregulation and liberalisation of the electricity market follow a New Public Management trend (Pollitt 1990; Hood 1995) and are also in the spirit of the "from government to governance" thesis. The traditional *government* approach highlights the chain of command of public organizations and top–down hierarchal decision-making by political actors. "Government" implies that control and direction takes place within governments and their formal institutions, and that the state's monopoly on the use of legitimate coercion is the focus (Stoker 1998). According to the governance concept, networks are self-organizing and cannot be fully accountable to governmental bodies; the matter of responsibility thus becomes fuzzy for the actors. (Rhodes 1997; Peters and Pierre 2004).

Actors participating in networks exchange resources and negotiate shared purposes. Co-operation and co-ordination in networks are regarded as the best ways to achieve common interests (Börzel, 1998). Powell (1991) explains that these resource exchanges are intangible, taking place over a long period of time through a general pattern of interaction. This happens between individuals in networks who take part in joint actions; with repeated interaction institutionalization occurs, that is, common views, participation patterns, and interaction rules evolve (Kickert et al., 1997). Kenis and Schneider (1991) view actors' inter-relationships as communication channels, and they emphasize that resources such as information, knowledge, and trust are exchanged between involved actors. MacNeil (1987) discusses 'entangling strings', which implies that actors are entangled with each other by means of rumours, friendship, interdependency, etc. These are all important parts of a relationship.

Lars Carlsson (2000) argues that policy network theory is not yet fully developed and thus cannot completely explain political inquiry. He means that a change in focus is needed whereby networks are regarded as *explaining* different outcomes instead of just being a dependent variable. He is arguing that network theory needs to be incorporated into a broader theoretical framework.

Notably, responsibility becomes more indeterminate and concealed in a governance rather than a government structure. In a government context, accountability is mainly regarded as a chain of command from the electorate to the politicians, and from them to public administration. In contrast, a governance structure makes accountability intertwined and multiple (Stoker, 2004). In networks and governance structures, contracts are thus vital in determining who is responsible for what (Wihlborg and Palm, 2008). Considine (2002) argues that in governance contexts, contracts are used as an alternative to legal mandates and hierarchical chains of command. Contracts thus represent a way of handling the lack of adjustments in legal frameworks to meet the development of partnership and networked governance (Pierre and Peters, 2000, p. 17).

Actors in a government structure are included or excluded according to their institutional given roles, defining a professional and/or democratic accountability. The broad distinction of government and governance is commonly used even if it is problematic since both approaches include several different models and they cannot be clearly separated.

Most researchers agree on the existence of a trend toward a more network structure in public policy making. The governance literature lacks however explanation power for how and why actors cooperate in networks. Governance theory needs to be complemented with empirical bottom-up studies on cooperating actors in network, which is an analytical purpose with this article. Resources are one such aspect that could be developed further in governance theory. Pierre Bourdieu has developed the resource concept and except from the well known "cultural capital", Bourdieu has also introduced 'social capital', which is defined as: "the sum of the resources, actual or virtual, that accrue to an individual or a group by virtue of possessing a durable network of more or less institutionalized relationships of mutual acquaintance and recognition" (Bourdieu & Wacquant, 1992 p 119, expanded from Bourdieu, 1980). Actors' social resources are in this perspective an important explanation to why actors cooperate with each other. Trust is close related to the social capital. If actors trust each other that will enhance cooperation. Powell (1991) emphasis that trust is more important for cooperation than both authority and negotiations. In the analysis I will discuss which actors cooperate and how and why they coordinate their work when it come to power outages. Which actors are included and which actors are excluded – and why?

The households are rather invisible in governance literature. If they are included they are so because they have gathered in for example consumer associations that have representation in a network. The household is also discussed as citizens and in connection to how the citizens can make decision-makers accountable for decision made in networks including private actors. The households are rather invisible in research on power outages as well. However, earlier research has shown that whether or not households prepare themselves for a crisis, such as a protracted power outage, depends on how they *perceive* the probability of the crisis. Tierney (1989) states that if households perceive the risk to be low, then they will feel it is not worthwhile to prepare for a possible crisis. Another reason for lack of preparedness is that households may feel that they lack information on how to prepare effectively (Tierney 1989). It is only when households believe that a crisis is real that they start taking precautionary action, and households continuously evaluate the information received to confirm the warnings. That is one explanation for the increased phone traffic during a crisis (Helslootn and Ruitenbergnn 2004).

Helslootn and Ruitenbergnn (2004) emphasize that citizens are usually not helpless or dependent on others during a crisis and seldom use public or volunteer resources for protection during a storm. Instead, they usually seek shelter with family or friends. Dynes (1994) discusses the existence of "situational altruism", which is a certain norm applicable during crises and that influences how citizens behave. To regard citizens as helpless victims is thus completely wrong-headed, and governments should instead treat them as the powerful actors they are. (Cited in Helslootn and Ruitenbergnn, 2004)

In the next section I will discuss cooperation and coordination in local networks in the county of Östergötland and analyze households' action space and actors responsibilities before and during an outage.

The Storm Per and Experience in the County of Östergötland

Sweden has shifted from an electricity system in which public actors take great responsibility for securing the electricity supply, to one in which private-sector legal arrangements and agreements determine the level of the supply security. All municipalities in Östergötland lacked formal agreements with the energy companies covering issues such as aggregate backup generating capacity, responsibilities in event of outage, and prioritizing users. According to representatives of both municipalities and energy companies, they have not needed to formalize their co-operation. The co-operation between municipalities and energy companies differs, however, between municipalities. Some municipalities have no contact with the grid companies, while others, often those that have long had an official responsible for emergency management issues, had established networks with ongoing meetings. For example, both Kinda and Norrköping municipalities had established informal co-operation with E.on, giving them influence on how E.on prioritized the grid repairs after Per. These municipalities also felt that they received enough information during the storms, contacting the energy company every night to discuss conditions and how they could limit damage for citizens.

The municipalities that owned an energy company, Mjölby and Linköping, experienced that they had both enough control and knowledge about the grid in their area:

> It is all about open communication and relations. And we own the grid, or parts of it. But we know all about the grid and the problems that can occur. (Interview chairman of the municipal executive committee, Linköping)

None of the municipalities that did not own an energy company (11 of the 13 municipalities) had any formal or informal agreement with the energy company concerning, for example, emergency reserves. The procurement of back-up energy sources was done based on personal contacts and networks. One municipality, Norrköping, had sold its electricity company to E.ON in 2001. At that time, the politicians and emergency manager in Norrköping believed that E.ON could handle the electricity supply because:

> As long as it is the same people working in E.ON who worked in NME [the former municipally owned electricity company] it will work. Because we know who to contact and they know who they will contact in the municipality. (Interview chair of the municipal executive board)

E.ON has concessions in several Östergötland municipalities and has had to develop working relationships with actors in each of them. E.ON's operations coordinator stated that cooperation and communication with the different municipalities differed greatly. In general, E.ON has forged better working relationships with larger municipalities, than with the smaller ones. For example, during the storms there were problems receiving updated contact lists regarding which municipal actors E.ON should inform of the occurrence of outages and of how long E.ON thought they would take to repair. In some municipalities, such information exchange simply did not happen. Both E.ON and the municipal representatives felt this indicated that some municipalities lacked updated emergency management systems.

The municipalities lacking cooperation with the grid company also felt they were powerless when it came to influence grid issues such as safety and investment plans. For example the administration manager in Boxholm meant that the only way to influence the grid company was through legislation. Vadstena lacked any contact with any grid company and did not even know who they should contact. The emergency management coordinator was at the same time critical to the grid companies and meant:

> They [the grid companies] think that we are just sitting here and think that we will pay the absolutely cheapest electricity supplier, and still got the best service at an outage They argue that other private actors pay a lot of money to have a secure electricity supply and then they shouldn't suffer because others are not willing to pay. They mean that the municipalities think that they still will have good service without paying for it, that the municipalities plan for that. But we have never had this in mind. (Interview emergency management coordinator, Vadstena).

The emergency management coordinator in Vadstena also said that supply and grid issues had not been on the municipal decision agenda very often. But recently, after the two storms, the municipality had been more aware of these issues and had started to look into for example their auxiliary generating capacity.

The unclear roles and relationship for all involved actors were also exemplified at a collaboration meeting held in Kinda, involving E.ON, politicians, officials, and local industry, to discuss how to coordinate efforts in the event of future power outages. During the meeting it became clear that politicians and officials in Kinda assumed that the municipal organization and services should be prioritized during an outage; however, the municipality had no agreement or contract with the energy company stating that they should be prioritized. Traditionally, however, municipal administrations have been prioritized, and municipal actors, and to some extent energy company actors as well, still held this to be appropriate.

Municipal officials and politicians returned to the question of how the municipality should act to influence which power lines E.ON restored first; the representatives of E.ON faltered in responding to this matter. In discussion, they alternated between emphasizing that they are a competitive company in a market that requires contracts, and that they must meet and agree to municipal requirements without demanding a contract.

Municipal contact with the grid companies is ultimately important for citizens. Municipalities that had a well-developed relationship with the grid company were better prepared and had considered how to prioritize between different users and activities and had drafted priority lists. These municipalities were also better informed during outages than those lacking such contacts. They had better knowledge of the grid and its weaknesses and better overall knowledge of where there were needs for auxiliary generating capacity. Where such involvement was lacking, priority lists were lacking and municipalities even thought it was not legally possible to prioritize one user over another:

> It exist no exceptions. According to the legislation the grid companies are supposed to deliver electricity to as many people as possible all the time and there are no possibilities to prioritize between users. The municipalities can express a desire that they should prioritize an elderly care centre or an industry, but that is nothing the grid company can take into consideration. (Interview Emergency manager in Boxholm)

Neither the municipalities nor grid companies had established regular contact with the households or their associations. Instead of networking, they developed one-way information dissemination, for example, via websites, and it was through these that the municipalities and companies communicated with citizens/customers, including regarding emergency management. This will be further discussed below.

The interviews with the householders instead highlighted the importance of informal neighbourhood networks. Neighbours were a common information source during power outages. Several householders took it for granted that "someone else" would phone the grid company; these householders did not, however, know who that other person was. There was a lack of any formal network or organized means of information dissemination, though information dissemination did seem to work very well.

> Yes, it is always someone here in the village that makes the call. Someone contacts Fortum [the grid company]. (Interview household O).

There was a tradition of common action in the villages, where the neighbours felt responsible for and looked after each other. Social altruism seemed to exist in Swedish rural areas.

Perception of Actors' Responsibilities during Outages

Municipal responsibilities during power outages were unclear to both households and grid companies. Both these actors, however, assumed that the municipalities would take care of their citizens and keep providing healthcare, schooling, preschooling, etc., during any outages. At the same time, care of the elderly was something the households said they would try to safeguard themselves. When municipal representatives were asked what their responsibilities were in the event of outages, they responded that they were healthcare and care of the elderly and children. They also said that they had a responsibility to keep local inhabitants informed. Municipalities and households believed that the grid company should prevent outages, repair the grid as fast as possible, and, together with the municipalities, prioritize between electricity users when possible. The grid companies agreed to this, but added that they also have a responsibility to keep customers informed and give them economic compensation.

The householders were more critical of the grid companies' preparedness for the storms than of their work repairing the damage. One important factor when it came to the level of household trust in the grid company was the company's awareness of the local context. If the household thought the company lacked such local awareness, trust in it quickly evaporated. One householder described how Vattenfall could not tell her when her electricity would be restored, because her house "did not exist" on its map. Vattenfall could not find her house on the grid map and said that they did not work in her area. In this situation, distrust in the company obviously arose, as well as a certain bitterness that the company did not regard rural customers as equally important as urban customers. In this case, trust was partly salvaged when Vattenfall's workers succeeded in repairing the grid. The woman in the household happened to run into Vattenfall's workers, who assured her that Vattenfall indeed knew that her house existed and that her electricity would be restored in a few days. The grid workers were mentioned by several of the respondents, who seemed to have great trust in them and their predictions of when their electricity service would be restored.

The respondents had reflected the least concerning the households' responsibility for preparedness for outages. Municipalities and grid companies expected households to have a certain level of emergency preparedness, and that rural households would be more prepared than urban ones. At the same time, nothing in the material indicated that the rural households received less support than any others during Gudrun or Per. However, this may be because it was mainly rural households that were affected by the storms, so all resources could be concentrated in rural areas.

The householders interviewed all lived in rural areas, and we asked them whether they believed they had a responsibility to be prepared for power outages and whether this responsibility was greater than that of urban households. Their answers were equally divided between the following three categories:

1) those who felt they had a responsibility and should be more prepared,
2) those who felt they had no responsibility to be prepared, and
3) those who felt that, ideally, they *should* have no responsibility and *should not need* to be more prepared, but who in practice said they needed to be prepared, otherwise they could not handle an outage.

Even though householders did not think they were responsible for preparing for outages, in practice they were all somewhat prepared. All householders said that they always kept a supply of, for example, batteries, candles, and extra heat sources at home. In the survey 17 of 62 households mentioned that they had portable generators at home.

Information as Replacement to Networking

For the municipalities and energy companies, better communication with customers and better information provision had high priority for the future.

> Seventy per cent of the crisis consists of a lack of information, so if we can handle the information issue, we can relieve a huge part of the suffering. In this regard, IT is an important tool. (Interview, Emergency Management Co-ordinator, Mjölby Municipality)

Both municipalities and grid companies worked a lot with and, for that reason, talked a lot about their websites and their future importance in disseminating information. It was through their websites they planned to communicate regarding emergency management with their citizens/customers in the future. In a survey, the Swedish Energy Agency asked 1072 households about their experiences during Per. Of these, 9% said that they had used the Internet to get information during the outages; half the respondents had Internet access during the storm (SEA 2007b). In interview, household members expressed their surprise at having been told to get their information from the Internet:

> And then they tell us to use the Internet for information, but how can I do that when I don't have any electricity? (Interview, household O)

According to both our and SEA's survey, most households use radio, followed by newspaper and neighbours to get information. Radio was a common way also for our interviewees when searching for information. At the same time, both Norrköping municipality and E.on., had experienced that it was hard to reach out by the media when the outages persisted over a long period which happened during the storm Per. When the number of power board customers are reduced than the outage is not news anymore and media stop their reports. In that stage both municipalities and the grid companies have experience difficulties to have the radio to give daily reports on where they are repairing the grid and which areas that must wait.

Other way of communication also existed. The municipality of Kinda used the postman to reach the households on the countryside. In this way all households except for five were reached. During the storm E.on also established information offices in many municipalities to facilitate communication and answer to questions from the public. This initiative was very appreciated by the municipalities because then they did not need to understand and mediate the information they got from the grid companies. No of our households had however been in contact with these information offices and according to E.on's representatives they had had more contact with people from the municipal administration than from the public.

In interview, several householders told us how they tried to contact their grid company by phone, but that it was nearly impossible because of the overloaded company switchboard. Several householders also stated that they did nothing during the outage, but relied on other people contacting the grid company to tell them about the outage. They also felt no need to get any information about the outage or when it would end. One householder put it this way:

> Yes, you can see that quite easily. When the light is on, then the electricity is back on. (Interview, household F)
>
> Another said:
>
> I never phoned. I don't think I would call anyone, even if it was long lasting ... I would only wait. It [i.e., the electricity] will eventually come back on, and they are doing what they can. That is what I would think. (Interview, household C)

Some phoned the company over and over again until they got an answer as to when the electricity would be restored. It is only that question – when the electricity will be restored – that the households wanted to get answered. Not knowing when power will be restored has a paralyzing effect on households. If they know how long an outage will last, households will also know what efforts to make; for example, they may need to move the groceries in the freezer to a "safer place" or start a portable generator.

Conclusions

The networks between municipalities and grid companies differ. The existing local networks are to a high degree built on social capital and on historical established informal relations. The result is rather stable but less flexible local networks. The actors include in a network is so because of personal contact and long-lasting relationship. As a result some municipalities totally lacked cooperation and coordination with the grid companies, while other had established personal contacts and discussed the outage more or less every day. Through the networks could the municipalities influence how the companies prioritize when they repaired the grid. The municipalities with well developed networks also had more knowledge and were better informed than municipalities lacking contacts.

In the end it is also the municipalities' networks that decide the citizens supply security; the municipality with informal contacts with the grid companies have more developed plans and better prepared for an outage, than municipalities without such contacts. Municipality engaging in a network with the energy company had also made priority lists.

All actors still seem to be seeking their appropriate roles in the deregulated market. This is why, for example, E.ON can emphasize both that they are a profit-maximizing company and that they feel responsibility for securing electricity supply to vital municipal activities, even though municipalities lacks a contract requiring this.

The households where not represented in any networks. The municipalities and the grid companies instead developed different tool to inform the households such as Internet. Internet will be more and more important and both municipalities and the grid companies develop their websites to communicate what causes the outage and how the actors deal with it. The use of internet is at the same time criticized by the households because internet will not be reached during an outage. For the users the phone was the most important tool used during black-outs. A restriction for the household was however the problem to get in contact with the energy company because of the overload on the companies' switchboard.

The households use their contacts with neighbours, friends and relatives during outages. In the country side the inhabitants' local knowledge about their neighbours are an important part in the emergency management. The neighbours look after each other and also coordinate the communication with grid companies and municipalities.

Not all households need however information and several households express trust to that the grid companies can manage and deal with the situation. An important factor for households trust to the companies' ability and willingness to repair the damages as fast as possible were the grid companies familiarity with the local conditions. The households were critical to that the national grid companies lacking local offices in the municipality also lacked an understanding of local conditions. The workers who repair the grid could in some cases up-weigh this criticism. The grid workers were spontaneously mentioned during the interviews as positive factors during the storms and the households had trust in their estimations of how long the outage would last.

When it comes to defining the grid companies', municipalities', and households' responsibility for being prepared for power outages, the expectations of each other differ. All actors stated that municipalities are responsible for the welfare of their citizens and must ensure that they do not suffer during outages. The grid companies' responsibilities, according to the actors, are to prevent power outages, repair damage, and pay their customers monetary compensation. Household responsibility for emergency management was less clear to the actors. Both municipalities and grid companies expected the citizens/customers to be somewhat prepared. The householders, however, believed that they were not formally responsible to be prepared for outages, even though they might *need* to be prepared to survive the outage.

Last, it is worth highlight the important role of municipalities for manage an outage. The municipalities have an obvious responsibility for the citizens' welfare services, even during blackouts. In those cases where the municipality not owns the local grid, the municipality has no influence over how to prioritize between users when that is possible. The municipalities must then also cooperate with private grid companies and need to play a more proactive role in the electricity market and for example contracting partners, which can provide emergency reserve power, or for securing electricity supply to specific municipal services. Some municipalities had established good relations and were cooperating in networks, while others lacked such contacts and also lacked formal contracts or agreements. This put both the municipality and their citizens in a vulnerable situation during an outage.

Acknowledgments

The research for this article forms part of the research programme, "Emergency management of technical infrastructure: regional and municipal strategies for coordination and implementation," funded by the Swedish Emergency Management Agency (SEMA) and of the "Reliable Electricity supply" project funded by ELAN.

References

Bladh, M. (2002). *Kraftbolagen, kommunerna och elavregleringen. Ägarförändringar i elbranschen 1991–2002*. Tema T Working-paper No 251. Linköping, Sweden: Linköping University.

Bourdieu, P. (1980). Le capital social. *Notes provisaires. Actes de la Recherche en Sciences Sociales 3*, 2-3.

Bourdieu, P. & Wacquant, L. (1992). *An Invitation to Reflexive Sociology*. Chicago: Univ. of Chicago Press.

Börzel, T.A. (1998). Organizing babylon – on the different conceptions of policy network, *Public Administration, 76*, 253-273.

Carlsson, L. (2000). Policy Network as Collective Action. *Policy Studies Journal, 28*, 502–520.

Considine, M. (2002). The end of the line? Accountable governance in the age of networks, partnerships, and joined-up services. *Governance: An International Journal of Policy, Administration, and Institutions 15*, 21–40.

Government bill 2004/05:62. *Genomförande av EG:s direktiv om gemensamma regler för de inre marknaderna för el och naturgas, m.m.*

Government Bill 2005/06:27. *Leveranssäkra elnät.*

Government Bill 2005/06:133. *Samverkan vid kris – för ett säkrare samhälle.*

Helslootn, I och Ruitenbergnn, A (2004). Citizen Response to Disasters: a Survey of Literature and Some Practical Implications, *Journal of Contingencies and Crises Management, 12*, 98-111.

Hood, C. (1995).Contemporary public management: A new global paradigm?. *Public Policy and Administration, 10*, 104-17.

Jacobsson K. (1999). *Att förändra sina omvärldsrelationer: En fallstudie av kommunala energibolag på den avreglerade elmarknaden.* Energy system Programme Working-paper no 9. Linköping, Sweden: Linköping University.

Kaijser, A.. (1986). *Stadens ljus. Etableringen av de första svenska gasverken.* Malmö, Sweden: Liber Förlag.

Kaijser, A., Mogren, A. och Steen, P. (1988). *Att ändra riktning. Villkor för ny energiteknik.* Stockholm, Sweden: Allmänna förlaget.

Kenis, P. and Schneider, V. (1991). Policy network and policy analyses: Scrutinizing a new analytical toolbox. In Marin, B. and Mayntz, R. (Eds.), *Policy networks, empirical evidence and theoretical consideration,* Campus Verlag, (25-62), Frankfurt am Main: Campus Verlag.

Kickert, W., Klijn, E-H. and Koppenjan, J. (Eds.) (1997). *Managing complex networks. Strategies for the public sector.* London: Sage Publications Ltd.

MacNeil, I.R. (1987). Relational contract theory as sociology: a reply to Professor Lindeberg and de Vos. *Journal of Institutional and Theoretical Economics, 143*, 272–290.

Palm, J. (2006). Development of Sustainable Energy Systems in Swedish Municipalities: A Matter of Path Dependency and Power Relations. *Local Environment, 11*, 445-457.

Palm, J (2008). Emergency management in the Swedish electricity market - the need to challenge the responsibility gap. *Energy Policy, 36*, 843-849.

Peters, G. & Pierre, J. (2004). Multi-level governance and democracy: a Faustian bargain? In: Bache, I., Flinders, M. (Eds.), *Multi-level Governance* (75-91). New York: University Press, New York.

Pierre, J. & Peters, G. (2000). *Governance, Politics and the State*. London: MacMillan.

Politt, C. (1990). *Managerialism and the Public Services*. Oxford: Blackwell.

Powell, W.W. (1991). 'Neither market nor hierarchy: network forms of organization. In Thompson, G. and Frances, J. et al. (Eds.). *Markets, hierarchies & networks. The coordination of life* (265-276). London: Sage publications.

Rhodes, R.A.W. (1997). *Understanding Governance: Policy networks, Governance, Reflexivity and Accountability.* Buckingham: Open University Press.

SEA, Swedish Energy Agency (2005), *Erfarenheter efter Gudrun,* ER2005:32. Eskilstuna, Sweden: SEA.

SEA, Swedish Energy Agency (2007a), *Prioritering av elanvändare vid elbrist – Slutrapport från Energimyndighetens Styrel-projekt åren 2004-2007,* ER 2007:38. Eskilstuna, Sweden: SEA.

SEA, Swedish Energy Agency (2007b), *Enkätundersökning Elanvändares förberedelser inför och hur de drabbades av stormarna Gudrun och Per, Energimyndigheten*, dnr 17-07-2831, not published.

SFS (Swedish code of statements) 2006:544, *om kommuners och landstings åtgärder inför och vid extraordinära händelser i fredstid och höjd beredskap.*

Stoker, G. (1998). Governance as theory: five propositions. *International Social Science Journal, 50*, 17-28.

Stoker, G. (2004). New Localism, Progressive Politics and Democracy?. *The Political Quarterly. 75*, 117-129.

Svenska Kraftnät et al. (2006). Svensk Elmarknadshandbok. Rutiner och informationsstruktur för handel och avräkning, version 2006-12-11, available at http://www.elmarknadshandboken.se/.

Tierney, K.J. (1989). The Social and Community Contexts of Disaster. In Gist, R. & Lubin R (Eds). *Psychosocial Aspects of Disaster* (11-39). New York: John Wiley and Sons.

Wihlborg, E. & Palm, J. (2008). Who is governing what? Governing local technical systems an issue of accountability. *Local Government Studies, 34*, 349-362.

In: Energy Policy: Economic Effects, Security Aspects... ISBN 978-1-60692-894-3
Editor: Noah B. Jacobs

Chapter 9

TRADE IN ENERGY GOODS AND SERVICES IN THE EARLY XXI CENTURY

Alberto Gabriele
United Nations Conference on Trade Development, Geneva, Switzerland

ABSTRACT

The world energy scenario in the late XX century was characterized by a widespread trend towards liberalization and privatization in the area of domestic policies. In the domain of international energy-related transactions, a naive belief in the omnipotence of technology and free market mechanisms contributed to consolidate uneven trade and geopolitical relations of forces strongly favoring advanced Western countries, and historically low prices prevailed in world energy markets.

This scenario underwent a sea-change around the year 2000, leading to a very different situation in the early XXI century. New large and middle-sized economic and geopolitical powers emerged in the developing world, from China to Venezuela. The South as a whole is less financially, economically, and politically and dependent on the North than it was the case in the past. Direct public control on energy resources are emerging again as a key factor of economic and geopolitical strength, with nation-states and SOEs increasingly dominating world markets. Multi-billion long term bilateral and multilateral agreements regulate an increasing share of world energy trade. The relative weight and role of material, non-reproducible resources is re-establishing itself vis-à-vis that of non-material, reproducible assets such as knowledge and technology.

Taking into account the profound change in geopolitical relations of strength worldwide, and the inadequate development of alternative energy sources, and notwithstanding the serious impact of the present international economic crisis, the persistent dependency of advanced and developing market economies alike on the availability of ultimately finite fuel reserves is bound to lead to an era of historically high energy prices in the medium-long term, and is not sustainable in the very long term.

"All around the world – from Kazakhstan to Alaska – oil-rich governments are reasserting that ownership as they squeeze international energy groups, raising taxes and royalties and taking a bigger share of projects. In Russia, the Kremlin has forced companies such as Royal Dutch Shell, BP and their peers to sell the majority share of their projects to Gazprom, the country's state-controlled gas monopoly. In Venezuela, almost every big international oil company has had to give up part or all of its oilfields... Oil production growth is slipping as many national oil companies prove themselves ill-equipped to manage the industries of which they are increasingly in control...The lack of growth means Opec's share of the market is projected to rise from 40 per cent to 50-60 per cent, shifting money and influence from western countries to oil exporters in the Middle East, Latin America and Africa...But even national oil companies within Opec are not investing as much as international oil companies...This lack of investment means oil prices are generally forecast to stay high unless a recession drastically erodes demand in large parts of the world " (C. Hoyos 2008)

1. The Collapse of Doha Negotiations on Trade in Energy Goods and Services

The Issues at Stake

Doha negotiations on trade in energy goods and services covered a broad range of activities relevant for energy companies, and spanned all energy sources, including renewables. Liberalizing commitments were sought on activities such as drilling; engineering; technical testing and analysis services; construction work for long distance and local pipelines, and for mining; wholesale trade services and retailing services of fuels. The negotiations addressed the establishment of commercial presence (Mode 3) and the intra-corporate transfers of specialists and professionals working for energy services companies (Mode 4). Some Members also proposed to negotiate additional disciplines on regulatory transparency, non-discriminatory third-party access to networks and grids, the need for an independent regulator, and requirements preventing certain anti-competitive practices. Other areas of negotiations focused on the liberalization of trade in environment-friendly goods and technologies, on trade facilitation (including the possible improvements and clarification of the "transit" obligation to allow passage of goods in transit across their territories), and other old and new issues such export taxes, subsidies, and bio-fuels trade-related policies (see Lamy 2008).

The last proposal relevant for energy services was put forward by the EU in 2007 in the Committee on specific commitments. The proposal focused on sub-sectoral classification issues in computer and related services and distribution services. According to the EU, energy distribution services have to be seen as part of a larger category, that of distribution services. In this framework, the W 120 classification list should be modified in order to allow (inter alia) a clear identification of energy distribution services, and member countries should undertake commitments in this sub-sector accordingly. However, the EU proposal was opposed by several countries, including the US, Switzerland, Brazil, Saudi Arabia, and Switzerland.

In July, 2008, in the last session held by the WTO Council for Trade in Services so far, the Chairman presented a Report on the "Elements required for the completion of the services negotiations" (WTO 2008). The Chairman reminded that members had been pursuing request-offer negotiations both on a bilateral and on a plurilateral basis, and noted that 21 collective requests on trade in services had been submitted by co-sponsors in various sectors and modes, including energy-related service. According to the Chairman, " the negotiations shall aim at a progressively higher level of liberalization of trade in services... There shall be no a priori exclusion of any service sector or mode of supply... . Respecting the existing structure and principles of the GATS[1], Members shall, to the maximum extent possible, respond to the bilateral and plurilateral requests by offering deeper and/or wider commitments. Such responses shall, where possible, substantially reflect current levels of market access and national treatment and provide new market access and national treatment in areas where significant impediments exist, in particular in sectors and modes of supply of export interest to developing countries, such as modes 1 and 4, in accordance with Article IV of the GATS. Commitments shall be commensurate with the levels of development, regulatory capacity and national policy objectives of individual developing countries."[2] The final drafts of commitments schedules were to be submitted by WTO members by end-2008.

The Hardening of Bargaining Stances on the Eve of the Last Doha Talks

In a pre-electoral period characterized by skyrocketing prices of oil and other energy sources, multilateral disciplines on trade in energy goods and services also became a hot important issue in the US political debate, leading some politicians to adopt the familiar tactic of playing the role of the tough guys in order to get the media's limelight In June 2008, eleven US senators called on the Bush Administration to take action in the WTO against eight OPEC countries. The senators were alleging that these eight countries, operating under the OPEC cartel, had refused to increase production, reduced the supply of oil on the market and driven up retail gasoline prices, claiming that "The very existence of the OPEC cartel violates GATT Article XI, which prohibits nations from maintaining quotas or any other quantitative restriction on exports. The refusal of OPEC nations who are members of the WTO to play by these rules is inexcusable, and they must be held accountable. I hope you will inform the President and direct your agency to look into these charges and work to directly confront OPEC's violations of international law." (Desta 2003)

From a technical point of view, OPEC countries could in fact be potentially vulnerable for challenge under the dispute settlement system.[3] In practice, however, there is little chance

[1] During the negotiations on energy services, some WTO members accused other ones of putting forward excessive liberalization requests, which went far beyond established GATS principles and thereby risked to undermine them. See Gabriele 2008 b (attached).

[2] As acknowledged by the Chairman, four Latin American developing countries (Cuba, Nicaragua, Bolivia and Venezuela), two of them energy exporters, expressed their disagreement with the formulation of the text.

[3] According to WTO rules, OPEC members can use export levies to achieve their price stabilization objectives, yet applying supply management techniques leaves them potentially vulnerable for challenge under the dispute settlement system. However, OPEC countries might successfully defend themselves on the basis of GATT Article XX(g) may however be quite a viable one even in this case" (see Desta 2003).

that this initiative (likely targeted mainly to the domestic electoral market) will lead to any substantial consequence in real-world negotiations on trade in energy services.

This initiative was representative of the aggressive stance displayed by important sectors of the globally energy-importing Western development establishment. However, their attitude was matched by a parallel hardening of rhetoric standings also on the part of some developing energy exporters. For instance, Evo Morales Ayma, president of Bolivia, declared on July 7, 2008 that "what must be done is definitively exclude the basic services of education, health, water, energy and telecommunications from the text of the WTO General Agreement on Trade in Services. These services are human rights that cannot be objects of private enterprise and the rules of liberalization that accompany privatization... So that this round of negotiations might truly be about development, anchored in humanity's present and future, they should... respect the interests of developing countries, without limiting their capacity to define and implement national policies at the agricultural, industrial and service levels...effectively reduce protectionist measures and subsidies by the developed countries...Assure developing countries the right to protect their nascent industries for whatever period necessary, in the same way that the developed countries did in the past...Guarantee the right of developing countries to regulate and define their policies regarding services, expressly excluding basic services from the WTO General Agreement on Trade in Services... (Morales 2008). Such a defiant declaration on the part of a small and poor energy-exporting developing country are representative of the confrontational climate which would lead to the collapse of the Doha negotiations.[4]

Energy Services and WTO Accession Negotiations

Energy is a key sector for various energy exporting countries engaged in WTO accession negotiations - such as Russia, Algeria, Iran, Yemen, Kazakhstan, and Azerbaijan. In the framework of accession negotiations, WTO member countries (most of them from developed countries) have been seeking seek to obtain GATS-plus liberalizing commitments from these acceding countries in various energy-related sectors and energy services.[5] Main areas where commitments are sought for are dual pricing (which implies trade distortions and can contribute to increase international energy prices) and demand barriers to commercial presence (Mode 3). With respect to the latter issue, acceding countries (like many other WTO member countries) presently allow foreign firms to invest in the energy sector only through production sharing agreements (PSAs), and/or impose other trade barriers constraining foreign investors' options in the realm of foreign ownership limitations and other areas. Demandeurs would like such barriers to be dismantled, along the lines and principles of GATS already agreed upon in other services sectors (i.e. telecom).

Access negotiations - taking into account the non-negligible weight of some of the countries involved in global energy markets - are likely to have had some form of indirect

[4] Morales' declarations are mentioned in this note because they explicitly relate to services and energy services in particular. However, small developing countries such as Bolivia were far from being the main "culprits" for the above-mentioned confrontational climate and the eventual collapse of the Doha negotiations.

[5] As is well known, energy services cannot be unambiguously identified on the basis of the W 120 list. Therefore, trade negotiations issues involving energy services proper cannot often be clearly distinguished from broader issues related to trade in energy.

impact also on overall WTO negotiations on energy services in the Doha Round. Tensions in the Middle East, largely related to the controversies on Iran's nuclear program, and between Russia and NATO were already mounting by the early summer of 2008.

Likely Consequences of the Doha Fiasco on Global Energy Trade

After the collapse of the Doha Round such tensions, far from abating, became more intense, leading to the Russian-Georgian war of August 2008 and to the return to a sort of cold war-like scenario. In this respect, it is likely that WTO accession negotiations for Russia and Iran, and possibly other energy-exporting countries, will be negatively affected. In turn, such a development might have a negative impact on any future perspective of resumption of multilateral negotiations on energy and energy services in particular. Conversely, such a setback of the movement towards multilateral liberalization of trade in energy goods and services can be expected to strengthen the now-prevailing trend towards bilateralization and the strengthening of the role of nation states in the energy sector[6] (see section 2).

In sum, Doha negotiations floundered in the midst of an ongoing energy crisis (which happened to coincide with a major financial crisis[7]). Not only this crisis is far from over. Quite to the contrary, it appears to be conducive to a structural change in the world energy scenario, rather than constituting a temporary perturbation which will be followed by a restoration of the status quo antes. This new scenario is characterized by a major market power shift towards energy exporters (with all its inevitable economic, political, and military implications), related to relatively contingent geopolitical factors , but also to the emergency of a long-term trend towards global energy scarcity. As a result, both developed and developing countries will be forced to change profoundly their present, energy-intensive economic growth patterns, and the presently-existing global market-oriented mode of production will undergo deep systemic changes (see Section 2).

These major historical changes are still poorly understood, yet they are already beginning to exert a relevant impact on overall economic and development ideology and on global trade. As a result, a resumption of WTO negotiations under the traditional market-fundamentalist agenda - which was supposed to eventually lead to a quasi-perfectly free, unbridled, unified global market for goods and services amidst the withering of the role of nation-states in the world economy - looks increasingly remote. This conclusion, a fortiori, holds for strategic and traditionally state-intervened and heavily regulated sectors such as that of energy.

---

[6] Similar trends are apparent also in other commodity-related sectors. Russia, for instance, announced in early August 2008 the reconstitution of a centralized and apparently monopolistic state trading agency to manage grain exports.

[7] Of course, such a "coincidence" is not due exclusively to casual random factors. There are a number of complex interrelationships between the global financial a and energy crises , which cannot be further explored in this brief note.

2. A Major Change in the Global Energy Trade Scenario

The World Energy Scenario in the 1990s and the Technological Omnipotence Fallacy

In the early 1990s it became apparent that developing countries' financing needs for the coming decade would be very high, and that these huge sums would not be available from public treasuries. Many developing countries were emerging financially exhausted from the long and hard experiences of the debt crises and falling commodity prices, and were still engaged in IFI-led structural adjustment programs. Consistently, the world energy scenario in the1990s was characterized by deep market-oriented reforms in major energy utilities in many developed and developing countries, and by a boom of private investment in FDI flows from developed to developing countries. A large share of these flows went towards the energy sector, mainly spurred by privatization processes. During the 1990s, in fact, private investment in infrastructure projects in developing countries reached over US$ 680 billion, out of which almost US$ 200 billion was accounted for by electricity and US$ 33 billion for natural gas. About half of these infrastructural investments were carried out in Latin America, and about one third in Asia.

However, due largely to the progressive realization that the liberalization process was more complex and fraught with difficulties than initially forecast, few developing countries implemented full-fledged structural reforms in the energy sector. Most developing countries, moreover, have not developed those emerging energy services that emanate from breaking up of integrated energy systems and the introduction of competition, especially in the gas and electricity segments.

The 1990s also witnessed a peak in the relative role of technology and institutions[8] (i.e., non-material, man-made, reproducible resources, including financial ones) with respect to that of natural resources in determining each country's economic well-being, competitiveness, and overall power ranking worldwide.

Conversely - and correspondently - during this period also the degree of dependency on non-OECD energy exporters on the part of most industrialized countries reached a nadir. This phenomenon was epitomized, in particular, by the political disarray and economic crisis in Russia and other former Soviet republics.

The above-mentioned trends led many observers to foresee a permanent technology-based worldwide hegemony of the most industrialized OECD countries, leading eventually to a progressive de-linking of global economic growth from its material base. Many also believed that the near- universal Washington consensus-inspired, neoliberal hegemony and the correspondent retreat of national states from the economic sphere would last forever, or at least for a long historical period. A by-product of this complacency was also a relatively scarce policy priority accorded to energy savings and the reduction of the energy intensity of economic growth in the major capitalist countries, and especially in the US. These trends

[8] A particularly crucial and strategically relevant set of institutions is that of domestic, international, and financial institutions, which was also dominated by the most industrialized countries.

contrasted markedly with those prevailing in China, where - starting from very high levels - the energy intensity of economic growth was undergoing a rapid reduction (see table 1).[9]

Problems with Market-oriented Energy Sector Reforms

In the energy area, the 1990s also witnessed a widespread tendency towards market-oriented reforms and privatization in the energy sectors in most developed countries and in many developing ones (see Gabriele 2004)[10]. The main goals were to achieve efficiency gains and cost reductions, to spur investment and energy diversification (thereby decreasing the systems' vulnerability), to foster technological progress and -especially in developing countries - to attract FDI inflows (see, for instance, OECD 2007).

Table 1. Energy intensity of economic growth

A. Tonnes per million $ GDP (Megajoules (mJ) per US$)				
	1980	1990	2000	2007
China	*564.5*	*359.8*	*191.5*	*181.4*
India	*110.7*	*113.1*	*111.3*	*na*
Japan	*196.4*	*161.3*	*168.3*	*140.2*
Republic of Korea	*224.9*	*219.7*	*260.1*	*212.4*
USA	*364.0*	*283.5*	*236.5*	*186.9*
B. Annual compound rate of energy reduction				
	1980-90	1990-2000	2000-2007	1980-2007
China	*4.40*	*6.11*	*0.54*	*4.12*
Japan	*1.95*	*-0.42*	*1.81*	*1.24*
Republic of Korea	*0.24*	*-1.70*	*2.00*	*0.21*
USA	*2.47*	*1.80*	*2.33*	*2.44*

Source: Author's elaboration from UN-DIESA 2001, and EIA Energy Outlook, various years.

Countries' overall experiences with energy sector reforms were very uneven. As a general rule, some goals (such as the realization of productivity increases and efficiency gains) proved easier to achieve than others. With respect, in particular, to the advantages expected from enhanced competition in energy markets, it was initially virtually taken for granted that the ultimate beneficiaries were going to be individual consumers, who would gain from more choice, better service and lower prices resulting from domestic and import competition. In practice, however, in many cases the true beneficiaries were industrial/financial lobbies, politicians, and TNCs. Prices for industrial consumers did decrease in many successful OECD and Latin American reforming countries during the first decade after energy market liberalization, but retail prices for residential consumers failed to follow this declining trend

[9] Under the push of climbing energy prices, the speed of reduction in the energy intensity of economic growth in the US accelerated again in the mid-2000s.

[10] Structural change in the energy sector among developing countries was pioneered by Chile and China, who enacted reform process inspired by two very different approaches already in the 1980s (see Gabriele 2004).

(see OLADE 2004, IEA 2005). Afterwards, retail prices have been increasing in many countries, under the pressure of rising fuel and pollution-control related costs.[11]

In fact, three major sets of problems with energy sector reforms eventually became apparent in many countries, and their effect was to undermine to varying degrees the hoped-for benefits for final consumers.[12]

The first set of problems stemmed from the intrinsic structure of energy markets and from the large size and limited number of energy traders, suppliers and distributors - and especially those of key infrastructural energy services such as electricity and gas.

The second one was the less-than-enthusiastic response of consumers themselves, who proved reluctant to demand more competition and to exploit in a theoretically rational way[13] the multiple market choices that reformed energy markets appeared to offer. This phenomenon, however, is not difficult to understand in political economy terms: consumers are dispersed and not organized, and suffer from asymmetry of information. Moreover, competition is a new concept which is unfamiliar in many developing countries.

The third set of problems was related to the difficulties encountered by many countries in the areas of legislation, institution-building, and implementation of regulatory and competition-preserving measures.

Under such circumstances, the task of national competition authorities turns out to be a far more difficult one than that once imagined by the less sophisticated advocates of market-oriented reforms, especially in the area of electricity and gas utilities. As expected, moreover, institutional and governance weaknesses render competition regulation in domestic energy markets particularly difficult in developing countries. These challenges make it imperative for national competition authorities to strive to compatibilize their actions with those of other public bodies in the overall framework of national energy policies, and to enhance information exchange and international coordination with other countries' sister institutions.

Even in the core, most industrialized EU member countries, which have constituted since a long time a very developed free-trade area where market-oriented energy reforms are particularly advanced, competition still does not exist in many member states and that it is doubtful that natural gas and electricity prices are the result of a truly competitive process rather than the outcome of decisions by companies with market power (EC (2007a,b)[14]). The exploitation of market power erodes the expected benefits to consumers from the introduction of competition. Of particular concern is ensuring that the market structure that develops during the transition from regulated to competitive markets does not prevent the development of effective competition.

In any case, competition law enforcement can make a significant contribution, but cannot by itself open markets and resolve all the shortcomings. A number of regulatory measures are

[11] In Asia and Africa, were market-oriented reforms and privatization were pursued only by some countries and usually not to an extent equal to that of OECD and Latin American countries, generalized trends are difficult to pinpoint. However, declining energy prices have been mostly limited to OECD countries. Many energy utilities in the developing world, moreover, keep performing poorly both in reforming and non-reforming countries (see Gabriele 2004 and Williams and Ghanadan 2006).

[12] Countries' overall experiences with energy sector reforms were very uneven As a general rule, some goals (such as the realization of productivity increases and efficiency gains) proved easier to achieve than others. , and in particular

[13] Actually, in many cases, consumers' passivity is fully warranted once factors such as risk, transaction costs and information-gathering costs are taken into account.

[14] EC 2007.. Prospects for the internal gas and electricity market. Communication from the Commission to the Council and the European Parliament. COM(2006)841 final. Brussels 10 January 2007.

also needed. So far, many developing and transition economies have adopted competition and regulatory laws as a follow up to their market oriented economic reforms. Yet, how important has this new form of economic governance been for growth and other developmental objectives?

The answer, broadly speaking, is: not a decisive one, at least so far. A few examples help to support this point - which, of course, would need a much more thorough analysis to be fully demonstrated. China, for instance, approved a competition law in June 2006, almost 30 years after it began economic reforms, yet the country has moved at extraordinary speed from low to middle-income status. Neither of the two major African success stories on growth, Botswana and Mauritius, had a formal competition law until Mauritius passed its Competition Act in April 2003. By comparison, Kenya passed the Restrictive Trade Practice, Monopolies and Price Control Act in 1989 and Zambia adopted the Competition and Fair Trading Act in 1994 but have been nowhere near the same economic success.

It is not surprising that both economic and political risks posed by energy reforms in developing countries are far more serious than in developed countries, taking into account the particularly strategic role of the energy sector. Ill-conceived and /or poorly implemented energy can have a strong negative impact on all sectors of the economy – such as reducing household disposable incomes due to retail tariff increases, to industrial operational cost increases that end up undermining export competitiveness and fuelling inflation, unemployment, and external debt. In turn, macroeconomic disorder can discourage in-bound FDI flows, thereby undermining one of the key requirements for successful energy reforms in capital-starved developing countries.

Employment losses in the electricity sector[15] and the unfair distribution of benefits stemming from efficiency gains (due to a lack of effective regulation mechanisms forcing quasi-monopolistic energy enterprises to adequately reduce retail tariffs) have proved long-lasing bones of contention even in the (relatively) most successful reforming countries, both developed and developing, such as the UK and Chile.

Domestic energy policy goals also keep conflicting with regional (let alone global) free trade objectives. Many European governments support consolidation and concentration in their national, thereby weakening the prospects for the creation of an authentic common EU market. By the same, severe problems with domestic energy supply recently led the South African government to renege on its pledge to honour long-term contracts to supply electricity to neighbouring countries.

The Sea-change at the Turn of the Millennium

Contrary to most observers' expectations, the aforementioned trends in the energy sector worldwide, both at the level of domestic policies and at that of global energy goods and energy services markets, proved short-lived, and peaked around the turn of the millennium. Around the year 2000, key elements of the global economic, trade, and geopolitical scenario started to undergo quite deep changes, leading to a very different picture in the early XXI century.

[15] In 1990-1995 employment in the energy sector fell by 17% in Western Europe and 42% in the UK (Hall 1999).

New, partly new, and old powers of very uneven strength and nature emerged, re-emerged, or kept growing steadily past a critical point where their weight and influence on global affairs became crucial, among them China, India, Russia and several medium-sized developing energy exporters, thereby profoundly modifying economic and political relation of power worldwide in a relatively short period of time.

China and India are following two very different but both very fast and rather environment-unfriendly growth patterns. Yet, such growth patterns appear to be sustainable - at least, in the medium term horizon). In the case of China, in particular, industrialization at an unprecedented scale has also been extremely energy-intensive in the past. Notwithstanding the remarkable gains in energy efficiency[16] (see table???) and a rapid expansion of renewable energies[17] (see???), economic growth in China is still rather energy-intensive now, when the sheer size of its manufacturing base is surpassing that of the US (according to a recent estimate[18]), and becoming the largest in the world. The two Asian giants - and, to lesser extent, other semi-industrialized countries - are exerting an increasing upward pressure on world energy demand. In the meantime, the newfound assertiveness of many fossil fuels exporters - along with a number of physical, technical and other objective constraints - allows them to exert an enhanced degree of control on global energy supply. In some cases, public energy enterprises' inefficiencies, as well as their financial and technical shortcomings, contribute to slow down the growth of energy supply, thereby further strengthening world energy prices.

Major Features of the Present Energy Scenario

The present historical period is characterized by a number of noticeable phenomena in the domain of international energy trade, and more broadly of geopolitical relations and the distribution of bargaining power between energy goods and services (and, to some extent, other commodities, especially non-reproducible ones) importers and exporters.

The new worldwide energy scenario is characterized by very volatile but historically and (probably) structurally high[19] (if compared to the previous two decades) energy prices (see Figure 1). In this framework, two specular, major trends are particularly worth mentioning:

---

[16] Since the 1970s China has reduced its intensity of energy use (tons OF WHAT??? Per US$ million GDP) at a rate far faster than that of any other major economic power. By the mid-1990s , China's intensity of energy use had become lower than that of the US. After a brief interruption in 2002-2004, China's its intensity of energy use resumed its declining trend in 2005.

[17] According to an estimate by the Climate Group, China invested US$ 12 bn in renewable energies in 2007, and is expected to overtake Germany as the largest investor in this area in 2009 (Kinver 2008).

[18] According to a study carried out by Global Insight, a US economics consultancy, in 2009 China will overtake the US as the world's largest producer of manufactures. China's manufacturing value-added output is expected to be over US$ 11.7 trillion in 2009, 17% of world total, against 16% for the US (see Marsch 2008).

[19] Of course, it is impossible to forecast future energy prices without any element of certainty. However, there are three major factors which appear to confirm the structural and relatively permanent character of recent increasing price trends. China, India, and other developing countries are growing rapidly, thereby pushing up global energy demand. Extraordinary technical advances, which could allow a strong increase in energy supply worldwide in a short- to-medium period of time, do not appear to be on sight. Political and cultural resistances against shifting towards significantly less energy-intensive growth and lifestyle patterns have proved to be very strong in OECD countries, and in particular in the US (see above, Section 1, and Table 1 on energy intensity).

- Direct state control on energy resources rises again as a key element of economic and geopolitical strength. Nation-states and large SOEs are increasingly becoming protagonists in the world market for energy services. Bilateral and plurilateral multi-billion long-term energy trade agreements are regulating an increasing share of world energy trade.
- The relative weight of knowledge-based, non-material endowments decreases vis a vis that of material, nature-based endowments.

The first trend leads to an enhanced role of the state in the economy and in the energy sector in particular. In turn, it magnifies the relevance of well-known structural characteristics of the energy sector (quasi-natural monopoly, economies of scale, etc.) that make energy particularly suitable for state intervention.

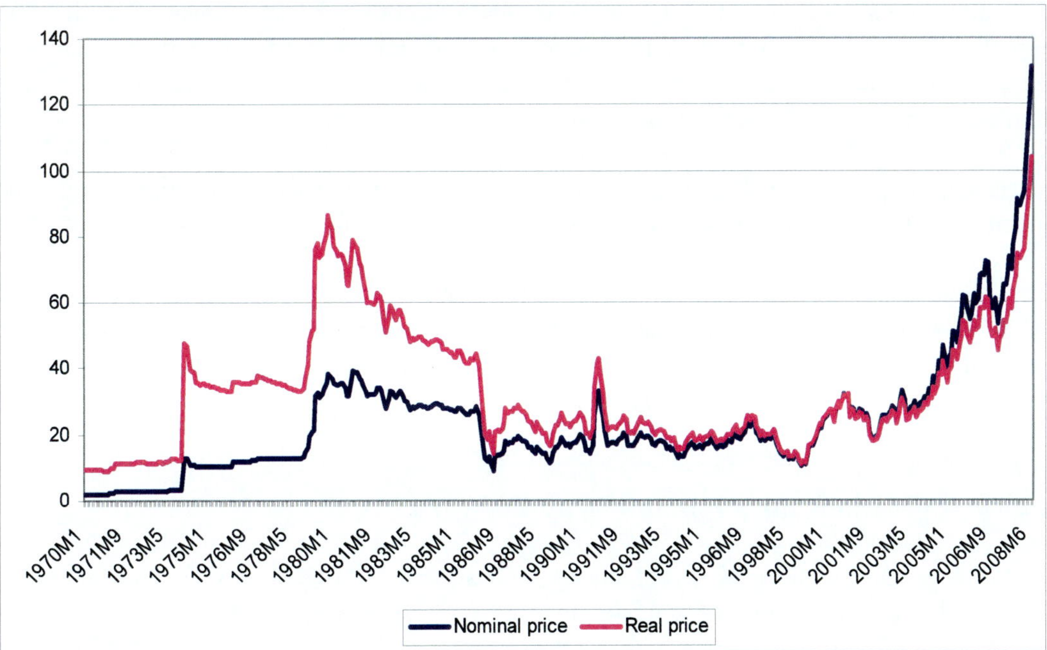

Source: UNCTAD secretariat calculations based on UNCTAD *Commodity Price Statistics* online; IMF, International Financial Statistics database; The World Bank Commodity Price Data; and Klein and Mak (2008).

Figure 1. Crude petroleum prices, nominal and realm monthly, 1970-2008.

Enhanced state control on energy resources and high international energy prices also mean that many developing energy exporters are now awash in cash. Therefore, they are less dependent on traditional OECD-dominated institutions such as the IMF, the WB , and large private financial institutions. They are also in a better position both to repay their external debt and to finance new bilateral and regional trade agreements with energy-importing countries.

The second trend suggests that, in spite of incessant technical progress, market-oriented economies worldwide (both developed and developing) appear to be inevitably heading towards generating an ever-increasing demand for a set of ultimately finite commodities (i.e., oil, gas, and other non-renewable sources of energy). This phenomenon, of course, adds to the already-existing tension in global energy markets, and is probably bound to persist also in the relatively long run, although - of course - it is not ultimately sustainable.

Shifting Priorities and Roles of Market Integration, Competition, and Planning

The energy sector worldwide, both in its goods- and services-producing segments, is witnessing a revamped trend towards concentration, vertical and horizontal integration, the scramble on the part of major private and public actors to achieve ever-increasing economies of scale and market power, and a generalized strengthening of the role of national states.

Contrary to the accepted wisdom of the previous period, when the progressive unbundling and atomization of energy markets was seen as the key pre-requisite for achieving the efficiency and reliability benefits stemming from competition, in many countries is emerging a trend towards vertical and horizontal integration and a long-term planning approach to energy production and trade policies.

M&As, including convergence mergers between natural gas and electricity firms, are increasingly popular, as well as large-scale, long-term, state-sponsored bilateral energy trade agreements. These phenomena appear to show that integration and market-compatible (but not simply market-determined) planning are seen as at least partly efficient risk management and welfare-maximizing strategies, both by TNCs and national states. Moreover, the sheer size of many of these operations renders their impact on world energy markets a is far from negligible one, thereby further strengthening the already-significant power of a reduced number of private and public global actors.

A Few Suggestions for a Possible Way Forward

More generally, the new world energy scenario and the mounting evidence on the increasingly binding nature of the supply constraint stemming from the limited global availability of fossil fuels show that market-oriented reforms are far from sufficient to withstand the present challenges, and can actually sometimes be part of the problem. To serve the ultimate interests and needs of both developed and developing countries the international community must fully realize that the world is in bad shape. Present trends in energy consumption in rich countries are neither economically nor environmentally sustainable in the medium/long term. Barring presently unforeseeable, amazing technological breakthroughs, an extension of now-industrialized energy consumption pattern to the bulk of the developing world is also not feasible. Still, billions of people are mired in abject poverty.

An indefinite (unplanned-uncoordinated-unreformed) continuation of the presently-existing market-oriented worldwide mode of production is neither possible nor auspicable. It is necessary to move towards a very different approach to global energy policies. As world prices are exceedingly volatile and do not adequately reflect the true value of energy resources, this approach should reduce (but not eliminate) the relative role of markets. Conversely, it should advance towards the establishment of a multilaterally planned, rational, environment-friendly and sustainable world system of producing and distributing energy.

Crucial elements of such a new systemic approach should be the transfer of energy and other resources from fostering luxury consumption in rich countries to basic consumption in poor countries, and radically enhanced priorities for environmental protection, energy efficiency and R&D investments, especially in the area of renewable energies. Such major, but badly needed changes in resources allocation and policy orientation can be favoured by a dismantling of the most blatant market distortions, such as those stemming from widespread subsidies to energy consumption. However, the weight of market failures, information gaps and asymmetries[20] , and a number of other factors constraining the effectiveness of energy and other world markets is so large that ultimately the solution can only be found in an enhanced and coordinated form of long-term, market-compatible global planning. Not an easy task, but a necessary condition for the survival and further development of the present form of human civilization.

References

Desta M., *OPEC and the WTO: Uneasy relation!?, Oil, Gas, and Law Intelligence,* 2003, vol.1, issue 1.

European Commission 2007a, (EC 2007a). *Prospects for the internal gas and electricity market. Communication from the Commission to the Council and the European Parliament.* COM(2006)841 final. Brussels 10 January 2007.

European Commission 2007b (EC 2007b)," Competition at National and International Levels: Energy", document presented to the Intergovernmental Group of Experts on Competition Law and Policy, Geneva, 17-19 July 2007

Farrell D. and . Remes J.K., *How the world should invest in energy efficiency,* The McKinsey Quarterly, Economic Studies, July.

Gabriele 2004, Policy Alternatives in Reforming Energy Utilities in Developing Countries, *Energy Policy*, Vol. 32/11, pp. 1319-1337, April.

Hall, D. 1999. Electricity restructuring, privatization and liberalization: Some international experiences. *PSIRU Reports.* 9910-E-U-Prob.doc.

Hoyos C., 2008, It is their oil – western energy groups yield to state producers, *Financial Times* July 9

IEA 2005. *Lessons from Liberalised Electricity Markets.* OECD/IEA

IEA Energy Outlook, various years.

Kinver M., 2008, China's 'rapid renewables surge',*BBC News*, August1.

Klein L.R. and Mak W. (2008). Current Quarter Model of the United States.Forecast Summary. Weekly Update on the U.S. E*conomy and Financial Markets.* University of Pennsylvania

Lamy P. 2008 , *Lamy on Doha on energy services* (attached)

Marsch P., 2008, China to overtake US as largest manufacturer, *Financial times,* August 10, London

[20] According to Farrell and Remes 2008, investing in enhancing energy efficiency is one of the most effective way to save energy resources and protect the environment. They estimate global investment requirements until 2020 at US$ 170 bn/year.

Morales E. 2008, in http://machetera.wordpress.com/2008/07/22/evo-morales-on-the-wto-negotiations-round/

OECD 2007, *Report of the Competition Committee Secretariat to the December 2007 Roundtable on "Freedom of Investment"*, Paris, December (available in http://www.oecd.org/dataoecd/3/19/40699061.pdf)

OLADE 2004. *Competencia En Mercados Energéticos: Una Evaluación De La Reestructuración De Los Mercados Energéticos En América Latina Y El Caribe.* Diciembre 2004

UN-DIESA 2001, Energy Uses per unit of GDP, in http://www.un.org/esa/sustdev/natlinfo/indicators/isdms2001/isd-ms2001economicB.htm

Williams and Ghanadan 2006. Electricity Reform in Developing and Transition Countries: A Reappraisal. *Elsevier. Energy 31* (2006) 815-844.

WTO 2008, Council for Trade in Services Special Session, *Elements Required for the Completion of the Services Negotiations*, Report by the Chairman, doc., TN/S/34, July 2008.

In: Energy Policy: Economic Effects, Security Aspects... ISBN 978-1-60692-894-3
Editor: Noah B. Jacobs

Chapter 10

DEVELOPING RENEWABLE ENERGY: THE POTENTIAL ROLE FOR ELECTRIFICATION COOPERATIVES

Dan van der Horst[1,*] and Peter Davis[2]
[1]School of Geography, Earth and Environmental Sciences, University of Birmingham, UK
[2]Director, Unit for Membership Based Organizations', School of Management, University of Leicester, UK

ABSTRACT

In the post-war period, energy (and especially electricity) provision was heavily regulated by the state which guaranteed long-term security of supply through a centrally planned predict-and-provide model. In the early 1990s, this was abandoned in favour of a neoliberal model which saw whole-sale privatisation of the sector. This resulted in lower electricity and gas prices for consumers in the short term, and the abandonment of investment and R&D in the energy sector. A few years later the political taste for intervention in the energy sector increased again because of environmental and (more recently) security-of-supply concerns. But despite a range of state incentives, including the setting of national targets and the provision of substantial subsidies, the private sector has been unable to deliver many of the renewable energy solutions which have been proven in other countries. The UK government has belatedly recognised the limitations of reliance on such large-scale private sector initiatives in the power sector and it has started to embrace the idea of supporting more community based initiatives. We examine the contrasting and ambiguous role of the state in the development of renewable energy provision in Denmark and the UK. Danish policy, based on the creation of an enabling environment for local initiatives –many of which are cooperatives-, has been highly successful (20% of national electricity supply). By contrast, UK policy for local-level initiatives has mainly sponsored social enterprise initiatives that are embedded in existing non-profit organisations such as schools and municipalities. The potential role of

---

* E-mail d.vanderhorst@bham.ac.uk Tel +44(0) 116 252 5517 Fax +44 (0)116 2525515

electrification cooperatives in developing the renewable energy sector in the UK (currently stuck at 3-4% of national electricity supply) is highlighted through a comparison with North American electricity co-operatives who are leading the US electricity sector with regards to renewable energy.

1. Oligopolistic Control and its Implications

The energy sector lies at the very heart of economic activity within what Averitt (1968) in his seminal institutional economic work dubbed the Dual Economy. It remains a sector of the economy where both extraction and distribution are very highly concentrated in terms of ownership. The recent transitions from public to private ownership even allowing for the establishment of state appointed regulatory institutions and opening up of markets to competition has done little to allay public unease concerning retail price levels nor academic concerns regarding the ethics of "confusion marketing" across utilities in general (Mitchell et al, 2005).

The OPEC cartel of oil extracting countries, the handful of giant multinational companies led by Esso, and the huge Russian former state energy enterprises ensure this sector is characterized by oligopolistic competition. It is a sector where superpowers like the USA and Russia along with the OPEC cartel can and do exert political, economic and military pressure to establish access, profitability and political hegemony. On the demand side for energy there is a rapidly rising curve as new superpowers like China and India emerge with aspirations for a life style comparable to that of the west.

There are serious challenges to maintaining energy supply in the face of increasing global demand. These challenges arise both in terms of the finite nature of fossil fuels and their environmental impacts. How quickly and how acutely this crisis may be upon us will depend on the growth and configuration of demand (Midttun, 2001). Here established co-operatives (such as the Co-operative Bank in the UK), alongside environmentalist pressure groups and NGOs like Greenpeace and Friends of the Earth may play an important role. Their direct engagement with the community may well be able to exert influence developing awareness of the benefits for domestic energy conservation and reduction in the so called domestic "carbon footprint".

In the UK the Government, using public concern at the country's growing dependence on overseas energy supplies, has put the nuclear option back on the agenda. Yet the nuclear option is for many too menacing and expensive to provide a viable alternative (Smil, 2006, p3). In addition to concerns about nuclear non-proliferation, accidental or deliberate (i.e. sabotage or terrorism) radioactive pollution, and the controversies surrounding the handling and long-term storage or disposal of radio-active waste, the nuclear option will maintain and possibly intensify the already high levels of concentration and centralization in the energy generation and distribution industry. Therefore, nuclear options seem to undermine rather than enhance the vision of free competitive market determining quality and price in energy generation and delivery even if one ignores their potential for environmental catastrophe:

"Nuclear technology presents even more formidable problems at it's present stage of development. Conventional reactors rely on uranium, yet another exhaustible fuel, and carry safety risks that are extremely costly to overcome - if, indeed, they ever can be. No one has convincingly solved the problems of nuclear waste disposal, and nuclear costs are so high that until now government subsidies have been essential to make atomic power remotely competitive with other sources. Fast breeder reactors are in a class by themselves. But while often presented to the uninformed public as perpetual motion machines be cause the plutonium they spew out can be used as a fuel, they, too, remain ultimately dependent upon the world's small and non-renewable supply of uranium. They are not only highly centralized, incredibly costly, volatile, and dangerous they also escalate the risks of nuclear war and terrorist capture of nuclear materials." (Toffler, 1995, pp3-4)

There are no serious economists writing today who have argued directly in favour of oligopolistic competition as a mechanism to optimize welfare. The reality of oligopolistic domination and control of the core economy - a reality for most industries for most of the last century has been intensified by the wave of privatisations, the general liberalisation of the global economy and the rolling back of communism since the 1980s. Those who espouse the current global concentration of economic power as in the public interest in the name of Adam Smiths call for liberalisation would do well to remember what he predicted concerning the behaviour of the business elite.

"People of the same trade seldom meet together, even for merriment and diversion, but the conversation ends in a conspiracy against the public, or in some contrivance to raise prices." (Smith, 1776, Bk. 1, Ch. X)

Energy provision exists in an oligopolistic market with varying degrees of institutional regulation, both informal and formal (Regibeau, 2000). The levels of influence through branding, the huge resources of capital and logistical and technological capacity and big business domination in market share terms may make the casual observer despair of there ever being any real prospect of the consumer gaining any leverage from the demand side.

Not withstanding the odds, different models of ownership have been able to defy and undermine the strategies of a concentrated and powerful oligopoly. The UK retail banking sector - one of the most concentrated service sectors in the world - was given a demonstration of the impact that the existence of even a small co-operative business can have. Here the UK Co-operative Bank and the Nationwide Building Society refused to charge anyone using their ATMs in the teeth of proposals from the biggest commercial banks to introduce such charges (Davis, 1999). This led to a climb-down by the big banks and years later the majority of the UK ATMs remain free of charge. This illustrates the importance of alternative ownership models for creating choice and competition (for ideas as well as for market share) in otherwise concentrated industrial and service sectors. We see no reason why in principle this mechanism for creating genuine competitive choices cannot be successfully transferred from the financial services context to the energy sector. Indeed we see cases that suggest it is already happening particularly in parts of Europe and North America.

2. Needs of Energy Conservation and Greater Consumer Awareness

There is today a formidable scientific consensus concerning the significance of fossil fuels as an important influence on climate change and the urgent need reduce its use through a combined strategy of energy efficiency measures and a switch to renewable energy sources. Although technological solutions are (or are becoming) available to generate certain types of renewable energy through large scale and centralized facilities (e.g. large scale hydro-electric dams; off-shore windfarms), many of these large scale technologies can only harness energy in areas that are relatively remote. Long-distance transportation of electricity over the grid creates significant energy losses whilst the price of this infrastructure shuts out smaller or local businesses and increases the risks for monopoly pricing. Unlike small-scale distributed renewable energy technologies, large scale renewables may not provide much consumer value or local economic benefits.

According to a recent Greenpeace report, 61.5 units of every 100 units of power generated centrally is currently lost through heat wastage and inefficient generation. With another 3.5 units lost in transmission, only 35% of the energy arrives at the end consumer where a further 13 units are lost through inefficient use. (Greenpeace, 2005; p24). The Greenpeace report also noted in part two how the British model of liberalizing wholesale electricity markets takes the centralized electricity model as given and in fact consolidates it.

One cannot help but draw the conclusion that the UK consumer is paying higher than necessary prices for environmentally damaging and high waste energy supplies. There is also the issue of the economic and social repercussions when there is a centralized power failure. The Greenpeace report quoted figures of $80 billion of losses due to centralized power failures in the US (Greenpeace, 2005; p30) and in the UK cited the case of the London Underground closure of its' independent power generator leading to their customers being left in total darkness as a result of the 2003 power cut (Greenpeace, 2005; p34).

Despite the existence of various comparative web-sites and government subsidies for domestic energy efficiency, the large companies who have taken part in the privatization of the UK electricity sector are failing to create and communicate choice in terms of price and (low-carbon) characteristics of the services they provide to their customers.

3. Recent Contrasting Experience in Europe. The UK and Denmark

Denmark has been widely admired for its achievement as the world's leading 'wind power'; 20% of electricity production is from wind farms and the strong level of domestic demand for wind turbines has created a vibrant national industry with many manufacturing and other related jobs. Technological advances and economies of scale are increasingly making electricity from windfarms price competitive with electricity from fossil fuel. The Danish success was based on state policies which created space for bottom-up co-operative efforts in which the owners of the wind turbine had to live within the same municipality (Christensen and Lund, 1998). High and guaranteed electricity prices (so-called feed-in tariffs) ensured the profitability of these projects, while local public support was enhanced by the ownership model; mostly small scale individual or cooperative, always locally owned. In recent years Denmark's right wing government has been reversing its policy away from locally owned and small scale towards large scale and owned by large companies and public opposition is subsequently rising (Loring, 2007).

The UK trajectory has been very much the opposite. The UK has Europe's best wind resources and in the time between the oil crisis of 1973 and the privatisation of the electricity sector, government funded research and development in wind energy was particularly advanced in the UK. When the Thatcher government closed this funding on the basis of short-term cost-benefit concerns, the Danes were to continue where the UK left off. The first significant funding for renewable energy in the UK came through an amended scheme that was originally designed to support nuclear (the Non Fossil Fuel Obligation or 'NFFO'; see Mitchell, 2000; Connor, 2003). However many projects were stalled as a result of opposition by NGOs and local residents. After a successful first phase supporting farm waste incinerators in core agricultural areas, almost all biomass energy projects stalled due to public opposition (see Upreti, 2004 and Upreti & van der Horst, 2004). Wind farm projects were larger in number, but also here about 50% failed to gain planning permission due to public opposition (Toke, 2005). Public opposition was due to a combination of factors, but distrust of large private corporations and a lack of local involvement (in terms of planning and decision making and in terms of ownership) played a major role. These experiences, and the UK's paltry record of 3% of electricity generated from renewables, stand in sharp contrast with the achievements of the renewable energy sector in Denmark. It slowly dawned on UK policy makers that despite high public support for renewable energy in principle, in practice public responses to top-down approaches could be very different. This resulted in two different strategies by the UK government. On the one hand they finally created schemes to support 'community level' renewables, and on the other hand they were looking for ways to 'streamline' the planning system to bypass local public opposition to certain developments.

This means that more decisions will be made at the national level, but also that new planning guidelines are developed to create a 'presumption in favour of wind' in certain delineated areas. What this effectively means is that although there is now more financial support for small scale and local developments, there is also more support for large-scale developments by major utilities. This is a dubious approach, especially since there is evidence to suggest that major utilities and the energy regulators are throwing up barriers for community initiatives.

In a recent special issue of *Journal of Cleaner Production*, Clark and Lund (2007) point out that there is a need for "agile energy systems" whereby regions, cities and communities can develop localized solutions that combine on-site and distributed energy generation, along with central grid power generation. They also recognize that under the currently dominant economic paradigm market forces do not provide level playing fields for communities and that better government and public involvement is required to develop such an energy system. Indeed a UK study shows that micro-grids are not more costly than the existing 'macro-grid' but that the way in which the (privatized) electricity market in the UK is organized, present considerable barriers to the large scale adoption of micro-generation (Abu-Sharkh et al., 2006). Similar barriers have been reported for heat and power (CHP) and for intermittent and small scale renewables.

It could be argued that some of these barriers are simply left-overs from a previous era of centralised planning by the state and the 'predict-and-provide' model, and that this large-scale approach still inadvertently pervades the culture and practices of the energy regulators. It could be that small scale and renewable is simply not part of their vision. However the fact that small local developments are threatening the comfortable oligopoly of the large utilities is something that is hard to overlook. Large firms can leverage a very significant amount of influence on government policy and many developers of renewable energy appear to receive state support for the purpose of creating 'export technologies' (van der Horst, 2005). Of course most policies are formulated as a compromise to different interests but the strongest interests often prevail in the details of the actual policy design. For example the UK's interest in off-shore wind farms is partially driven by the interest to avoid local public opposition, but also looming large is the UK's oil engineering industry that has the know-how to build such large structures on the sea bed – at a time when north-sea oil is declining rapidly and this industry is looking for a new lease of life.

Several hundred projects have been funded by the UK government under the community energy initiative. A full analysis is ongoing, but it is clear that many projects were initiated by municipalities and schools, to provide heat or electricity for on-site consumption (i.e. no access to the grid required and no competition with existing large scale utilities). Indeed very few of these projects were co-operatives that were selling excess power to the grid. We came across two exceptions.

In both cases much of the public opposition to wind farms was turned around through the sense of community ownership. The Baywind Energy Co-operative Ltd. was launched in 1996 and now has 1,300 shareholders. The return on investment when tax allowances are taken into account is between 7% and 8% (Block, 2003, p46) The Baywind co-operative has subsequently set up a subsidiary to promote community ownership in renewable energy generation called Energy4All Ltd. Initiatives are underway in Scotland and in Oxfordshire but institutional resistance remains significant with co-operatives actually excluded from the governments community grant scheme. (Block, 2003, p47)

The importance of community ownership was emphasised in a recent comparative study of attitudes towards wind power in the case of the community wind farm co-operative on the Isle of Gigha in the South – West of Scotland and an adjacent project under private ownership in the Kintyre Peninsular, (McFadyen and Warren, 2007). These researchers found their results pointed to the need for a change of development model towards community ownership as the latter had a positive effect on public attitudes. The smaller scale and the economic advantages in addition to simply the fact of local ownership are also likely to form part of the underlying reason for more positive attitudes in the Gigha community (McFadyen and Warren, 2007, p10). The Baywind Co-operative experience also argues strongly that the economic case for community ownership as a means of retaining economic benefits within the locality is an important factor in addition to the social and environmental case for local ownership (Block, 2003, p46).

These two examples demonstrate the potential value of electricity cooperatives, to the public (who are broadly in favour of renewable energy), the government (who have set targets for renewable energy), and obviously to the members - as both cooperatives have been successful sofar. It is therefore all the more remarkable that the UK government has made available significant economic resources for renewable energy without considering policy designs that would be more supportive for the creation of new electricity cooperatives. Other modes of promoting renewable energy, such as school projects or municipal projects, can clearly be beneficial in terms of education or demonstration of a particular technology but the structure and governance of such projects is less likely to guarantee self-maintenance, or have a potential for innovation or growth. We will examine these potentials through a case study of existing electricity coops in the US.

4. Some Co-operative Examples of Community Ownership of Energy Generation and Distribution in North America (1)

Across wide swaths of the global economy co-operative businesses provide leverage for the small business, farmer or fisherman, the individual worker and the consumer.

Many of these co-operatives, mainly in the Americas, supply and generate energy. There is evidence that where electrical co-operatives exist they are taking the lead in developing sustainable power generation. Here the co-operatives close proximity to members drawn from farming communities has meant that they have been able to take a lead in applying renewable energy technologies relating to biomass sources of energy. So far over a hundred Co-operatives in twenty two states are utilizing biomass in their power supply (NRECA, 2007 p1). Of the top ten utilities ranked for their commitment and performance in the provision of green power by the US Gov Energy Department five were Electric Co-operatives (Stuteville, 2007, p17).

Wind is proving to be a valuable renewable energy resource and North Dakota ranks highest in the USA in wind resource potential. Employing that potential, Basin Electric Power Co-operative in Bismarck, North Dakota, has added approximately 136 megawatts (MW) of wind energy to its portfolio over the past several years through joint projects and power purchase agreements (Blowers, 2007). In Missouri the Electric Co-operatives are building the states first wind energy farm (McCarty, 2007, p1). In South Carolina Electric Co-operatives and Santee Cooper, the state-owned utility, are the first and only electric service providers to offer renewable energy to South Carolina consumers (Bruce, 2007).

Co-operatives are also active in the R&D of energy conservation, distribution and generation. This is important because,

> "....wind energy has its barriers. The congested and limited transmission infrastructure inhibits strong continued growth of energy development, and adding to the dilemma is the intermittent nature of wind. It can't provide firm, continuous power like a base-load production facility and any additional power generated could not be stored for future use, although, for rural areas at least, this problem appears to have been overcome........The objective then is to develop an alternative use for wind energy," says Ron Rebenitsch, Basin Electric manager of member marketing speaking of the idea of converting wind to hydrogen. This innovative project will convert wind power from its intermittent form to an energy resource we can store and use as needed." (Bowers, 2007, p1)

With grants from the U.S. Department of Energy (DOE) secured by U.S. Sen. Byron Dorgan and in-kind contributions from Basin Electric and a consortium of partners, a research wind-to-hydrogen production facility and accompanying hydrogen use demonstration project are currently under construction (Bowers, 2007, p1).

Basin Electric wind resources plan for the project include two wind turbines located approximately 10 miles south of the research center and two larger wind farms located near Edgeley, ND, and Wilton, ND. These wind farms will provide the electricity required for the hydrogen production process. As electricity is produced by the wind turbines, it will be dynamically scheduled over the local transmission system in "near real time" from the turbines to the hydrogen production site. The dynamic scheduling will be done using software currently being developed. Meters at the production facility and each wind farm will use telecommunication paths to achieve the dynamic scheduling interconnection. (Bowers, 2007, p2)

According to Rebenitsch, the wind energy will power an electrolyzer, which is a commercial hydrogen generator that uses electricity to separate the oxygen and hydrogen contained in water. "The hydrogen produced at the facility will be stored. It will then be used mainly as a transportation fuel, but also for the creation of firm schedulable power from a small generator or possibly, a fuel cell, he says…….The electrolyzer will be one of the nation's first production sources of hydrogen from a renewable resource.." (Bowers, 2007, p2). In this case, the hydrogen will be produced using electricity generated by wind resources either owned or contracted by Basin Electric. The hydrogen produced at the production site will primarily be used to refuel hydrogen-powered vehicles. For example, General Motors 2006 Flexfuel pickups will be converted to run on hydrogen fuel to demonstrate the new link between wind power and vehicle transportation. "Even with the conversion to hydrogen, these vehicles will still be able to run on gasoline or E85 ethanol," Rebenitsch says." (Blowers, 2007, pp1-2)

We saw in the section above concerning the UK that local community based resistance to green energy projects had often acted as a barrier to developing green energy initiatives. Yet in America it is member education and volunteer actions by segments of their co-op membership that is raising awareness of the significance of the threat of climate change. The rural electric co-operative press has many examples of member campaigns on energy saving as well as green energy generating strategies.

Members of the Delta Montrose Electric Association for example, ran a campaign and sold 3,000 compact fluorescent lights reducing green house gas emissions by 700 metric tons in the process. In the process they also raised $6000 for local not for profit groups participating in the activity. The co-operative, in promoting energy saving equipment claims their members measured as a typical household can save around $450 pa as well as reduce their carbon footprint. (NRECA, 19.06. 2007, pp1-2) Another example, this time from the State of Southern Carolina, also illustrates the way co-operative membership activities provide awareness raising contexts and encourage good intentions to turn into practical actions.

"…..most electric cooperatives have begun offering their members the option of purchasing Green Power, sold in blocks of 100 kilowatt-hours. The response has turned out to be a pleasant surprise that despite the higher cost more members have signed up than expected. A surprising number have committed to 100-percent participation, meaning they have requested the maximum amount of Green Power blocks possible. The Green Power Program is a great way for people to do something concrete for the environment." (Bruce, 2007, p1)

It is not only in the area of prevention of climate change that co-operatives can be a valuable model. They can respond to disasters that may be the result of climate change because of their membership culture of embed community volunteerism which can mobilize assistance and raise community consciousness to the issues that face them. In the case of Hurricane Katrina the Electric Co-operatives put the power back in New Orleans and members of their staff were honored for their courageous work during the aftermath of the hurricane. (Stuteville, 2007a) In "…..Mississippi, 11,000 workers came from 22 states to repair damaged systems…Power was restored to almost every meter in the state that could accept it within three weeks…" (Johnson, 2007, p1)

5. Case Study: Wind Co-operative of the Year 2005

In 2005 the Department of Energy (DOE), in conjunction with National Rural Electric Co-operatives Association (NRECA) and the Cooperative Research Network, honored Illinois Rural Electric Co-op with its 2005 Wind Cooperative of the Year Award, in recognition of the co-op's leadership, success and innovation in wind power. (Cunnigham, 2007)

Illinois Rural Electric has "demonstrated that wind power can contribute to a cleaner environment, a stronger local economy and act as a hedge against rising fuel costs," said Douglas Fowler, DOE acting assistant secretary for energy efficiency and renewable energy, as he announced the award. (Cunningham, 2007)

The Winchester-based co-op in May 2005 was the first in the state to install wind capacity, completing a 1.65 MW turbine on a site 35 miles east of the Mississippi River. According to the US DOE, the co-op's service territory is one of the state's best sites for wind energy development. (Cunningham, 2007) Founded in 1936, the co-operative commenced providing electricity to farmers in rural Illinois in 1938. Today it serves 10,250 members in some of the most sparsely populated parts of the State.

Accepting the award, Bruce Giffin, Illinois REC's general manager, thanked DOE and other federal and state agencies that extended financial support for construction of the $1.89 million project. "If we can utilize renewable energy to generate electricity, and do it at the right price, that is the right thing to do", added Sean Middleton, the co-op's manager of engineering. Wind levels in Pike County could support as many as 100 turbines, adding as much as $7 million to the local tax base, according to representatives of the rural electric co-op. (Cunningham, 2007) However, Sean Middleton acknowledges such an investment would provide enough for about half a million average homes at current usage and cannot be justified for a co-operative serving a rural community unless the institutional environment enabled the co-operative to be able to sell on its surpluses. (Interview with PD 27th June, 2007)

The Pike County Wind Project's turbine, with a hub 235 feet above the ground and a height of 365 feet with the blades straight up, transmits its power into the co-op's distribution lines for use by members, offsetting wholesale power purchase. (Illinois Rural Electric Co-op, *Pike County Wind Power*, 2007). Monitoring the turbines at this location over the last two years Sean Middleton confirms predictions that the turbines are achieving an annual average utilization at around the 30% level. (Interview with PD dated 27th June, 2007) This generates enough power for about 500 homes which leaves the co-operative currently still heavily dependent on coal generation for its energy provision.

What is interesting about the co-operative business model in America is not that it has very different technical problems to solve or that there solution does not require an equally radical rethink about role and function but rather that there is less of a vested interest in the status quo. Consider the statement by Gregory Boyce President and CEO of the giant coal multinational Peabody which has posted five consecutive years of record earnings with 2007 forecast to be another record year. (Powers, 2007, p30)

"Each time we have regulated emission in this country, we have had commercial technologies to enable the reductions at known costs. That doesn't exist today for carbon dioxide and it must before carbon dioxide is capped. Technology first, controls later, to protect our energy security, keep jobs in the United States and fuel our growing economy as the environment continues to improve." (Powers, 2007, p30)

Contrast this with the statement by Bruce Giffin General Manager of the Illinois Rural Electrical Co-operative.

"Our members are really enthusiastic about green energy and want us to move faster in this direction. Many have already taken steps to generate green power themselves. We are organizing our next away day policy development meeting around the topic of converting to green energy." (Interview with PD dated 27th June, 2007)

The co-operative business model is better able to respond to consumer pressure because it is owned by consumers not investors. The co-operative has recently started to experiment with solar panels at its head offices at Winchester as one of a number of green energy options it is exploring. For many of the co-operatives farmer members' bio-mass is another viable option as is methane in dairy producing regions. Sean Middleton, the co-operatives technical manager, is also taking an interest in experiments with the latest giant ammonia batteries being experimented with by one of their neighboring co-operatives. The batteries may make it feasible to convert wind power into storable Hydrogen as a reserve energy source particularly for the powering of motor cars and other vehicles.

"These giant storage units would be unacceptable in urban areas but look promising for sparsely populated rural areas where the large size of the storage units would not be a problem." (Sean Middleton interview with PD 27th June, 2007)

The Illinois Rural Electric Co-operative would require considerable amounts of external funding to make economic sense of central investment in wind or any other green energy source. Its current wind turbine project was approximately half funded by governmental and other external agencies. When such finance is available there is no question that the co-operative is anxious take up the opportunity. Bruce Giffin notes that going green attracts tax breaks for the private sector not available to co-operatives so the co-operatives need alternatives such as the direct financial incentives up front provided for the Pike County Wind Turbine Project. (Interview 27th June, 2007)

However as with other rural co-operatives for marketing and procurement of agricultural produce rural electric co-operatives may be able to transform their function into one of purchaser of members' surplus energy and provider of energy of last resort when domestic generation cannot fulfill domestic requirements. In addition there would be the possibility of equipment procurement and servicing as additional products and services they could provide their membership.

Without the signing by President Roosevelt of the executive order setting up the Rural Electrification Administration (REA) on May 11th, 1935 (Pence et al, 1984, p63) it would have been impossible to establish rural electrical co-operatives in America and without the co-operatives most of rural America would have remained to all intents a third world country. Even with the Presidents Rural Electrification Administration order many private companies still tried to prevent the economic success of co-operatives by actions known locally as skimming the cream. This was the tactic of private operators responding to rural co-operatives by "spite-lining" into one or two of the more densely populated of the rural areas to undermine the economic viability of the whole co-operative project.

> "A 1940 REA survey showed there had been some 200 cases of spite-line or cream-skimming activities in 38 states since 1936. Eight newly organized co-operatives were wiped out entirely; many others were considerably weakened." (Pence, 1984, p88)

As one of the first wave of rural electric co-operatives following Roosevelt's order the Illinois Rural Electric Co-operative finds itself fifty years later once again a pioneer in the development of progressive community based energy as it struggles to meet its memberships aspirations for clean energy and the autonomy of self-generation.

6. Research Agenda

It is doubtful that in the UK or any other densely populated country context that the political leadership will ever exist to transform energy generation and distribution into the co-operative model even though American experience shows it to be the most socially progressive business model. The lobby for big power solutions which only the existing business structures can finance and operate remains dominant in the UK. This is not to detract from the relevance globally of the American Rural Electric Co-ops or the encouragement they give to those arguing for the viability of the co-operative business model in the context of green energy generation, distribution and conservation.

The problem is three fold. First there is the need to determine an appropriate ownership and business model that can implement the establishment of sustainable energy generation as an arena for small scale enterprise. Here we think the co-operative model given the right institutional framework has demonstrated real merits over the last half century and is more flexible and responsive to the imperatives of climate change.

The second issue is that of knowledge transfer and dissemination. This is a social question of networking, communicating and mobilization. Here the question is how to bring together technologists and environmentalists with co-operative and social economy development agencies in an interface with local community needs.

Thirdly, we conclude that, given the power of oligopoly in this sector and the ambivalence of the state, research into the domestic or micro power generation options should be stepped up. The technologies are developing at a cost profile that is increasingly bringing them within reach of the average householder or small co-operative of householders. Judging by previous experiences, such initiatives will initially face barriers in terms of access to the grid but they could, given sufficient numbers of consumers adopting them, create a new political climate and undermine the market power of the established energy suppliers by reducing levels of demand for centralized energy.

References

Abu-Sharkh S., Arnold R.J., Kohler J., Li R., Markvart T., Ross J.N., Steemers K., Wilson P., Yao R. (2006). Can microgrids make a major contribution to UK energy supply? *Renewable and Sustainable Energy Reviews* 10, 78–127.

Averitt R.T. (1968), *The Dual Economy. The Dynamics of American Industrial Structure*, W.W. Norton and Co. Ltd. New York.

Bruce J. (2007). The Greening of Power in South Carolina, Adapted from *Electric Co-op Today*, 2007, National Rural Electric Co-operative Association. www.nreca.org/main/NRECA/About Us/CooperativeDifference/

Blowers A. (2007), Basin Electric prepares to launch pilot project, Adapted from *Electric Co-op Today,* 2007, National Rural Electric Co-operative Association. www.nreca.org/main/NRECA/About Us/CooperativeDifference/

Christensen, P., Lund, H., 1998. Conflicting view of sustainability: the case of wind power and nature conservation in Denmark. *European Environment* 1, 1–6.

Clark W. and Lund H. (2007). Sustainable development in practice. *Journal of Cleaner Production* 15, 253-258.

Connor, P.M., 2003. UK renewable energy policy: a review. *Renewable and Sustainable Energy Reviews* 7 (1), 65–82.

Cunningham T. (2007), Winning is a Breeze for Illinois Co-op, Adapted from *Electric Co-op Today*, 2007, National Rural Electric Co-operative Association. www.nreca.org/main/NRECA/About Us/CooperativeDifference/

Davis P. (1999), Ch.7 "The Co-operative Bank Case Study" in Managing *the Co-operative Difference*, Co-op Branch, International Labour Office, Geneva.

Electric Co-op Today, National Association of Electric Co-operatives, USA. http://www.nreca.org/AboutUs/CooperativeDifference.htm

Friends of the Earth, *Briefing. New research on the economics of climate change. A briefing ahead of the Stern Review*, London, October, 2006 www.foe.co.uk

Greenpeace, *Decentralising Power: An Energy Revolution for the 21st Century*, London, July, 2005 www.greenpeace.org.uk

Johnson S. (2007). On the long Road to Recovery, Co-ops Lead the Way, Adapted from *Electric Co-op Today*, 2007, National Rural Electric Co-operative Association. www.nreca.org/main/NRECA/About Us/CooperativeDifference/

McCarty J. (2007), Missouri Electric Cooperative Members to Benefit from Wind Energy, Adapted from *Electric Co-op Today,* 2007, National Rural Electric Co-operative Association. www.nreca.org/main/NRECA/About Us/CooperativeDifference/

McLaren Loring J. 2007. Wind energy planning in England, Wales and Denmark: Factors influencing project success. *Energy Policy* 35, 2648–2660.

Midttun A. (2001). (Editor), *European Energy Industry Business Strategies*, Elsevier Science.

Mitchell, C., 2000. The England and Wales non-fossil fuel obligation: history and lessons. *Annual Review of Energy and the Environment* 25, 285–312.

Mitchell V.W., Walsh G., and Yamin M. (2005), "Towards a Conceptual Model of Consumer Confusion" in *Advances in Consumer Research*, 32(1), 143-150.

NRECA (2007). Electric Co-ops Squeeze Power form Unlikely Sources for Renewable Energy, Adapted from *Electric Co-op Today*, 2007, National Rural Electric Co-operative Association. www.nreca.org/main/NRECA/About Us/CooperativeDifference/

Pence R.A. (1984). *The Next Greatest Thing. 50 Years of Rural Electrification in America,* The National Rural Electric Cooperative Association, Washington DC.

Powers J.M. (2007). "No Doubts About Coal. Interview with the President and Chief Executive Officer of the worlds largest private sector coal company Peabody Energy", *Electric Light and Power* 85(3), 28-34.

Regibeau P. (2000), "The global energy industry: is competition among suppliers ensured?" in *International Journal of Global Energy Issues* 13(4), 378-399.

Smil V. (2006), "Energy at the Crossroads". Background notes. Global Science Forum Conference, *Scientific Challenges for Energy Research*, Paris, May 17- 18, 2006.

Smith A. (1776). *An Inquiry into the Nature and Causes of the Wealth of Nations*, Book 1, Chapter X.

Stuteville G. (2007a), Co-ops Move onto DEO's Green Power Top ten Lists, Adapted from *Electric Co-op Today*, 2007, National Rural Electric Co-operative Association. www.nreca.org/main/NRECA/About Us/CooperativeDifference/

Stuteville G. (2007b), Electric Co-op Linemen Honoured for Hurricane Restoration, Adapted from *Electric Co-op Today*, 2007, National Rural Electric Co-operative Association. www.nreca.org/main/NRECA/About Us/CooperativeDifference/

Toffler A. (1995). "The Third Wave, Selected Excerpts", from Alvin and Heidi Toffler, *Creating a New Civilization: The Politics of the Third Wave*, Andrews and McMeel, 1995. http://www.zpower.com/fr/documents/Paper_TheThirdWave.pdf

Toke, D., 2005. Explaining wind power planning outcomes: some findings from a study in England and Wales. *Energy Policy* 33 (12), 1527–1539.

Upreti, B. R. 2004. Conflict over biomass energy development in the United Kingdom: some observations and lessons from England and Wales. *Energy Policy* 32(6), 785-800.

Upreti B.R. and van der Horst D. (2004). National Renewable energy policy and local opposition in the UK: the failed development of a biomass electricity plant. *Biomass and Bioenergy* 26 (1), 60-69.

Van der Horst D. (2005). UK biomass energy since 1990; the mismatch between project types and policy objectives. *Energy Policy* 33(5), 705-716.

In: Energy Policy: Economic Effects, Security Aspects... ISBN 978-1-60692-894-3
Editor: Noah B. Jacobs

Chapter 11

REGULATORY CHALLENGES: COMPETITION DEFENSE IN THE BRAZILIAN ELECTRICITY SECTOR

Nivalde José de Castro
Institute of Economics, Federal University of Rio de Janeiro, Brazil
André Luís da Silva Leite
University of Southern Santa Catarina, Brazil

ABSTRACT

The main goal of this paper is to show that in the Brazilian Electricity Sector the Regulatory Authority (ANEEL) faces a significant challenge. We show that due to the physical features of the Brazilian Electricity Sector coordination is more important than competition, so bundled transactions are more efficient than transactions through verticalized firms. We concluded that firms strategies are to be organized as holding companies, and this lead them to have a high market power, which demands from ANEEL a strong position against the abuse of market power.

1. INTRODUCTION

The very intense dynamics of recent electricity sector development worldwide have, with few exceptions, followed a common path. Reforms in the industry structure in the 1980s and 90s reduced quite substantially the State's participation as a direct stakeholder in production and financing. The main goal was to enhance social welfare and introduce competition in generation segment. The financial crisis of the 80s acted as the main factor driving those changes, which were directed essentially to attracting private investment to the infrastructure sector. Given the sector's intrinsic features, the State built structures to enable a regulatory framework to be formatted to prevent market concentration, which would be prejudicial to the competition that it was intended to introduce.

From the late 90s, these reforms began to show a series of problems resulting from errors of design, planning and execution, which led to a new process of restructuring that is now being intensified by the world energy crisis and efforts to achieve security of supply. The aim of this chapter is to examine this process in relation to Brazil. To that end, the text is divided into three sections. The first examines – in general terms and highlighting the most important and decisive aspects – the reform and restructuring process in Brazil's electricity sector. The second analyses what are regarded as the major problems and the adjustments needed to bring better balance to the sector. The third addresses the issue of the trend towards economic concentration in favour of companies structured in the holding format. Lastly, we present our conclusions, indicating in general terms the need for the sector regulatory agent (ANEEL) to examine the issue of holdings, so as to prevent market capture and greater difficulties for the operations of the agency itself.

2. Brazil's Electricity Sector: Reform and Restructuring

Since the 1980´s the electricity industry, in many countries, has being restructured. So, electric energy is being considered and traded like a commodity. But, energy is not an ordinary commodity, because it cannot be stored (Collins, 2002).

The guarantee of energy supply is crucial to economic development. The cost of a shortage of energy can be measured by its effect on society. Therefore, reliability is the main goal in electricity markets. In a competitive industry, the demand should be responsive to changes in the spot price, so, consumers would choose to consume or not. To assure an optimal level of capacity reserve, installed capacity (ICAP) of generation must be greater than the forecasted demand. Especially, in a competitive market there must be market mechanisms to stimulate firms to invest in installed capacity.

In this sense, reliability can be achieved through market mechanisms (Stoft, 2002). One important characteristic of a capacity market (ICAP Market) is the establishment of a penalty to assure that all generators will deliver power if required. If a generator is required by the System Operator to generate power, and it does not deliver it, the generator will be punished by the regulator.

In line with a trend observed in numerous countries, the central purpose of Brazil's electricity sector reform beginning in 1990 was to introduce a competitive environment and increase private participation in sector investment, after the crisis of the 1980s had rendered the public sector-based pattern of investment a thing of the past. Initially the purpose was to set up a system where competition and prices defined by the wholesale market would encourage new investment to expand capacity – from generation through to distribution. In practice one can define the new Brazilian model as a mixed model with wholesale and retail. In these cases, especially wholesale, free entry and free access to the transmission network (or essential facilities as in Armstrong (2001)) must be compulsory instruments of regulatory policy.

However, it is well known that the Brazilian electricity is mainly hydro electrical and, in this case, important gains of energy efficiency can be obtained with the joint operation of hydroelectric plants. So, the transmission lines are essential elements to optimize the use of the plants. Actually, in a system like this, in which the plants are located far from the load centers and different regions have complementary hydrological regimes, the "transfer of water" through the network is one of the factors that determine the expansion and the operation of the system at minimum costs. In other words, free access to the network in a system like this is not just an instrument to facilitate competition. It is, maybe, a mechanism to optimize the use of energy resources.

From 1950 until the mid-80s, the electricity sector was financed from public capital and grew at significantly high rates, accompanying and outstripping electric power demand deriving from the expansion of Brazil's economy. In the 1980s, however, the financial crisis that befell public finances left the State with insufficient capital to effect necessary investments.

The financial crisis in the State led to dwindling levels of investment in Brazil's electricity sector, which paralysed construction of generating plants and left transmission and distribution systems inadequate. In parallel with this, public utility tariffs were used as a tool to control inflation, eroding electric utilities' self-financing capability still further, as signalled by Castro & Francescutti (1998). The proposal to reform Brazil's electricity sector thus gained force on the political agenda in the 1990s in the wake of State reform.

One important feature of electricity sector reform in Brazil, as in other countries, was the change in the structure of electricity industry firms. Companies were unbundled, i.e. the three segments of their operations – generation, transmission and distribution – were separated. Unbundling was designed to prevent predatory behaviour and, given free access to the grid, to increase the number of competitors in the generating industry. In that regard, and as compared with other international experiences, the expression "sector reform and restructuring" does not necessarily refer to a programme of privatisation although in the Brazilian case, sale of public assets was on the agenda of the State reform that began in 1990.

Brazil's electricity sector restructuring in the 1990s was designed along lines similar to reform of the UK system, particularly as regards mechanisms to encourage competition and efficient production (Leite, 2003). Monetary stabilization from 1994 onwards paved the way for infrastructure enterprises to be included in the privatisation process. Four factors contributed to this situation.

Firstly, a constitutional amendment abolished the public monopoly over infrastructure industries. Secondly, differential treatment for domestic and foreign capital was eliminated[1], enabling foreign firms to dispute auctions to purchase Brazilian companies. Thirdly, Law No. 8.987/95, the law of concessions, laid down the basic conditions for entering, leaving and operating in infrastructure sectors. Lastly, negotiation of states' debts to the federal government was made subject to a certain level of amortisation, which state governments could only meet by divesting assets.

In order to attain the level of competition desired in the electricity sector, it was proposed to:

a. Unbundle the enterprises by activity, i.e. power generation, transmission, distribution and sale;
b. Set up a National System Operator (*Operador Nacional do Sistema*, ONS), acting independently of the owners of transmission assets;

[1] Constitutional amendment No. 6/95.

c. Set up a spot market (*Mercado Atacadista de Energia*, MAE), where short-term power sale and purchase transactions would take place;
d. Set up an independent National Electric Power Agency (*Agência Nacional de Energia Elétrica*[2], ANEEL), responsible for regulating and overseeing electricity services;
e. Establish new economic agents – independent producers and free consumers – in the electricity sector; and
f. Lay down specific rules under which to monitor the market and ensure free competition. These include horizontal segregation of the major generators and limits on market shares, restrictions on cross ownership and a ceiling on self-supply.

As proposed and implemented, the reform suffered from significant failings of both planning and execution. Some of the deficiencies deserve specific analytical attention here. The first was the sale – privatisation – of two distribution companies (Escelsa and Light) owned by the Eletrobrás Group, before the regulatory agency, ANEEL, was in place.

Both the reform and the privatisation process fell behind schedule. There was no consensus, either in society or in political circles, as to whether the privatisations were really necessary, as described by Peci (2007). The privatisations were accompanied by significant shortcomings in the new contract rules. Clauses of the concession contracts and the initial contracts between generators and distributors comprised potential sources of conflict of interest between the parties (Correa et alli, 2006). The contracts tended to contribute to inflating the value of the assets to be privatised, so as to boost competition at auction and, most importantly, ensure higher premiums. This policy indicated that the process of privatising state enterprises, i.e. the reform overall, was subordinated to the endeavour to overcome macroeconomic imbalances by obtaining revenue to reduce the public deficit.

In addition to this hasty start marred by inconsistencies, another error in the implementation of the new model was the priority given to privatising distribution enterprises, an activity typically classified as a natural monopoly. It would have been logical, given the aim of increasing competition and economic efficiency, to privatise the generators – especially the federal ones. That idea was abandoned, however, for lack of political consensus on the matter. The distributors, meanwhile, were defaulting (as a result of the macroeconomic policy to curb inflation by controlling public tariffs in 1970 and 80). This deterred private investment in generation, because there were no concrete guarantees that distributors would be able to pay for power supplied, which would depress the value of the generators and consequently the extent to which revenues from their sale could contribute to the macroeconomic adjustment (Carvalho, 2001). Accordingly, the solution was to give priority to selling off the federal distributors and encouraging the state governments, via public debt settlements, to sell their distributors.

The complexity of Brazil's electricity sector – in terms of the volume of investments, its continental scale and installed capacity – prevented the reform process from being pursued to the limit (unlike in Argentina, for instance, where all the assets came under the control of private capital). In Brazil, generation remained predominantly under federal government ownership, while distribution – a natural monopoly with captive consumers – was 80% privatised.

[2] Law No. 9.427, of 31 November 1997.

The main – and crucial – evidence of faulty institutional design was given by the supply crisis that led to electricity rationing programme adopted in May 2001, which resulted in a 20% reduction in nationwide consumption (Pinto Jr. et al., 2007). Pires, Giambiagi & Sales (2002) list four major, interrelated reasons for the 2001 power supply crisis. They are:

a. exhaustion of the State model;
b. faulty planning of the transition from the former model to the private model;
c. contractual and regulatory problems; and
d. a lack of coordination among government departments.

The risk of a supply shortage forced consumers to reduce demand in 20%, excluding the three states in the South of Brazil. The main reasons of this crisis were the scarcity of rain and, most important, the lack of investments in generation and transmission. The lack of investments was the consequence of the hybrid market structure, where the government was also an investor and the regulator. This fact created unstable rules to private investor, especially rules concerning the rate of return, pricing process and the uncompleted privatization process. The financial consequences for utilities were severe, because consumers demanded less energy and they were exposed to debts in foreign currency, which value has raised since then (Von der Feher and Wolak, 2003).

Note that in the post-rationing period the utility companies faced a serious liquidity crisis and operating losses (Pires et al., 2002). In Brazil's electricity sector, the price of electric power is a function of conditions in the electricity industry, i.e., the availability of water in reservoirs and the level of rainfall. In predominantly hydro systems, the price of power tends not to be very volatile in the short term, but more so in the medium term. That is because, in the short term, reservoirs transfer power from low load times to peak times, modulating supply and buffering price volatility. In the medium term, meanwhile, the price of power is more volatile, because the hydro systems are designed to guarantee load under adverse hydrological conditions.

From 2003 onwards, Brazil embarked on a process of electricity sector adjustment and restructuring determined by the new authorities produced by national elections. The Ministry of Mines and Energy (MME) conducted a series of studies in order to formulate and implement a new model for Brazil's electricity sector, the legal and institutional bases for which were approved by the National Congress in Laws 10.847 and 10.848 of 2004. The two aims of the new model can be summarised as follows: to guarantee electricity supply and to ensure moderate tariffs.

As regards secure supply, Brazil's new (present) model incorporated:

a. a reversal in the focus of electric power contracts, from the short to the long term, with a view to reducing price volatility and creating a market of long-term contracts (Power Purchase Agreements, PPAs) that could be used as firm guarantees for financing;
b. compulsory contractual coverage, by distributors and free consumers, of 100% of their electric power consumption;
c. an Electric Sector Monitoring Committee (Comitê de Monitoramento de Setor Elétrico, CMSE) with the function of monitoring the balance between supply and demand to short-, medium- and long-term time horizons;

d. prior environmental licensing as a requirement for any new project to participate in the tender process; and
e. a return to integrated, centralised sector planning by the State, through the Power Research Enterprise (Empresa de Pesquisa Energética, EPE), set up by the new law.

In order to secure moderation in tariffs, one of the key aims of the new model, a virtual pool was formed among the distributors to act as sole purchaser on the electricity market. The distributors individually forward their electric power demand estimates to 3- and 5-year time horizons to the EPE. The EPE, in a monopsonistic environment, effects the purchases via auction at the lowest price and issues the sale contracts for thermoelectric and hydroelectric supply for 15 and 30 years, respectively.

The 2003 model divided the Brazilian electric power market into two trading environments, with clearly distinct logics and structures. The first was designed to meet demand from captive – primarily residential, service and industrial – consumers with lower levels of consumption: this is termed the Regulated Contract Environment (*Ambiente de Contratação Regulada*, ACR). The second is directed exclusively to firms, referred to as free consumers, consuming larger volumes. This market, denominated the Free Contract Environment (*Ambiente de Contratação Livre*, ACL), seeks to ensure that its participants compete and operate freely. In the ACL, players can enter freely into bilateral contracts setting prices, volumes, timeframes and hedge clauses with minimal government intervention.

On the captive market (ACR), power supply contracts can be of two types:

a. Contracts for a set quantity of electricity, where the generators assume all risks (of loss or gain) of integrated power operation and all costs relating to supplying the power contracted; and
b. Power availability (capacity) contracts, where the risk of supply varying from assured power are allocated to the pool and transferred to regulated consumers.

In addition, electricity purchase auctions are differentiated into old power auctions (power from existing projects) and new power auctions (from projects to be built). Contracts for new generation projects are decided at auctions five and three years ahead of initial delivery (known as *A*-5 and *A*-3 auctions). The former are designed for construction of hydroelectric-based generating plants able to come on-line within five years and the latter for projects for conclusion in three years, directed to thermoelectric-based generating plants. Contracts signed for the purchase of new power should cover terms of from 15 to 35 years (depending on the time necessary to amortise the investments) and should contain provisions to encourage moderation in tariffs. This, it is hoped, will significantly reduce the investment risk for generators, which will be able to start the project with a guaranteed revenue flow at acceptable rates of return and be able to raised inexpensive financing, particularly via project finance and by granting guarantees (receivables) to financial institutions (Correia et alli, 2006).

The results obtained to date warrant the conclusion that the present model is managing to ensure expansion and moderation in tariff rates. Under the present model, Brazil's electricity sector is taking shape as more stable, sound and predictable than under the model in place in the 1990s.

However, there are still numerous challenges to be met by Brazil's electricity sector. Although there is a substantial – 160 GW estimated – hydro potential to be harnessed, the environmental legislation issue has to be resolved more precisely. This is currently one of the major restrictions on expansion of the hydroelectric matrix. Another imperative is to refine certain points in the regulatory framework, which would mitigate uncertainties in the sector.

From the foregoing, given an environment still gaining in maturity and with challenges ahead, it can be inferred that the firms, especially the larger ones (which in the case of this study are those structured into holdings), tend to set strategies designed to maximise their reliability, contract compliance and larger profit margins. These strategies can lead, in the medium term, to levels of market concentration that could jeopardise sector stability and, on the other hand, hinder the activities of the regulatory agency. That is the issue to be examined in the next section.

3. Regulatory Challenges: Emphasis on Defending Competition

ANEEL's various duties include defending competition. It is the agency's duty to set down restrictions on players' participation in the industries that make up the electricity sector production chain. It also oversees concentration in Brazil's electricity sector, under Resolution 278/2000, which sets limits on electricity sector firms' market-share.

The limits specified by this Resolution are:

a. no generator may hold more than 20% of national installed capacity or of the national distribution market;
b. participation in the South/Southeast/Mid-West System may not exceed 25% of installed generation capacity (or the distribution market) and, in the North/Northeast System, 35% of installed generation capacity (or the distribution market);
c. no player may hold a share greater than 20% of final or intermediate brokerage in the national system or greater than 25% the sum of final and intermediate brokerage in the system.

In principle, these limits constitute *per se* prohibitions against any act that would raise a firm's share above the allowances (Pinto Jr. & Borges, 2002).

However, when the new electricity sector model – and the new trading rules – came into force in 2004, it became possible to modify Resolution 278/2000, which was not properly suited to the institutional arrangements of the new model.

In the Regulated Contract Environment (ACR), where long-term electric power purchase transactions are carried out by way of regulated auctions, the focus in appraising practices conducive to market power shifts from the market environment as such to the stages prior to the auction. This viewpoint stems from the fact that firms compete for *the* market, but not *on* the market. That is to say that, at the time of the auction, the process of competition has taken place *ex-ante*. In this case, the distributor companies are obliged to close 100% of the contracts (PPAs) and are required to inform the EPE of their load needs, which form in aggregate the marginal market demand for 3 or 5 years ahead. After the auction, distributors that estimate demand growth in their concession areas close contracts with several generators. As the distributors do not know in advance which generators they will be signing contracts with, they cannot discriminate among sellers, which minimises any incentive to bundling or self-dealing.

The auctions are held by the ANEEL on the basis of studies conducted by the EPE. One such study is crucial to ensuring a basic minimum of tariff moderation. This is the setting of an upper limit on the monetary value of the MW, on the basis of which the auction will be disputed. This upper limit is referred to as the "ceiling price". This minimises any likelihood of gaming. Thus any move towards concentration in the distribution industry has little or no anti-competitive effect, because the distributors operate in a natural monopoly environment in their concession areas. Even an increase in concentration in the generation industry will have little effect, given the nature of the ACR.

In the Free Contract Environment (ACL), meanwhile, generators and brokers compete for contracts with the group of free-consumer firms. In this case, the competition takes place on the market. Free-consumer firms have autonomy to celebrate contracts with generators and/or brokers. On this market, the volumes, delivery dates and prices are settled between the agents, either directly or by way of sale/purchase auctions.

Adjustment and settlement transactions are conducted at a spot price, by way of the Difference Settlement Price (*Preço de Liquidação das Diferenças*, PLD). PLD is a *proxy* to the spot price. The PLD is an extremely volatile indicator, because it is calculated essentially on the basis of water volumes in reservoirs and rainfall. As the electricity sector involves large volumes of water dammed in very small areas and the rainfall regime is tropical, the PLD is not a good price indicator for electric power. Its volatility – in the course of a year it can vary from R$17 to R$570 – indicates that the PLD does not provide grounds for long-term investment decision-making processes. That deficiency imposes very high – and as yet not precified – financial risks on generator and distributors.

On the free market, negotiation is open and unrestricted, and there the ANEEL must be more alert to any increase in market concentration in the generation industry. This calls for more thorough studies of possible consequences and trends. Even in the distribution industry, there is a possible loophole in the unbundling provisions. This is the possibility, apparently not prohibited by the regulations, that a generator forming part of a holding may successfully bid for a plant that is to offer power for marginal demand from a group of distributors. There may then be a firm from the same group among the distributors named in the power sale contract.

The problem is how to anticipate and comprehend such outcomes beforehand with a view to taking the proper measures. For example, depending on whether the action towards concentration occurs in the same Submarket or in different Submarkets, the regulatory agency will proceed differently in its appraisal. If the action occurs in two different Submarkets, the analysis of a possible increase in a firm's market power will be framed mainly by restrictions on transmission between these Submarkets. If on the other hand, it occurs in a single Submarket, where the restrictions on transmission are lesser, although not negligible, the ANEEL must ascertain the possibility of market power being exerted at peak load times.

Thus, the market share ceiling imposed on the electricity generation, transmission and distribution industries may represent a constraint on system expansion, given the trend towards concentration inherent in capitalist economies. However, the regulatory agency should centre its efforts on minimising the market power of sector firms, especially in the free contract environment (ACL).

4. The Market Power of the Holdings

One important concern arises, however, when one considers the increasing numbers of firms in holdings subordinated to a single controller. In the model for Brazil's electricity sector, firms were unbundled. However, the law places no restriction in terms of market share on firms structured into holdings. In this regard, the law – i.e. the electricity sector and antitrust regulatory agencies – assume in fact that a holding plays a neutral role. That interpretation is mistaken and if not recognised and corrected can lead to firms of the holding type gaining considerable market power.

There is empirical evidence, although for a relatively short time period, indicates that the trend is for the holdings operating in the four industries to gain in strength, particularly in generation, distribution and trading. This market power stems not just from the holding's financial capacity, which makes for greater leverage. It also enjoys greater competitive advantage in terms of market knowledge, especially as regards the market in the concession area of the distributor in the group. This fact points to an information asymmetry that, in practice, acts as a barrier to entry by other competitors in the generation and trading industries. The competitive advantage consists in the opportunity to capture the distributor's customers with the potential to enter the free contract market. Thus the regulatory agency's role, which ultimately is to eliminate the power that the natural monopoly structure gives the distributor, is surmounted by the bundled operation of the holding. This argument gives grounds for saying that the larger the distributor's market, and the larger the influence of industrial consumption, the greater will be the potential for it to bring billing gains into the holding by creating barriers to entry. Thus, according to the line of reasoning presented here, there is a strong likelihood of a process of full re-bundling taking place through contracts.

Table 1 shows the firms structured into holdings in Brazil. The firms showed in the table are responsible for, approximately, 75% of generation and 65% of distribution capacities in the Brazilian system. The main tendency is for a generator company to integrate with a trader.

Shortly after the rationing period, the PLD was very low. In this sense, the ACL became more attractive for free consumers. So, generators companies created their own traders to attract free consumers to have contracts with one – or more than one – generator of the group.

So, the holdings show their market power in two ways: First, in financial terms they can cross-subsidize controlled firms. Secondly, they have a larger competitive advantage derived from the knowledge of the market, specially their geographical area, which is concession from public authorities. In this case, specially when a distributor company is part of the group, there's a significant information asymmetry, which, in turn, act as a barrier to entry. So, our hypothesis is that holdings use their market power to attract free consumers through their distributor company.

Table 1. Holdings in the Brazilian Electricity Sector

2007				
Holding	**Companies**			
	Generation	**Transmission**	**Distribution**	**Trader**
Companhia Brasiliana de Energia	AES Tietê, AES Uruguaiana		Eletropaulo Metropolitana	
CEMIG	CEMIG, UTE Barreiro, UTE Ipatinga, Cogeração S.A.	Transleste S.A., Centrooeste Minas, ETEP, ENTE, ERTE, EATE	CEMIG,	CEMIG Trading S.A.
DUKE	Duke Energy Brasil			Duke comercializadora
ENERGISA	UTE-JF, Zona da Mata Geração		CFLCL, CENF, Energipe, CELB, SAELPA	Cat-Leo
NEONERGIA	Rio PCH, Termope, afluente, UHE Baguari, Termoaçu, Itapebi, Goiás Sul		COELBA, CELPE, COSERN	NC Energia
COPEL	17 UHE, 1 UTE	Copel Transmissão	Copel Distribuição	Copel comercialização
REDE	Tangará Energia, Juruena Energia, Rede Lajeado S.A.		CAIUA, CELPA, CEMAT, CELTINS, CFLO, CNEE, EEB, EDE VP	REDECOM
ENERGIAS DO BRASIL	Energest, EDP Lajeado, Enerpeixe, Enernova		Bandeirante, Escelsa, Enersul	ENERTRADE
VBC/CPFL Energia SA	CPFL Geração		CPFL Piratininga, RGE	CPFL Brasil Comercialização S.A.
ELETROBRÁS	Furnas, Eletronuclear, CHESF, CGTEE, Eletronorte.	Eletrosul, Eletronorte, CHESF		

Source: SÉRIES Econômico Financeira das Empresas do Setor Elétrico. 2007.

Therefore, the regulatory Agency faces a significant challenge in order to minimize the market power of large firms. So, we argument that, in this case, the larger the market and the number of industrial plants in the geographical area, the greater the possibility for the holding company to internalize gains from this strategy.

This process could lead to even greater growth of the short-term market, given the larger number of free consumers. This growth would tend to raise the average rate charged to captive consumers, because the tariffs include certain costs, such as the Energy Development Account (*Conta de Desenvolvimento Energético*, CDE) and the pro-rata cost of power from Itaipu. If more captive consumers become free, this account will have to be shared among a smaller number of regulated consumers, thus increasing their rates. On the other hand, the groups structured into holdings will come to command greater market power, which will hinder the regulatory agency's activities in the future.

5. Conclusions

From the mid-1990s, the electricity sector in various countries underwent substantial modifications designed to yield a more competitive environment. This unprecedented process, for lack of economies of learning, led to errors of design, execution and planning in several countries. Brazil was no exception, as underscored by the supply crisis of 2001.

From that crisis onwards, Brazil went on to encounter a solution with the State playing a significant role, and introduced a regulatory framework for the sector. Even given this sound framework, there are important challenges to be met as regards sector regulation. This paper thus set out to discuss one of the regulatory challenges facing the sector: defending competition.

For competition to be introduced successfully into the sector, unbundling is one essential condition. It can thus be seen that, as in other sectors of the economy, there is a trend towards greater concentration, especially by way of mergers and buyouts.

The paper has shown that, in the electricity sector, however, unlike other sectors, concentration indices are inadequate to specify degree of market power precisely. This makes the regulatory agency's task even more challenging, as its appraisal of any move towards concentration has to be made *ex ante*.

Another important point raised in the paper is the role of holdings, which can strategically attract captive consumers through broker firms. This issue becomes even more important given that – because of how tariffs for captive consumers are calculated in Brazil – an increase in the number of free consumers would lead to a rise in rates.

References

Armstrong, M. *The theory of access pricing and interconnection.* Nuffield College Working paper, February, 2001.

Carvalho, M. *Privatização, dívida e déficit públicos no Brasil.* Texto para discussão 847, IPEA, Rio de Janeiro, 2001. [In portuguese]

Castro, N J. & Francescutti, F. G. *A few consideration on the recent transformation of the Brazilian Electricity Sector*. In: "III Encontro de Economistas de Língua Portuguesa", Macau: jun 1998. [In portuguese]

Collins, R. A. *The economics of electricity hedging and a proposed modification for the future contract for electricity*. "IEEE Transactions on Power System", vol. 17, n.1, February 2002.

Correia, T,; Melo, E.; Costa, A.; Silva, A. *Trajetória das reformas institucionais da indústria elétrica brasileira e novas perspectivas de mercado*. "Revista de Economia". V.7, n.3, set/dez, 2006 (disponível em www.anpec.org.br, acesso em 02/fevereiro/2007).

Leite, A. L. *A capacity market model for the Brazilian electricity industry*. Florianópolis-SC: UFSC, Brazil, (Doctoral Dissertation), 2003, 146p. [In portuguese]

PECI, A. *Reforma regulatória brasileira dos anos 90 à luz do modelo de Kleber Nascimento*. "Revista de Administração Contemporânea". V. 11, n.1, jan/mar 2007, p. 11-30.

Pinto JR, H. & Borges, H. *Concentration in the Brazilian Electricity Industry*. Rio de Janeiro: UFRJ (mimeo), 2002.

Pinto JR, H. (org.). *Economia da energia*. Rio de Janeiro: Elsevier, 2007.

Santana, E.A. E Oliveira, C.A.N.V. *Transaction costs economics and th reform in the Brazilian Electricity Industry*. "Estudos Econômicos", v. 29, n. 3, 1999.

Stoft, S. *Power System Economics*: Designing markets for electricity. IEEE/ Wiley Inter-Science: Pistacaway, 2002.

von der Fher, N. & Wolak, F. *Power sector reform in Brazil: some issues*. Stanford University (draft), 2003.

In: Energy Policy: Economic Effects, Security Aspects... ISBN 978-1-60692-894-3
Editor: Noah B. Jacobs

Chapter 12

IDENTIFYING THE SOURCES OF ENERGY CONSUMPTIONAND CO_2 EMISSIONS IN THE TRANSPORT SECTOR: AN EMPIRICAL ANALYSIS OF THE SPANISH CASE

Pablo del Río**
Institute for Public Goods and Policies, Centro de Ciencias Humanas y Sociales
C/Albasanz 26-28, 28037 Madrid, Spain
Miguel Angel Tarancón
Faculty of Law and Social Sciences, University of Castilla – La Mancha
C/ Ronda de Toledo, s/n, 13071 Ciudad Real, Spain

ABSTRACT

Energy consumption in the transport sector is significantly growing at the global level, with negative socioeconomic and environmental implications, such as concerns over security of energy supply and CO_2 emissions. According to the Intergovernmental Panel of Climate Change (IPCC) energy consumption in this sector has doubled between 1971 and 2000 (from almost 1000 to 2000 million tonnes of oil equivalent), despite the significant energy efficient improvements in automobiles. This trend is expected to continue in the short, medium and long terms. According to the same source, energy consumption is expected to more than double between 2000 and 2050. Although developing countries are expected to account for the largest increase in energy consumption in this sector, some developed countries will probably increase their energy consumption in the transport sector significantly, such as Spain. This is so because scale effects are likely to offset technological effects and other factors, leading to a net increase in energy consumption and CO_2 emissions. This paper aims to provide a diagnostic on the main factors influencing energy consumption and CO_2 emissions in the passenger transport sector in Spain. The paper will also identify those policy measures which are likely to be more effective to reduce this energy consumption.

* E-mail: pablo.delrio@cchs.csic.es

1. Introduction

The transport sector is receiving a considerable attention in the context of compliance with the greenhouse (GHG) emissions reduction targets set in the Kyoto Protocol. This focus is certainly due to the high share of this sector in GHG emissions, both at a global, European and national level, as well as to the high growth rates experienced by these emissions. In addition, energy consumption in the transport sector, mostly based on oil, is a significant source of concern, especially for those countries which are oil importers, in a context characterised by increasing oil prices. Both aspects are interrelated: an increase in energy consumption in the transport sector entails an increase in GHG emissions. Both of these have negative economic consequences for countries where this happens: an increase in GHG emissions will oblige those countries to buy permits in the international market and, in addition, an increase in energy consumption will make them more vulnerable to oil price increases. This paper focuses on the GHG aspect, however.

At the world level, CO_2 emissions from the transport sector amounted to 6.3 Gt of CO_2 in 2004, representing 23% of energy-related CO2 emissions. Its growth rate was the highest within the final energy use sectors. The subsector of road transport accounts for 74% of overall CO_2 emissions from transport. The largest growth rates in emissions are expected to occur in the developing and least developed countries (IPCC 2007).

In Europe, transport was responsible for 28% of total GHG emissions and its growth rate was 24% in the period from 1990-2005, although significantly different growth rates could be observed among European countries (EEA 2008). It is estimated that, if transport emissions had shown the same decreasing trend as in the rest of sectors, overall CO_2 emissions in the EU would have been reduced by 14% instead of 7.9%. In turn, the weight of passenger transport by road in overall CO_2 emissions in the EU is quite significant (12% n 2004). Road represents 84% of passenger transport.

The difficulties in reducing GHG emissions at the world, European and national (Spanish) levels as well as the particularities of the transport sector require the identification of the complex combination of economic, political, social and technological factors which lead to GHG emissions.

Therefore, the aim of this paper is to identify the factors behind the situation and trends of energy consumption and GHG emissions in the transport sector in Spain and to propose some policy measures which could be effective to reduce energy consumption and its associated GHG emissions in this sector.

Accordingly, this chapter is structured as follows. The theoretical framework is provided in section 2, which discusses the main factors influencing energy consumption and CO_2 emissions in the transport sector in general. In section 3, the main factors affecting energy consumption and CO_2 emissions in the Spanish transport sector, with a special emphasis on passenger and road transport, are discussed. Section 4 identifies the most relevant policy measures to tackle those factors. The chapter closes with a conclusion section.

2. Drivers of Energy Consumption and CO_2 Emissions in the Transport Sector[1]

There are several studies which decompose the CO_2 emissions in the transport sector according to the factors causing them (see IEA 2000, Halnaes et al 1999, Schmutzler 2005, Holden and Hoyer 2005, Tarancón and del Río 2007, among others). These factors can be summarised as follows:

- *Scale*. The scale (or absolute use of transport services) depends on the number of passengers per kilometre (km). In turn, this depends on three variables: the number of trips, the distance of each trip and the occupancy levels of vehicles travelling. The use of vehicles depends on the income of individuals. In addition, structural and geographical factors play a very relevant role in the demand for transport services and, thus, on the scale of transport activity. For example, more densely populated cities (i.e., less urban sprawl) generally lead to a lower need for transport (fewer number of trips and shorter distances travelled). Thus, urban planning leading to more compact cities could be expected to result in a lower level of transport demand. A widespread problem is that the increase in transport demand has more than offset energy and CO_2 emissions reductions resulting from technological improvements.
- *Modal split*. Different modes of passenger transport have different energy consumption levels and, thus, distinct CO_2 emissions per kilometre. In general, travelling by road with a car leads to higher CO_2 emissions than other transport modes, including train and the non-motorised alternatives (bicycle and walking).
- *Technological change I: energy consumption*. Within a specific transport mode, different types of vehicles can have different levels of fuel consumption per kilometre and, thus, different emissions per kilometre. In this context, it is necessary to distinguish between new and existing vehicles. Their specific energy consumption depends on factors related to the energy efficiency of the engine as well as the vehicle weight or size. Although engine improvements have led to a substantial energy savings (for example, in the case of TDI engines), this has been offset by a higher horsepower of the vehicles and the additional installation of energy-consuming devices (such as air conditioning).
- *Technological change II: carbon-intensity of fuels*. Road transport of passengers has a large dependency on oil. The alternatives are less carbon-intensive fuels (such as biofuels) and hybrid and electric vehicles.
- *Drivers' behaviour*. Driving patterns and other decisions of drivers influence the level of energy consumption of vehicles and their CO_2 emissions. A less aggressive driving pattern (in the context of "ecodriving") at moderate speeds and an appropriate maintenance of the vehicle, with periodic inspections, are associated with lower energy consumption levels[2].

---

[1] The discussion in this section is partly based on del Río and Hernández (2008).

[2] Maintaining an appropriate pressure of the tyres, avoiding excessive accelerations and keeping car speeds within moderate limits could improve the fuel economy of cars between 5% and 20% (Gallagher et al 2007).

In addition to technological and behavioural changes, changes in infrastructures might be needed to tackle those factors.

3. ANALYSING THE MAIN FACTORS AFFECTING ENERGY CONSUMPTION AND CO_2 EMISSIONS IN THE SPANISH TRANSPORT SECTOR

The aim of this section is to identify what factors are particularly relevant to explain the situation and trends in energy consumption and CO_2 emissions in the Spanish transport sector, with a special emphasis on road passenger transport. The reason we focus on passenger sector is that its emissions are greater and their growth rate higher than the emissions from freight transport. Within passenger transport, road is, both, relative energy intensive. In addition, road transport is the main mode of passenger transport, and the one which is expected to experience, together with aviation, the highest growth rates in the future (see data below).

In order to analyse the situation of the Spanish transport sector in this context, we need a yardstick to which we can compare it. Therefore, we compare the Spanish situation and trends with data for the European Union. In order to carry out this analysis, we use official data sources from the European Environmental Agency, Eurostat and the Spanish Ministries of the Environment and Promotion, among others[3].

Table 1. Final energy consumption in the transport sector in Spain

Year	Total		Transport		Share
	Mill. Toe*	Annual change (%)	Mill. Toe*	Annual change (%)	%
1990	60.67		22.72		37.45
1991	62.75	3.43	23.20	2.11	36.97
1992	62.63	-0.19	23.90	3.02	38.16
1993	62.83	0.32	23.75	-0.63	37.80
1994	65.45	4.17	25.23	6.23	38.55
1995	68.67	4.92	26.59	5.39	38.72
1996	70.72	2.99	27.46	3.27	38.83
1997	73.94	4.55	29.10	5.97	39.36
1998	80.21	8.48	30.31	4.16	37.79
1999	82.64	3.03	31.52	3.99	38.14
2000	86.77	5.00	32.28	2.41	37.20
2001	90.30	4.07	33.79	4.68	37.42
2002	92.14	2.04	34.38	1.75	37.31
2003	97.16	5.45	36.20	5.29	37.26
2004	100.73	3.67	37.84	4.53	37.57
2005	103.13	2.38	38.70	2.27	37.53

* Million tones of oil equivalent

Source: Ministry of Environment; I.D.A.E

[3] The Spanish Ministry of Fomento ("Promotion") is the one responsible for public works.

Taking into account the discussion provided in the previous section, the following paragraphs include several indicators regarding the different factors influencing the situation and trends in energy consumption and/or GHG emissions in the transport sector in Spain. This will be compared to the case of the EU. We start with a discussion of the situation and trends of overall and transport sector GHG emissions in Spain.

3.1. Energy Consumption in the Spanish Transport Sector

Final energy consumption in the transport sector has increased significantly in the last 15 years, leading to a high share in total final energy consumption. This share has remained constant during the period (Table 1).

This share is comparatively high in Spain, when the same data for the EU are considered. The following table shows that, according to Eurostat services, transport accounts for 40.5% of final energy consumption in Spain, whereas this percentage is 31.2% in the EU. Road transport accounts for one third of final energy consumption in this country.

Table 2. Final energy consumption per sector in the EU and Spain

			Transport					Household and Services		
	All sectors	Industry	Total	Road	Railways	Air	Inland Navigation	Total	Households	Services
EU-27 (Mtoe)	**1170**	**327**	**361**	296	9	50	5	**481**	309	171
Share (%)	**100**	**28**	**31**	82	2.6	13.8	1.5	**41**	64.4	35.6
SPAIN (Mtoe)	**97.2**	**31**	**39.4**	31.4	1.2	5.3	1.5	**26.7**	15.2	11.6
Share (%)	**100**	**31.8**	**40.5**	79.7	3.0	13.5	3.8	**27.4**	57	43

Source: Eurostat.

Table 3. Final energy consumption by fuel/product

	1990	1995	2000	2005
Solid fuels	3.5	2.2	1.6	1.7
Oil	33.4	38.9	45.7	52.7
Gas	4.9	6.8	12.1	17.9
Electricity	10.8	12.1	16.2	20.8
Renewables	3.9	3.2	3.4	3.8

Source: Eurostat.

As a result of the increasing consumption of energy in the transport sector, and its dependence on oil, this fuel has experienced a substantial increase in its share in final energy consumption (1990 = 100, 2005 = 175). This has been due especially to the increase in the consumption of diesel in road transport.

The dependence on oil in this sector is clearly shown in the next table.

Table 4. Final energy consumption in the transport sector by fuel (ktoe)

	Petrol and diesel*	Motor spirit	Gas Diesel Oil	Share of biofuels (%)**	Biofuels	Bio-gasoline	Bio-diesel	Other liquid biofuels
EU-27	296258	114302	181955	1.1%	3211	690	2508	13
SPAIN	33254	7640	25614	0.8%	259	113	146	

Source: Eurostat.

* Total final consumption of petrol and diesel for transport.

** Share of biofuels in total final consumption of petrol and diesel for transport.

This represents a serious problem for the Spanish economy, whose imports of oil amounted to 78 million tonnes of oil equivalent (MTOE) in 2005, almost two thirds of total energy imports. This is especially worrying in a context of increasing oil prices.

3.2. Situation and Trends of Overall GHG Emissions

Total emissions of GHG had increased by 50% in 2006 with respect to 1990 levels, significantly more than in the other large economies of the EU (Table 5). This means that Spain is far from its Kyoto Protocol target (a 15% increase between 2008-2012 and 1990). One of the reasons for this situation is related to the substantial increase of GHG emissions in the transport sector, as shown in the next section.

Table 5. GHG emissions in the EU-27 and selected countries (1000 Tons CO_2e)

Country	1990	2006	%1995-1990	%2000-1990	%2006-1990
Germany	1227688	1004794	-10.81	-16.96	-18.16
U. Kingdom	768481	652290	-8.05	-12.83	-15.12
Italy	516898	567922	2.62	6.84	9.87
France	563288	541308	-1.42	-1.37	-3.90
Spain	287687	433339	10.81	33.82	50.63
Poland	453575	400459	-2.86	-14.13	-11.71
Netherlands	211651	207477	5.83	0.94	-1.97
Romania	247698	156680	-25.68	-44.00	-36.75
Czech Rep.	194244	148204	-21.28	-24.34	-23.70
Others EU-27	1100834	1030336	-8.58	-9.68	-6.40
EU-27	5572044	5464800	5142809	-9.10	-7.70

Source: Eurostat.

Table 6. Evolution of GHG emissions from transport (million tones of CO_2e)

	1990	1995	2000	2005	% change (2005-1990)*
EU-27	960	1037	1174	1277	33%
Spain	72.6	83.5	114.6	140.3	93%

Source: Eurostat.

* Accumulated percent change.

3.3. Situation and trends of transport GHG emissions.

Emissions in the transport sector in Spain have increased by 93% between 1990 and 2005. This is in sharp contrast with the EU-27 increase of 33% (Table 6).

This trend has led to an increase of the share of the transport sector in overall GHG emissions in this country, which in 2006 reached 25% (Figure 1).

Figure 2 shows that there is an intense divergence between transport and total GHG emissions in Spain.

The large share of road transport, which accounts for 90% of total GHG emissions, can be observed in the next table. However, the share per transport mode is quite similar to the case of other EU countries.

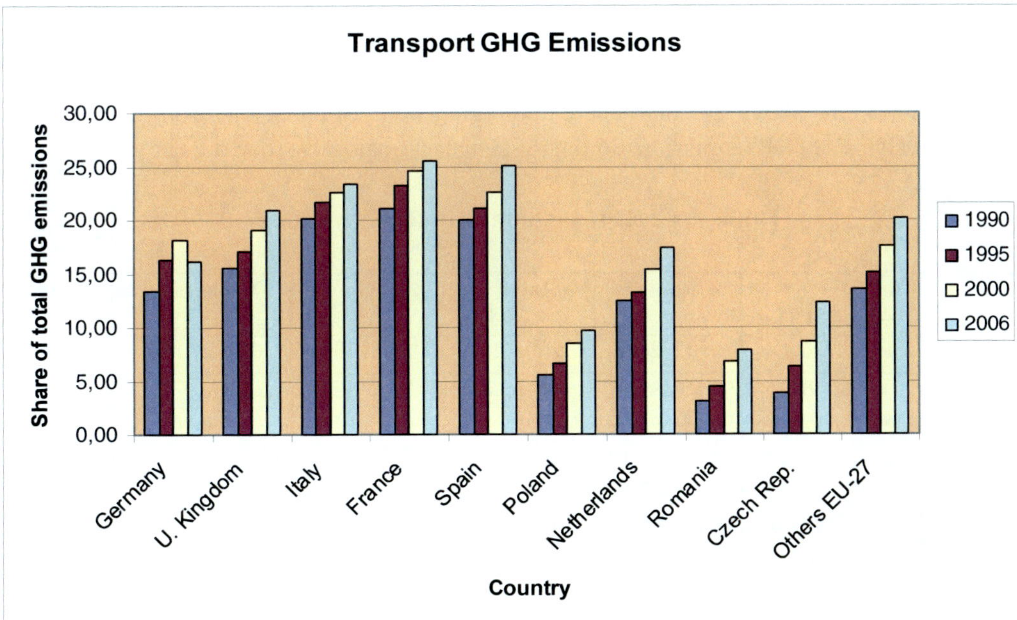

Source: Eurostat.

Figure 1. Share of the transport sector in GHG emissions (%).

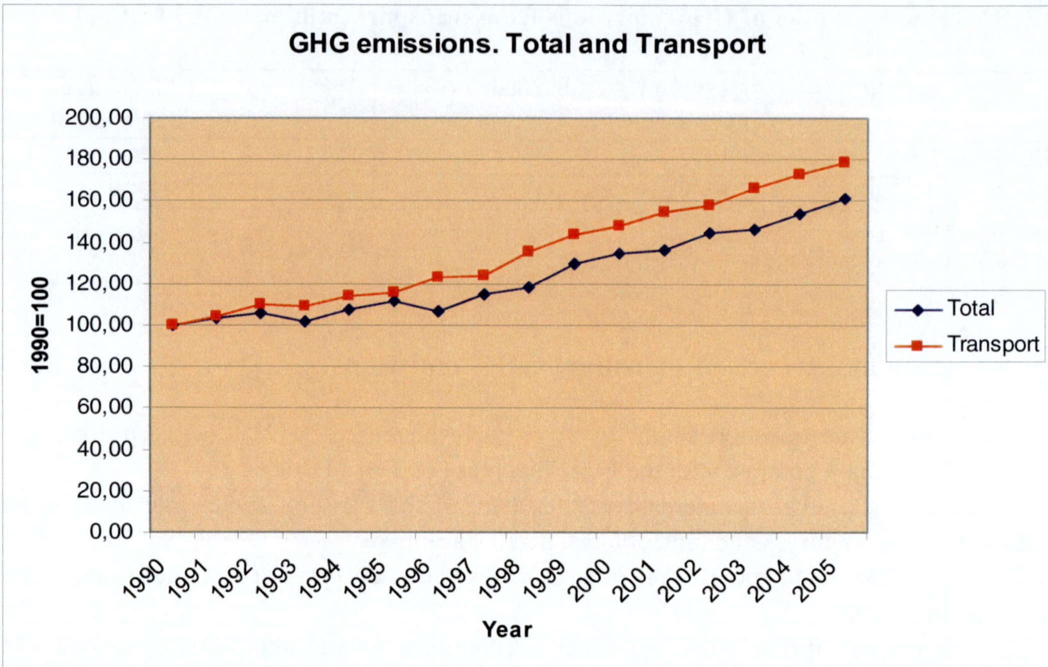

Source: Ministry of Environment; I.D.A.E.

Figure 2. Trends of transport and total GHG emissions in Spain (1990-2005).

The growth rate of GHG emissions in the passenger transport is higher than in the freight transport, in sharp contrast to the EU (Figure 3).

In addition to the above, the high share and high growth of emissions from road transport makes the analysis of GHG emissions in the passenger transport by road a clear priority.

Table 7. GHG transport emissions by mode

Country	**Total 1000 tons CO_2e**		**% Road**		**% Rail**		**% Civil Aviation**		**%Navigatio n**		**% Others**	
	1990	2006	1990	2006	1990	2006	1990	2006	1990	2006	1990	2006
Germany	164418	162011	92.59	92.67	1.76	0.79	1.76	3.31	1.25	0.53	2.64	2.70
U.K.	118889	136715	93.64	92.07	1.58	1.78	1.03	1.73	3.49	4.06	0.26	0.37
Italy	103952	133198	92.32	92.18	0.48	0.30	1.55	2.10	5.26	4.64	0.40	0.79
France	118817	138604	93.86	93.68	0.91	0.45	3.61	3.42	1.44	2.02	0.18	0.43
Spain	57530	108619	89.27	90.25	0.81	0.31	7.25	6.70	2.63	2.57	0.04	0.17
Poland	25373	38620	85.19	96.87	8.18	1.20	0.22	0.20	0.86	0.04	5.54	1.70
Netherlands	26439	36147	97.96	97.88	0.34	0.27	0.16	0.11	1.54	1.74	0.00	0.00
Romania	7692	12352	84.67	97.33	12.50	1.81	0.32	0.10	2.41	0.32	0.09	0.44
Czech Rep.	7453	18197	81.73	97.49	8.80	1.44	2.06	0.09	0.76	0.10	6.64	0.87

Source: Eurostat.

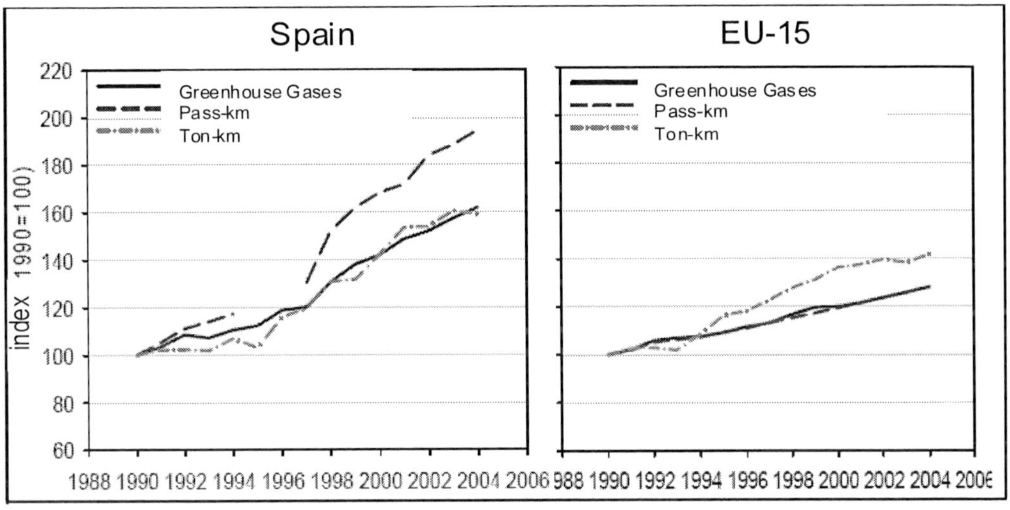

Source: Monzón (2007).

Figure 3. Trends of passenger transport and freight transport GHG-related emissions in Spain and the EU-15

3.4. Technology Effects

When analysing the factors behind the rapid growth in emissions, technology is a key feature. Technology has two sector-specific aspects: energy efficiency (i.e., consumption of energy per kilometre or per person travelling) and types of fuels used (oil or less carbon-intensive fuels, such as biofuels).

Regarding the first aspect, the Table 8 shows that there are significant differences with respect to the energy consumption of different types of modes and vehicles.

The energy efficiency of the different transport modes has increased (the energy intensity has been reduced), which has led to a reduction of the CO_2 emissions intensity of those modes, although still significant differences exist (Table 9).

Table 8. Energy specific consumption per vehicle

	Energy consumption (kilograms of oil equivalent/passenger/kilometre							
	0.00-0.01	**0.01-0.02**	**0.02-0.03**	**0.03-0.04**	**0.04-0.05**	**0.05-0.06**	**0.06-0.07**	**0.07-0.08**
Interurban		High-speed train, long-distance trains, standard buses, microbuses		Cars	4x4			
Urban		Urban trains	Buses			Cars		4X4

Source: own elaboration based on Monzón (2007).

Table 9. Energy intensity and emissions intensity per mode

	Energy Intensity (megajoules/passenger-km)		Emissions Intensity (gCO_2/passenger-km)	
	1990	**2005**	**1990**	**2004**
Road	2.34	1.98	162	136
Rail	0.45	0.37	31	26
Airplane	13.81	9.24	986	693

Source: Own elaboration from Monzón (2007).

Therefore, generally, technological change has had an important and positive impact on the energy consumption and the emissions reductions per mode. In the more particular case of passenger cars, there have been important improvements in energy efficiency, although these improvements have slowed down in the last years. Table 10 shows the data for the European (ACEA), Japanese (JAMA) and Korean (KAMA) automobile manufacturers

The above data suggests that, since energy and emissions intensity differ per vehicle and mode, fiscal instruments should be applied to encourage the least energy and emissions intensive modes and penalise the most intensive ones. For example, a tax on the carbon intensity of fuels could be applied to the former and the revenues used either to support the later or for other social objectives, in the context of the so-called ecological tax reform.

Concerning the use of less carbon-intensive fuels, the share of biofuels in total final consumption of petrol and diesel for transport was only 0.8% in 2005 in Spain (146 ktoe of biodiesel and 113 ktoe of bioethanol), compared to 1.1% in the EU-27.

Table 10. Targets for passenger car energy efficiency

Year	**g CO_2/km**			
	ACEA	**JAMA**	**KAMA**	**Total**
1995	185	196	197	186
2000	169	183	191	172
2004	161	170	168	162
2005	160	166	167	161
2006	160	161	164	160
Target	**140 g CO_2/km by 2008 and 2009**			

Source: EEA (2008, p.23).

3.5. Scale Effects

Scale effects refer to the use of transport services. In turn, they are affected by several factors, including price, income and weather. Some of them will be analysed in this section.

What does the data tell us regarding the scale factor? Table 11 and Figure 4 show the evolution of passenger transport activity in total and by modes between 1995 and 2005. The increase in transport activity is above the growth in GDP and is mostly related to the road and civil aviation sectors. The later shows a substantial increase in its share.

Table 11. Passenger transport activity

Passengers transport in Spain															
Year	Road			Rail			Navigation			Civil Aviation			Total		
	1000 mill. p-km.	% variation	share	1000 mill. p-km.	% variation	share	1000 mill. p-km.	% variation	share	1000 mill. p-km.	% variation	share	1000 mill. p-km.	% variation	share
1995	296.97		91.49	16.58		5.11	1.03		0.32	10.03		3.09	324.61		100.00
1996	304.30	2.47	91.31	16.80	1.33	5.04	1.10	6.80	0.33	11.05	10.17	3.32	333.25	2.66	100.00
1997	322.30	5.92	90.90	17.88	6.43	5.04	1.20	9.09	0.34	13.20	19.46	3.72	354.58	6.40	100.00
1998	320.97	-0.41	90.60	18.88	5.59	5.33	1.30	8.33	0.37	13.12	-0.61	3.70	354.27	-0.09	100.00
1999	341.33	6.34	90.59	19.66	4.13	5.22	1.30	0.00	0.35	14.49	10.44	3.85	376.78	6.35	100.00
2000	352.89	3.39	90.16	20.15	2.49	5.15	1.34	3.08	0.34	17.02	17.46	4.35	391.40	3.88	100.00
2001	359.67	1.92	90.01	20.83	3.37	5.21	1.31	-2.24	0.33	17.77	4.41	4.45	399.58	2.09	100.00
2002	387.65	7.78	90.69	21.21	1.82	4.96	1.29	-1.53	0.30	17.31	-2.59	4.05	427.46	6.98	100.00
2003	397.12	2.44	90.60	21.15	-0.28	4.83	1.28	-0.78	0.29	18.78	8.49	4.28	438.33	2.54	100.00
2004	410.16	3.28	90.57	20.79	-1.70	4.59	1.44	12.50	0.32	20.46	8.95	4.52	452.85	3.31	100.00
2005	417.83	1.87	90.02	21.60	3.90	4.65	1.48	2.78	0.32	23.24	13.59	5.01	464.15	2.50	100.00

Source: Ministry of Promotion (2005).

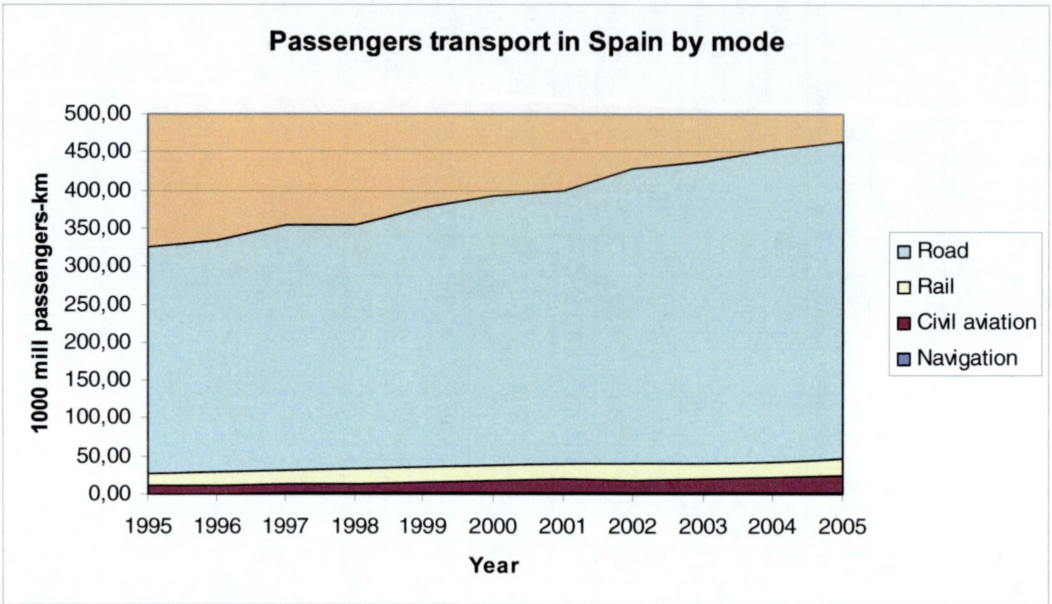

Source: Ministry of Promotion (2005).

Figure 4. Trends in passenger transport activity per mode.

When analysing the factors behind this growth, several aspects should be considered: population growth, economic growth, price increases, public expenditures on infrastructures etc.

The population increase between 2007 and 1990 has been above the EU average, as shown in the Table 12. Therefore, this factor is useful to explain the increase in transport activity since 1990.

Given its high income elasticity of demand, the use of transport is closely related to GDP growth rates[1]. These have been higher in Spain than in the EU (Table 13). However, although increasing, the expenditure on transport by Spanish households is still below the EU average (Table 14).

Table 12. Population trends in Spain and the EU (million)

	1970	1980	1990	2000	2006	2007	% change (2007-1990)*
EU27	435.4	457.0	470.3	480.5	491.1	493.2	4.8
EU15	339.9	354.5	363.4	375.5	387.7	389.9	7.2
SPAIN	33.5	37.2	38.8	40.0	43.7	44.4	14.4

Source: Eurostat.

*Accumulated percent change.

[1] As people get richer, they tend to travel more (make more trips and travel greater distances). In addition, the increase in income leads to a shift to more energy and carbon-intensive modes of transport (changes from train to cars and from smaller to bigger cars)(Stern 2006).

Table 13. Growth in GDP (real growth at constant 1995 prices)(%)

	2000	2003	2006
EU-27	3.9	1.3	3.0
EU-15	3.8	1.2	2.8
SPAIN	5.0	3.1	3.9

Source: Eurostat.

Table 14. Final consumption of households for transport (by type of expenditure, 2006)

	Final consumption for transport	Of which			Transport as a % of total final consumption	Expenditure per head on transport
		Purchase of personal transport equipment	Operation of personal transport equipment	Purchased transport services		
	Million €				%	€
EU27	892 563	297236	440260	155068	13.6	1800
EU15	841625	283610	415057	142958	13.7	2200
SPAIN	64184	22854	30922	10408	11.8	1500

Source: Eurostat.

3.6. Modal Factors

The above factors, in addition to other more "psychological effects" lead to the aforementioned dependence on private cars, although this dependence is similar to the EU one (Table 15).

This is the result of the increase in passenger cars, compared to other modes (Table 16).

This trend has led to a high growth rate in the number of passenger cars per thousand inhabitants since 1970 and the share is closer to the EU-27 average, although still lower than the EU-15 average (indeed, the growth rate between 2005 and 2006 was the lowest in the EU-27).

Table 15. Modal split of land passenger transport (2006, %)

	Passenger cars	Bus and coach	railway	Tram&metro
EU27	82.3	9.3	6.9	1.5
EU15	83.3	8.6	6.9	1.3
SPAIN	81.4	11.8	5.3	1.5

Source: Eurostat.

Table 16. Trends in different transport modes (thousand mio-pkm)

	1990	1995	2000	2005	2006	% change (2006-1990)*
PASSENGER CARS						
EU15	3101	3520	3859	4012	4052	30
ES	174.4	250.4	302.6	337.8	340.9	95
BUSES AND COACHES						
EU15	367	377	401	416	416	13
ES	33.4	39.6	50.3	53.2	49.4	47
TRAM AND METRO						
EU15	49.0	48.8	54.7	60.0	61.5	25
ES	4.4	4.3	5.2	6.0	6.2	40
RAILWAYS						
EU15	268	273	306	324	334	24
ES	15.5	16.6	20.1	21.6	22.1	42
HIGH SPEED RAIL TRANSPORT						
EU27	15.2	32.8	58.8	80.1	89.9	491
ES	0	1.2	1.9	2.3	8.4	-

Source: Eurostat.

* Accumulated percent change.

Table 17. Number of passenger cars per thousand inhabitants

	1970	1980	1990	1995	2000	2005	2006	% change (2006-1990)*
EU-27			345	381	427	459	466	35
EU-15	183	293	405	435	478	503	508	25
SPAIN	70	201	309	360	431	463	464	50

Source: Eurostat.

* Accumulated percent change.

3.7. Behaviour of Drivers

In addition to the above, transport-related energy consumption by private cars depends to some extent on the behaviour of drivers. For example, it is well known that speeds of 100km/hour lead to significant lower energy consumption levels (around 20% lower) than speeds above 130km/hour and fuel consumption increases exponentially after that speed level is surpassed. In addition, appropriate maintenance activities and the right tire pressure increase the fuel efficiency of cars. However, there are no public data on this behaviour and, thus, this factor is very difficult to tackle from an analytical perspective. This should be the focus of further research.

4. Policy Implications

What policy measures are out there? There is already a quite voluminous literature on policy instruments to reduce energy consumption (and CO_2 emissions) in the transport sector. The following box provides a classification of these instruments, which are usually grouped in three categories: command-and-control (CAC) regulation, economic instruments and other instruments. IPCC (2007), among others, provides a full description and assessment of those policies.

Box. Classification of mitigation instruments in the transport sector

Regulatory Instruments

- User restrictions
- Speed limits.
- Vehicle maintenance requirements.
- High occupancy requirements for vehicles
- Standards.
- Biofuel oblligations (quota).

Economic Instruments

- Emissions trading
- Subsidies
- Price-related measures (taxes):

Information Instruments

- Information/awareness raising campaigns
- Ecodriving campaigns.
- Education.

Other Instruments

- Voluntary agreements
- Land-use planning and transport planning
- Public procurement
- Public investments in R&DD.
- Promotion of public transport.

Source: del Río and Hernández (2008).

According to the data in the previous section, scale and (lack of) modal shift seems to be the major factors behind energy consumption and, thus, GHG emissions in the Spanish transport sector. These two aspects should thus be tackled by public policies. In contrast, the adoption of technological improvements by manufacturers leading to an increase of the energy (and CO_2) efficiency of cars, has been a norm, and a (relatively) effective policy (the agreements between the European Commission and the European, Japanese and Korean manufacturers, i.e., ACEA, JAMA and KAMA) has already been implemented. There is room for further adoption of technological measures, however, regarding use of information and communication technologies (for example, better logistics and regulation of transport flows). In addition, a greater effort has to be made to encourage the adoption of cleaner cars. The hybrid option is worth mentioning in this regard. Its greater fuel efficiency compared to conventional cars does not seem to be enough to offset its larger up-front costs and lack of familiarity by users. An initial subsidy covering the cost difference (and justified due to the negative externalities avoided in terms of oil dependence and environmental pollution) could be quite effective in this regard. However, this should be complemented by information provision to the general public on the advantages of hybrids.

However, although more effort should be devoted to encourage technological changes, behavioural changes should be the major priority. These measures should be adopted to tackle two aspects: scale and modal shifts, although both issues are interrelated. Instruments should try to encourage a reduction in the number and length of trips, but the priority should rather be towards facilitating the shift from the most energy-intensive forms of passenger transport (such as cars) to more energy-efficient types, including rail and the non-motorised alternatives (walking and cycling). Rather than a single instrument, a combination of measures is likely to be more effective.

In this context, economic instruments have been proposed by economists. Since the implementation of emissions trading schemes is particularly problematic in this sector (see Stern 2006), taxes on the most energy-intensive and polluting forms of transport are a clear alternative at hand[2]. They also raise revenue, which can be used by the government to stimulate more energy-efficient forms of transport or be used for other social objectives, including the reduction of pre-existing distortionary taxes, in the context of the so-called double-dividend approach.

However, economic instruments are unlikely to be effective if they are not combined with other instruments. The reason is quite simple: income effects predominate over price effects. As people get richer (something that has been occurring in Spain, whose GDP growth rates have been more than 1 percentage points above the EU average), they tend to use transport services more intensively, regardless of its price. This results in a highly inelastic demand curve for transport (i.e., low price elasticity of demand). This means that higher transport prices as a result of the application of taxes are unlikely to significantly reduce transport demand, at least in the short to medium terms. Therefore, in some cases, these instruments should be complemented by regulatory, information provision and land-planning measures. These all have different time horizons.

It is interesting to see that many of these "hard" measures have been applied in the Spanish context. Instead, the implementation of soft measures has been the norm (or the lack of implementation of instruments at all). The recent increase of the purchasing tax for the most energy-inefficient automobiles seems to be a step in the right direction, but other less popular measures (including fuel taxes and regulations restricting the use of private cars) are unlikely to be politically feasible, especially in a context of increasing oil prices. The trend towards urban sprawl in the major Spanish cities is not a positive trend either in this context. Personal mobility in private cars seems to be a social value in itself and this cultural-behavioural aspect will be difficult to bend, at least in the short to medium term.

In general combinations of instruments belonging to all the categories should be adopted. Some provide negative incentives to act (i.e., sticks), mandating some type of behaviour (i.e.,

[2] Taxes can be applied to the purchase of a vehicle, but there are also circulation taxes (paid annually) and fuel taxes. The later are more likely to affect the use of cars, although the former can discourage purchasing a more energy-intensive car.

user restrictions) or financially penalising certain conducts (i.e., taxes on fuel use). Others rely on positive incentives (i.e., carrots), financially encouraging the realisation of certain activities (i.e., subsidies for public transport or for the purchase of cleaner vehicles) or using information or education measures to change behaviour.

5. Conclusion

The aim of this chapter has been to illustrate the situation and trends of Spanish GHG emissions in the transport sector and the drivers for the growth in those emissions. Some policy implications have been derived from this analysis. Further research should apply more formal methodologies to assess the factors behind the growth in emissions and their respective relevance. Since it seems that the rise in emissions is mostly related to the scale factor, this should be tackled with appropriate policy measures. What of those instruments are more effective in this regard and how they interact between each other also deserves special attention.

References

European Environmental Agency (EEA)(2008). *Climate for a transport change. TERM 2007: indicators tracking transport and environment in the European Union*, Copenhague, EEA Report No 1/2008.

Gallagher, K.S., Collantes, G., Holdren, J.P, Lee, H., Frosch, R. (2007). *Policy Options for Reducing Oil Consumption and Greenhouse-Gas Emissions from the U.S.*, Harvard (Massachussets), Transportation Sector. Discussion Paper July 27, 2007. John F. Kennedy School of Government, Harvard University.

Halsnaes, K., Callaway, J.M., Meyer, H.J. (1999). *Methodological Guidelines. Main Reports: Economics of Greenhouse Gas Limitations*. UNEP. Riso National Laboratory, Denmark.

Holden, E., Høyer, K.G. (2005). The ecological footprints of fuels. *Transportation Research Part* D10, 395–403

International Energy Agency (IEA)(2000). *The road from Kyoto, Current CO_2 and transport policies in the IEA*. IEA, Paris.

Intergovernment Panel on Climate Change (IPCC)(2007). *Fourth Assessment Report. Working Group III*. Geneve (Switzerland).

Ministerio de Fomento (Ministry of Promotion)(2005). Los transportes y los servicios postales. Informe anual 2005. Madrid

Monzón, A. (2007). La contribución del transporte a la emisión de gases de efecto invernadero en España. VII Seminario Internacional sobre Cambio Climático. Spanish Ministry of the Environment. 20th November 2007. Madrid.

Del Río and Hernández (2008). Análisis de los procedimientos de reducción de emisiones de CO_2 en el sector del transporte de pasajeros por carretera. *Ekonomiaz* (in press).

Schmutzler, A. (2005). Transportation and the Environment-Perspective for Future Research. In: Böhringer, C., Lange, A. (Eds.), *Applied Research in Environmental Economics*, vol. 31. Physica Verlag, Springer, Heidelberg.

Stern, N. (2006). *Stern Review on the economics of climate change*. Cambridge (Reino Unido), Cambridge University Press.

Tarancón, M.A., del Río, P. (2007). Structural factors affecting land-transport CO_2 emissions: a European comparison. Transportation Research D: Transport and the Environment, 12, pp.239-253.

In: Energy Policy: Economic Effects, Security Aspects... ISBN 978-1-60692-894-3
Editor: Noah B. Jacobs

Chapter 13

DILEMMA OF ENERGY EFFICIENCY IMPROVEMENT: MARKET FAILURE, GOVERNMENT FAILURE AND REBOUND EFFECT

Sang-Hyeon Jin*

Economic and Industrial Policy Research Division,
Gangwon Development Research Institute, Gangwon-do, Korea

ABSTRACT

There are two important challenges: new oil shock and climate change where energy efficiency improvement is considered as a measure of them. In fact, IPCC and EEA suggested that energy efficiency improvement policy can be the most effective and economic measure in the short term. According to these suggestions, organizations and governments are planning for efficiency improvement in the international and national levels. But this argument has been refuted in the point where goal of this policy will not be accomplished if the market mechanism is not considered seriously. The main issue is who will take responsibility for this policy. Will it be the government or the market? This paper checks the roles of the market and government in respect to energy efficiency improvement. This paper also reviews the controversies of market failure, government failure and rebound effect theoretically and empirically. In conclusion, it shows the policy implication of energy efficiency improvement in the context of climate change. Also it suggests that smart government, which hopes to reflect market mechanism in this policy, has to reform traditional tax system ecologically.

1. CHALLENGES OF ENERGY POLICY

There are two important challenges to overcome energy policy: One is a new oil shock and the other is climate change. While a new oil shock is a problem of oil supply and demand,

* E-mail: upperhm@gdri.re.kr

climate change originated from global warming of greenhouse gases. They can be very different issues, but their fundamental reasons are the same, namely, modern industrial society has grown by using fossil fuel. Fossil fuel is not only bringing about climate change but also being depleted rapidly. Therefore, energy efficiency improvement has recently been spotlighted by many researchers and governments.

There are two reasons why so many people have an interest in energy efficiency improvement. First, if we use fossil fuel efficiently, then we can reduce demand for it. Resource economists and engineers argue that governments have to regulate extravagant energy consumption behavior. Actually, energy efficiency improvement programs by governments started after the first and second global oil shocks. Second, if we reduce consumption of fossil fuel, then we can mitigate climate change. International organizations and environmentalists argue that emissions of greenhouse gases have to be restricted globally. In 2007, the International Panel on Climate Change (IPCC) also suggested that energy efficiency improvement can be the most effective and economic measure in the short term. Likewise, the European Environment Agency (EEA) argued in 2006 that the EU has to improve energy efficiency because it is the most cost-effective measure for climate change.

Based on these suggestions, international organizations and governments are making plans for energy efficiency improvement at the international and national level. But some researchers think that energy efficiency improvement may not be as effective as expected if governments do not consider market mechanisms and consumer behavior of energy use. Saunders (2000b) even argues that energy efficiency improvement can increase energy use instead of reducing it because it decreases the price of energy services. It is important to estimate how the effect of efficiency improvement measures can be reduced by market mechanisms and to think how this effect can be reflected in climate change policy.

It is necessary to review the role of the government and market in improving energy efficiency. In fact, energy efficiency improvement programs were started by governments after the first and second oil shocks because they thought there was market failure in respect to the energy policy. They began to regulate energy consuming behavior of not only companies but also general consumers because efficient technologies were not spontaneously accepted in the market. But now energy efficiency improvement policy is substantially problematic because it is guessed that there is government failure in energy efficiency improvement too. Rebound effect is the clearest evidence of government failure without understanding market mechanism. Therefore market failure, government failure, and rebound effect in energy efficiency improvement will be reviewed theoretically and empirically in this paper. Finally, in the context of climate change and a new oil shock, a policy implication of energy efficiency improvement will be suggested.

2. Energy Efficiency Improvement and Climate Change

Many countries have had an interest in energy efficiency improvement since the first and second oil shocks of the 1970s. Energy security was the most important national issue in the world. Most governments implemented an energy efficiency improvement policy to buffer against the oil shocks and to reduce energy use. While energy efficiency improvement policy has a long history, it is becoming more important than during the oil shocks of the 1970s because the context of policy is rapidly changing. First, a new oil shock would be more serious and fundamental than previous oil shocks. While in the 1970s it was a temporary imbalance between oil supply and demand, depletion of oil is the main factor for the current rise in oil prices. Second, since the Kyoto Protocol was declared, the international society has come to a consensus to try to mitigate climate change together. Likewise, the combination of a new oil shock and climate change makes it a mission of governments to reduce energy use and demand. Therefore policy makers have an interest in energy efficiency improvement because it is the most effective and viable way to reduce energy use and greenhouse gas emissions.

Lovins and Hennicke (1999) argued that energy efficiency improvement policy is feasible from an economic standpoint. They suggested that the 'factor four scenario' should be adopted for a sustainable energy system. They presented about fifty cases at the Rome Club and showed that the factor four scenario could reduce resource consumption by half while at the same time doubling welfare. They thought that there are a lot of factors which obstruct performance of energy efficiency improvement in spite of sufficient economic feasibility. For example, exclusive possession by a government of the energy supply system, biased subsidies, unfair tax schemes are causes to obstructing the implementation of the factor four scenario.

Other reports which urge that energy efficiency improvement policy should be accepted by governments are increasing worldwide. The 3rd Assessment Report of the International Panel of Climate Change (IPCC, 2001) estimated that efficient energy technology can reduce carbon dioxide emissions by 30%. The report of the European Environmental Agency (EEA, 2006) predicted that total energy use can be curtailed by 20%, if only existing energy efficiency improvement programs are implemented in Europe. Reduction of greenhouse gas emissions using efficient energy technology is getting support from various researchers and policymakers (Howarth et al., 2000).

3. Market Failure of Energy Efficiency Improvement Policy

Energy conservationists argue that energy efficiency improvement is a 'no regret policy' because it is environmentally sound and economically profitable. Much research has shown empirically that there are lots of efficient energy technologies that will be profitable not only for consumers but also producers. But these efficient energy technologies are not spreading naturally through the market. This phenomenon is called 'efficiency paradox' because economic efficiency technologies are not selected spontaneously by companies (DeCanio, 1997). Generally efficient energy technology is known as profitable technology because it is not risky and usually makes a profit of 30%. Then what is the real reason why companies do not choose economic and efficient energy technologies?

Economists are skeptical that there are profitable but unselected technologies systematically in the market like efficient energy technologies. They think that only price change can make enough efficiency improvement, if the market mechanism works normally.

It is the modern technology change hypothesis of Hicks' 'induced innovation theory.'[1] Hicks argued that "relative price change can make innovation by itself and expensive production elements be used less than before" (Birol and Keppler, 2000). In a competitive market, technology change depends on the relative price of energy, material, and labor. Thus energy price is always opposed to technology innovation. Namely, if energy is expensive, efficient technologies can be selected in the market. On the other hand, if energy is cheap, inefficient technology can be selected. But sometimes energy efficiency improvement can progress when the energy price is decreasing; this is a kind of ratchet effect[2] in energy efficiency improvement.

Economists who believe in the profit maximization hypothesis deny market failure of economically feasible energy efficiency improvement policy. But profit maximization hypothesis cannot explain the existence of undeveloped profits and is an unscientific argument. It can even be refuted by the fact that a company exists. If profit maximization hypothesis is right, management is not necessary because the cost for executive ability is superfluous.[3] So they explain 'efficiency paradox' as a 'transaction cost' or 'hidden cost.' Namely, they explain that these transaction costs and hidden costs can make the effectiveness of energy efficiency improvement decrease. But transaction cost is not well defined to estimate and has not been supported by empirical foundations. Moreover Koomey and Sanstad have shown that there is no hidden cost to obstruct the spreading of efficient energy technologies (DeCanio, 1997).

On the one hand, some people say that efficient energy technology is not the most profitable technology. This is the same argument that best practice is the enemy of the second best practice. But this is based just on inference. According to the Pareto efficiency theory,[4] the second efficient improvement also has to be realized. The company which does not accept the second efficient technology for the best efficient technology is not reasonable. In other words, if any reasonable company can make profits from efficiency improvement, it has to seek an even smaller profit. The reason is because the second efficiency improvement also is able to induce optimal efficiency improvement. For this reason many researchers have had an interest in Porter's hypothesis that environmental regulation can be helpful in finding profit which is left undeveloped for best efficient improvement.

---

[1] Studies on the mechanism of technology spread and transfer process are based on so-called 'induced innovation theory. Habakkuk is a pioneer of this field and argued that the United States introduced labor-saving technology from the United Kingdom in the industrialization process because labor was insufficient. In other words, the United States had to induce labor-saving technology because the United States had enough land but more expensive labor than the United Kingdom. As a result, expensive labor transformed the direction of technology development. While Habakkuk's induced innovation theory has been debated in academic society, the meaning of his argument is the explanation of technology progress using economic factor. Especially induced innovation theory explains technology progress from the viewpoint of demand factor and this theory has been applied to other technology innovation processes.

[2] Ratchet is a toothed wheel that can rotate toward only one side. Therefore, ratchet effect means the phenomenon that people maintain a consumption pattern of high income even when income is decreasing. In respect to energy, the ratchet effect means that people still consume energy less and improve efficiency, while the energy price is decreasing.

[3] Modern enterprise theory is generated to investigate the role of management for getting over agency, insufficient information, and incentive problems.

[4] The Pareto efficiency theory is based on the principle that improving someone's utility without hampering another's utility is socially desirable. Here, the second best efficient improvement is also a reasonable choice because it doesn't hamper anyone but improves total utility.

Porter's 1990 hypothesis has aroused controversy in academic circles because it argued that environmental regulation can even increase the profit of a company. It was a very interesting argument which broke the rules. Porter studied some efficiency improvement cases of enterprise in respect to energy and environmental policy. He then argued that accepting government regulation was sometimes able to make a company check and review management's inefficiency and extravagancy. As a result, it could develop new efficient technology and reform the management process and organization. The case study of Porter shows that a regulation is able to help a company check inefficient elements, improve itself, and increase profits.

But neoclassical economists have refuted Porter's hypothesis. In 1995, Palmer argued that Porter ignored environmental regulation costs of 135 billion dollars and his study was based on not a strict statistical analysis but just a case study. But Porter's study is still an interesting hypothesis because it gives a base for justifying government intervention when efficient energy technology is not chosen voluntarily by the market.

Clinch and Healy (2000) argued that there are lots of reasons that efficient energy technology is not naturally chosen in the market. For instance, the interest rates a company must consider as a criterion for investment is different from the social discount rate that a government considers a criterion for energy efficiency improvement policy. Insufficient information is also another obstructing factor for energy efficiency improvement. Likewise, efficient energy technology is not selected in the market for both of these reasons. Thus they argue that governments have to inform companies, reduce the opportunity cost for energy efficiency improvement, prepare funds for technology development, curtail transaction costs in the process of investment, and internalize social benefits in the private benefit. In other words, they argue that governments should intervene actively to correct market failure in respect to energy efficiency improvement.

4. Government Failure of Energy Efficiency Improvement Policy

Until now government intervention has been justified for market failure of energy efficiency improvement. A lot of efficiency improvement policies have been executed and an enormous sum of public money has been spent for efficiency improvement. Despite these expenditures, the World Wildlife Fund (WWF) estimated that if investment for efficiency improvement had been maintained at the 1993 level when fund for energy demand management reached its peak, the electricity fee for consumers would be reduced by 1 billion dollars, greenhouse gases by 11 million tons, and air pollutants by 79 thousands tons. In conclusion, it argued that governments should continue support for efficiency improvement and utilities have to invest 5% of revenue to efficiency improvement obligatorily (Loughran and Kulick, 2004).

But there is not only market failure but also government failure of energy efficiency improvement policy. In particular, economists have criticized government intervention because it can reduce social utility and make additional costs. Hausman and Joskow said "even though insufficient information and failure of capital market make consumers choose incorrectly, any government policy restricting their choice cannot be helpful to them" (Fischer, 2005).

Brookes (2000) also opposes government intervention for efficiency improvement. He thinks that government intervention is justified only when it corresponds to public interest. He objects to efficiency improvement policy because efficiency improvement is a behavior for a producer to increase productivity of input. In other words, he insists it is just a behavior not to reduce national energy use but to increase personal profit.[5] So if the government wants to intervene, it has to prove that this intervention is helpful to the public interest. The reason is because governments do not intervene to improve productivity of any other resource, e.g., water and iron. He argues that if governments want to intervene in energy efficiency improvement for climate change and the main goal of its policy is reduction of greenhouse gases, they have to redirect the policy goal to not efficiency improvement but reduction of greenhouse gases.

Efficiency improvement policies of governments can also make equity problems since taxes collected from the people are mostly given to a company consuming lots of energy. Sutherland (2003) also argues that reinforcement of a minimum efficiency standard can be a burden on the poor. He pointed out that the U.S. Department of Energy ignored time preference of the poor when it carries out a cost-benefit analysis. The poor prefer inefficient products with a lower initial cost and a higher operational cost because the present is more important than the future to them. Therefore if a government increases the minimum efficiency standard, it will exclude the preference of the poor and make the initial cost increase. He warns that a minimum efficiency standard policy can increase the price of cheap home appliances and decrease utility of the poor's welfare.

Some researchers argue that a government's efficient improvement policy will certainly fail. Brookes (2000) thinks efficiency improvement policy is surely necessary for greenhouse gas reduction but there is a dilemma. If efficiency improvement policy succeeds and there is a decrease in the energy demand, the price of the energy will also decrease. But if the price of energy decreases, there will be a loss of interest in efficiency improvement policy. On the other hand, if the policy fails and prices increase, there will be an interest in efficiency improvement policy. Therefore absolute energy use keeps increasing and only relative efficiency, namely energy use per production value, keeps improving. In other words, efficiency policy always succeeds relatively but fails absolutely. Brookes calls it 'the principle of the unidirectionality of energy efficiency savings.' Actually, this principle can be identified worldwide. In the United States, the energy use per production value has decreased, but total energy use has increased (Brookes, 2000). South Korea also has made relative energy intensity (energy use/production value) improve but cannot reduce absolute energy use (Luukkanen and Kaivo-oja, 2002).

Some research has shown the failure of efficiency improvement policy through empirical analysis. In 1988, Tran estimated that even though there had been no energy demand management program, 70% of the total energy reduction could be achieved. Furthermore Loughran and Kulick (2004) pointed out that there was selection bias[6] in demand

[5] Brookes (2000) also insists that only improvement of productivity can induce energy efficiency improvement naturally; he calls it 'the principle of indivisibility of economic productivity.'

[6] Selection bias is a statistical error generating in the process of selection. In other words, bias of selection itself can distort statistical results. Generally selection bias can make statistical significance more exaggerated than real significance, so it shows spurious results.

management program. Namely, subsidies were given to large enterprises which didn't need them. In the United States, 14.7 billion dollars has been invested toward efficiency improvement policy between 1989 and 1999. It was assessed that 1.8% of total electricity use has been reduced. But they objected that the actual reduction was only 0.3% when considering selection bias. In other words, even though it had not been a financial subsidy, a large enterprise would improve efficiency. So they insisted that while actual effect of efficiency improvement policy was only 0.3%, it was exaggerated as 1.8%.

5. Energy Efficiency Improvement and Rebound Effect

In this context, a government has tried energy efficiency improvement policy directly because of market failure. But some researchers have objected that energy efficiency improvement policy led by a government could be another 'government failure' if the market mechanism is not properly considered. Rebound effect is a typical phenomenon of a government failure when a government which does not understand the market mechanism can even increase energy use far from achieving the goal of energy reduction (Khazzoom, 1987).

5.1. Definition of Rebound Effect

Sometimes technical measures to improve energy efficiency can even increase energy use rather than reducing it, if the market mechanism works. Generally it is expected that improving efficiency will reduce energy use, but if the energy service price decreases with improved efficiency, then people have a tendency to use more energy and increase their utility. This tendency is called 'rebound effect,' and it is related to consumers' tendency to consume more energy due to economic benefit from efficiency improvement (Berkhout et al., 2000). That is, the unexpected increase of energy use is the rebounded part because energy use increases again after decreasing through efficiency improvement.

If all conditions are the same without efficiency improvement through technology development, a so-called *ceteris paribus*, producers can produce the same amount of product with less energy use. Likewise consumers can also get the same amount of service with less energy use. But efficiency improvement tends to change energy consuming behavior. That is, if the price of energy service decreases, people want to get more energy service than before.

For example, if a fuel efficient car has better gas mileage than others, the cost of fuel per kilometer will fall. Generally speaking, a price decrease leads to increased consumption. People may want to go on a picnic every weekend or commute from the suburbs because the energy service price is so cheap. In other words the precondition of *ceteris paribus* changes because people want to get more energy service and it means increasing energy use. Therefore, some part of the expected energy savings can be lost because people tend to get more energy services at a decreased price. This lost part is denoted as 'rebound effect.'[7] A rebound effect of 10% means that 10% of the expected savings are offset by increased consumption. For instance, if expected energy savings are 20% and the rebound effect is

[7] Rebound effect is sometimes called 'take-back effect' because it means the lost of the expected energy savings.

10%, then the actual savings is 18% because 2% is offset. Rebound effect can be defined using the simple equation (Haas and Biermayr, 2000):

$$\text{rebound effect} = \frac{\text{expected savings - actual savings}}{\text{expected savings}}$$

A rebound effect of 0% means full achievement of energy reduction because expected savings is the same as actual savings. On the other hand, a rebound of 100% means complete failure because actual savings is zero. Also, if rebound effect is greater than 100%, efficiency improvement measures can even increase energy use, which is called 'backfire effect.' If energy use increases after efficiency improvement, actual savings is minus and rebound effect is greater than 100%. For instance, if a person changes the old inefficient car for a new efficient car, fuel economy will be better than before. Then he can move into a new house in the suburbs and commute from there because transportation costs decrease. Namely, an efficient car bought for reduction energy use can even make energy use increase. Eventually, if a backfire effect is noted, efficiency improvement measures are of little use as energy reduction policies.

5.2. Arguments on Rebound Effect

Discussions on rebound effect have aroused various responses from researchers and policymakers. At first, there was an objection to the existence of rebound effect itself. Before rebound effect was discovered by Khazzoom, there were a lot of papers and reports suggesting energy efficiency improvement programs as a measure for the first and second oil shocks (Laitner, 2000). In 1988, Amory Lovins, who advocated energy efficiency improvement, denied the existence of rebound effect in *The Energy Journal*. But Khazzoom pointed out that Lovins ignored the economic side of energy consumption in 1989. Eventually it was proven that Lovins was mistaken and had no academic basis for his denial (Grepperud and Rasmussen, 2004). Later, researchers developed the theory of rebound effect into the 'Khazzoom Brookes Postulate.' Although there is consensus about the existence of rebound effect, Howarth (1997) argued that backfire effect is impossible after conducting studies that used a microeconomic model. But Saunders (2000b) countered that Howarth had used a restricted production function and confirmed that backfire effect is possible, at least theoretically.

While some arguments were resolved, rebound effect theory still has a few issues. It is the size of rebound effect that continues to be the most contentious issue. In other words, the degree of rebound effect is still debated by researchers and policymakers because the effectiveness and necessity of energy efficiency improvement policy may depend on the degree of rebound effect. Some researchers argue that the impact of rebound effect is so negligible that it can and has to be ignored (Bentzen, 2004; Berkhout et al., 2000; Haas and Biermayr, 2000; Hertwich, 2005; Howarth et al., 2000; Laitner, 2000); others argue that the impact of rebound effect is so substantial that it has to be taken seriously (Brännlund et al., 2007; Grepperud and Rasmussen, 2004; Jin, 2007; Roy, 2000; Saunders, 2000a). Greening et al. (2000) reviewed all the articles on rebound effect estimation in circulation in the United

States. They showed that there was no general rule and the size of rebound effect was very wide (See Table 1). From this, the conclusion can be drawn that the impact of rebound effect has to be gauged individually, because each country and sector has very different consumption characteristics and patterns.

6. Smart Government Tuning Market Mechanism

Many papers and reports have shown that energy efficiency improvement policy is the best economic and technically feasible measure for reducing energy use and greenhouse gas emissions. But there is not only market failure but also government failure in energy efficiency improvement. In particular, rebound effect is very important in relation to this issue because it shows the failure of a government quantitatively which does not consider market mechanism.

While there have been various arguments on this issue in academia, it is not reflected in the real policy of governments and international organizations. Actually there are only a few rebound effect estimations in the world. But as Greening et al. (2000) showed, each country and sector has very different consumption characteristics and rebound effects. Each country and international organization has to estimate the rebound effect by sector and reflect it into real policy. A government has to invest and implement energy efficiency improvement policy to the sector which has the least rebound effect. Then a government can get the best effect of policy and reduce not only energy use but also greenhouse gas emissions. Considering the rebound effect by a government can be helpful for measuring a new oil shock and climate change.

If a government propels an economic subsidy policy or regulation of energy use without considering market failure and government failure, this policy will certainly fail. Therefore a government has to be smart with understanding market mechanism. For example it will be better to subsidize the sector which rebound effect is less and to regulate energy use of the sector which has a bigger rebound effect. If there is a backfire effect, then regulation or voluntary agreement will be desirable without economic incentives or subsidy. Basically, a smart government has to introduce a carbon tax because it is very useful for transforming a carbon intensive social structure at a national level. It also can be effective in offsetting rebound effect.

At this time, it is an argumentative issue whether government regulation would be more effective than using market mechanism. In particular, it is critical in respect of climate change. Since the Kyoto Protocol, a lot of economic incentive systems have been introduced by governments and international organizations. Clean development mechanism, joint implementation, and emissions trading system are just a few. A smart government can be the answer. At first, a smart government has to recognize the necessity and effectiveness of its efficiency improvement policy. After that, it has to understand market failure, government failure, and especially rebound effect. Then it can make a no regret policy for climate change and new oil shock.

Table 1. Summary of empirical evidence for rebound effect

Economic actor	End use	Potential size of the	Comments	Number of studies
Consumers	Space heating	10~30%	The unmeasured part of this effect includes an increase in space conditioned and an increase in comfort.	26++
	Space cooling	0~50%	The unmeasured part of this effect includes an increase in space conditioned and an increase in comfort.	9+
	Water heating	10~40%	Reports of increased shower length or the purchase of increased water heating unit size indicate some indirect effects, which cannot be measured.	5-
	Residential lightning	5~12%	An indirect effect in terms of an increase in operating hours was reported.	4-
	Appliances ("White Goods")	0%	Indirect effects in terms of the purchase of larger units with more features were reported.	2-
	Automotive transport	10~30%	The unmeasured part of this effect includes changes in automotive attributes, particularly the shifts toward attributes such as increases in weight, horsepower	22-
Firms	Process uses (Short-run)	0~20%	Although increases in output occurred for less than 20% of the study participants, no values were reported.	1-
	Lighting (Short-run)	0~2%	Changes in output were not reported. However, labor productivity probably improved.	4-
	Long-run aggregate impacts	0~100%	Changes in output show a great deal of variability in the literature.	Any number of studies with a variety of conclusions.
Economy-wide effects	Change in total output growth	0.48%	Postulated effects include an increase in standard of living and consumption of more energy-consuming "luxury" goods.	1-

* Grading system used for the quality of estimate:

++ These studies are done with a number of methods that provide good correspondence of estimates.

+ These studies are done with only a moderate number of different methods that show some variability in estimates.

- These studies are done with only one or two methods and are inconclusive in results.

Note: All estimates assume a 10% increase in efficiency of fuel consumption.

Source: Greening et al., 2000.

REFERENCES

Bentzen, Jan. (2004). Estimating the rebound effect in US manufacturing energy consumption. *Energy Economics*, vol.26, pp.123-134.

Berkhout, Peter H.G. et al. (2000). Defining the rebound effect. *Energy Policy*, vol.28, pp.425-432.

Birol, Fatih and Jan Hors Keppler. (2000). Prices, technology development and the rebound effect. *Energy Policy*, vol.28, pp.457-469.

Brännlund, Runar, et al. (2007). Increased energy efficiency and the rebound effect: Effects on consumption and emissions. *Energy Economics, Vo*l.29, pp.1-17.

Brookes, Leonard. (2000). Energy efficiency fallacies revisited. *Energy Policy*, vol.28, pp.355-366.

Clinch, J. Peter and John D. Healy. (2000). Domestic energy efficiency in Ireland: correcting market failure. *Energy Policy*, vol.28, pp.1-8.

DeCanio, Stephen J. (1997). The efficiency paradox: bureaucratic and organizational barriers to profitable energy-saving investments. *Energy Policy*, vol.26, pp.441-454.

European Environment Agency. (2006). *Energy and environment in the European Union: Tracking progress towards integration*, EEA Report.

Fischer, Carolyn. (2005) On the importance of the supply side in demand side management. *Energy Economics*, vol.27, pp.165-180.

Greening, Lorna A. et al. (2000). Energy efficiency and consumption: the rebound effect: a survey. *Energy Policy*, vol.28, pp.389-401.

Grepperud, Sverre and Ingeborg Rasmussen. (2004) A general equilibrium assessment of rebound effects. *Energy Economics*, vol.26, pp.261-282.

Haas, Reinhard and Peter Biermayr. (2000). "The rebound effect for space heating Empirical evidence from Austria. *Energy Policy*, vol.28, pp.403-410.

Haas, Reinhard, et al. (1998). The impact of consumer behavior on residential energy demand for space heating. *Energy and Buildings*, vol.27, pp.195-205.

Hertwich, Edgar G. (2005). Consumption and the Rebound Effect: An Industrial Ecology Perspective. *Journal of Industrial Ecology*, vol.9, pp.85-97.

Howarth, Richard B. (1997). Energy efficiency and economic growth. *Contemporary Economic Policy*, vol.15, pp.1-8.

Howarth, Richard B. et al. (2000). The economics of energy efficiency: insights from voluntary participation programs. *Energy Policy*, vol.28, pp.477-486.

Intergovernmental Panel on Climate Change. (2001). *Climate Change 2001: IPCC Third Assessment Report*, IPCC.

Jin, Sang-Hyeon. (2007). The effectiveness of energy efficiency improvement in a developing country: Rebound effect of residential electricity use in South Korea. *Energy Policy*, vol.35, pp.5622-5629.

Khazzoom, D.J. (1987). Energy saving resulting from the adoption of more efficient appliances. *Energy Journal*, vol.8, pp.85-89.

Laitner, J.A.S. (2000). Energy efficiency: rebounding to a sound analytical perspective. *Energy Policy*, vol.28, pp.471-475.

Loughran, David S. and Jonathan Kulick. (2004). Demand-Side Management and Energy Efficiency in the United States. *The Energy Jounal*, vol.25, pp.19-45.

Lovins, Amory and Peter Hennicke. (1999). *Voller Energie*, Campus Verlag Gmbh.

Luukkanen, J. and, Kaivo-oja, J. (2002). Meaningful participation in global climate policy: comparative analysis of the energy and CO_2 efficiency dynamics of key developing countries. *Global Environmental Change*, vol.12, pp.117-126.

Roy, Joyashree. (2000). The rebound effect: some empirical evidence from India. *Energy Policy*, vol.28, pp.433-438.

Saunders, Harry D. (2000a). A view from the macro side: rebound, backfire, and Khazzoom-Brookes. *Energy Policy*, vol.28, pp.439-449.

Saunders, Harry D. (2000b). Does predicted rebound depend on distinguishing between energy and energy services. *Energy Policy*, vol.28, pp.497-500.

Sutherland, Ronald J. (2003). The high costs of federal energy efficiency standards for residential appliances. *Policy Analysis*, vol.504. Cato Institute, Washington, DC.

SHORT COMMUNICATION

In: Energy Policy: Economic Effects, Security Aspects… ISBN 978-1-60692-894-3
Editor: Noah B. Jacobs

Short Communication

Energy Policy Options for Swedish Non-energy-intensive and Small- and Medium-Sized Manufacturing Industries

P. Thollander
Institute of Technology, 581 83 Linköping, Sweden

Abstract

Since the early 1990s, industrial energy end-use policy activity in Sweden has been scarce, in particular towards small- and medium-sized industrial enterprises (SMEs) and non-energy-intensive manufacturers. While the other Nordic states have taken considerable actions towards the industrial sector, Sweden has not. Rising electricity prices in Sweden and the Energy End-use Efficiency and Energy Services Directive (ESD), among other things, address the need for the Swedish authorities to design and adopt energy policies towards the non-energy-intensive and SME industrial sectors. This chapter aims to explore this by discussing plausible policy solutions, based on current research on barriers to energy efficiency and industrial energy programs. Initially, results from the three available Swedish studies on barriers to energy efficiency for these sectors are outlined. Then, a few examples of successful and less successful actions towards the industry and in particular SMEs are presented and finally, a discussion is held regarding plausible energy end-use polices, based on the presented research. In conclusion, a highly useful approach towards non-energy-intensive manufacturers and SMEs in particular is to provide energy audits free of charge and involve the local authority energy consultants.

Introduction

Since the early 1990s, industrial energy end-use policy activity in Sweden has been scarce, in particular towards small and medium sized industrial enterprises (SMEs) and non-energy-intensive manufacturers (Thollander et al., 2007). While the other Nordic states have taken considerable actions towards the industrial sector, Sweden has not. Rising electricity prices in Sweden and the Energy End-use Efficiency and Energy Services Directive (ESD), among other things, however address the need for Swedish authorities to design and adopt energy policies towards the non-energy-intensive and SME industrial sectors.

THE ESD

The ESD came into force in 2006 and proposes a reduction in energy use of 9% in each member state, to be achieved by the ninth year of application of the directive (EC, 2006). The directive addresses a number of policy activities and energy efficiency services, such as the availability of energy auditing for SME industries. It also highlights the availability of energy efficiency funds to all market actors and promotes energy audits and financial incentives for the adoption of energy efficiency measures and energy services (EC, 2006). The directive stresses, among other things, the need to find possible energy end-use policy initiatives directed at SMEs in a national context: "*In order to enable final consumers to make better informed decisions as regards their individual energy consumption, they should be provided with a reasonable amount of information thereon and with other relevant information, such as information on available energy efficiency improvement measures.*" (EC, 2006). The purpose of the ESD is to enhance the cost-effective improvement of energy end-use efficiency in Member States by: *"(a) providing the necessary indicative targets as well as mechanisms, incentives and institutional, financial and legal frameworks to remove existing market barriers and imperfections (market failures) that impede the efficient end use of energy and (b) creating the conditions for the development and promotion of a market for energy services and for the delivery of other energy efficiency improvement measures to final consumers."* (EC, 2006). Companies participating in the EU ETS (EU Emission Trading Scheme) are however not included in the directive (EC, 2006).

It has been left out to each member state to design and adopt National Energy Efficiency Action Plans (NEEAPs) in order to fulfill the ESD. In the ESD as well as in the reports written in regard to the Swedish NEEAP, on behalf of the directive, it has been stated to be of major importance that the authorities are acting as role models. Moreover, the Swedish authorities have considered to include even actors participating in the ETS, but which energy use is not included in the directive, e.g. electricity (EEC, 2008).

OVERALL AIM

The introductory part of this chapter raises the question how one could formulate and adopt accurate policy/policies towards Swedish industrial non-energy-intensive manufacturers and SMEs? This chapter aims to address this question by discussing plausible policy solutions, based on current research on barriers to energy efficiency and industrial energy programs. Initially, results from the three available Swedish studies on barriers to energy efficiency for the regarded sectors are outlined. Then, a few examples of successful and less successful actions towards the industry and in particular SMEs are presented and finally, a discussion is held regarding plausible Swedish industrial energy end-use polices, based on the presented research.

Barriers to Energy Efficiency and Industrial Energy Programs

As regards studies on barriers to energy efficiency, there are a few international studies (Sorrell et al., 2000, Schleich, 2004, Schleich and Gruber, 2008). In Sweden there have been four scientific studies on the issue (Rohdin and Thollander, 2006, Rohdin et al., 2007, Thollander et al., 2007, and Thollander and Ottosson, 2008). In Table 1, the largest barriers to energy efficiency in the different studies regarding non-energy-intensive and small- and medium sized industries are outlined. For a more detailed presentation, it is recommended to read each study separately.

Energy information programs have internationally been an important means of stressing industrial energy end-use efficiency. The American Information Assessment Centers (IAC) program began in 1976 and has provided energy audits to more than 10 000 small- and medium-sized manufacturers (Andersson and Newell, 2004), where more than half of the recommended projects were adopted, the main reason for non-adoption being that it was economically undesirable. The authors conclude however that: "*other, more costly policies targeted at increasing the financial attractiveness may be needed to further promote energy efficiency in this sector.*" (Anderson and Newell, 2004). The Australian Commonwealth Government's Enterprise Energy Audit Program (EEAP) was another large-scale energy efficiency program covering some 1 200 firms, offering energy audits to a 50% discount between 1991 and 1997 where the adoption rate was more than 80% (Harris et al., 2000). Evaluation results from the two programs revealed that the higher the average cost of the energy efficiency investment, the less likely it was that a recommendation was actually implemented (Harris et al., 2000, Anderson and Newell, 2004).

Another policy instrument, the Dutch Energy-Bonus, offered investment subsidies from 1980 to the end of 1990. An evaluation of the latter concluded that it did not seem very effective in stimulating investments in energy efficiency, and the evaluation estimated that the free-rider effect was above 85%, i.e. less than 15% of the investments were carried out as a consequence of the Dutch Energy Bonus (Farla and Blok, 1995).

Yet another study by Gruber and Brand (1991) found that a grant for energy audits in German SMEs only showed limited success. Many SMEs were not aware of the grant and a large portion were reluctant to use it as the outcome of future benefits was uncertain, i.e. the costs of the audit (with the grant) may exceed future potential energy savings at the firm (Gruber and Brand, 1991).

Another widely used instrument within many European countries is Long-Term Agreements (LTAs) or Voluntary Agreements (VAs). Such instruments have been given increased attention; Bertoldi (1999) estimated the saving potential of LTAs within the EU to be about 20% if at least 70% of the non-energy-intensive industry is included. In Sweden, the two industrial energy programs since the 1990s, launched from the Swedish government, EKO-energi and PFE (Program for improving energy efficiency in energy-intensive industries), both aimed towards the energy-intensive industry, have consisted of such an instrument. EKO-Energi, a VA-concept, began in 1994 and lasted until 2001 and engaged some 72 large energy-intensive customers. An evaluation of the program found that in particular the audits had given increased priority to energy efficiency and environmental issues. (Uggla and Avasoo, 2001). The other program PFE, and which is still running, involves some 120 companies (Ottosson and Peterson, 2007) where the firms in the program are offered a tax discount on the recently introduced tax on electricity for the manufacturing industry, 0.55 Euro/MWh if the company fulfil the requirements (Ottosson and Peterson, 2007).

Table 1. A summary of the largest barriers to energy efficiency spotted in the different case studies (Based on Rohdin and Thollander (2006), Rohdin et al. (2007), and Thollander et al. (2007))

	Non-energy-intensive manufacturers		Swedish foundries		Swedish Highland SMEs
1	Cost of prod. disruption/hassle/ inconvenience	1	Access to capital	1	Lack of time or other priorities
2	Lack of time or other priorities	2	Technical risks such as risk of prod. disruptions	2	Other priorities for capital inv.
3	Difficulty/Cost of obtaining info. on the energy consumption of purchased equipment	3	Lack of budget funding	3	Access to capital
4	Technical risks such as risk of prod. disruptions	4	Difficulty/Cost of obtaining info. on the energy consumption of purchased equipment	4	Cost of prod. disruption/hassle/ inconvenience
5	Other priorities for capital inv.	5	Other priorities for capital inv.	5	Lack of budget funding
6	Technology is inappropriate at this site	6	Possible poor performance of equipment	6	Lack of sub-metering
7	Lack of staff awareness	7	Lack of sub-metering	7	Difficulty/Cost of obtaining info. on the energy consumption of purchased equipment
8	Lack of technical skills	8	Poor information quality regarding energy efficiency opportunities	8	Lack of technical skills
9	Access to capital	9	Cost of identifying opportunities analyzing cost effectiveness and tendering	9	Low priority given to energy management
10	Poor information quality regarding energy efficiency opportunities	10	Low priority given to energy management	10	Lack of staff awareness
11	Possible poor performance of equipment	11	Lack of time or other priorities	11	Technical risks such as risk of prod. disruptions
12	Cost of identifying opportunities analyzing cost effectiveness and tendering	12	Technology is inappropriate at this site	12	Slim organization

Table 2. Swedish industrial energy efficiency programs (Thollander et al., 2007)

Energy program, year	Type of program	Industries	Quantitative evaluation	Qualitative evaluation	Sub-sidiaries	Calculated energy efficiency potential
EKO-Energi, 1994-2001	Voluntary agreements,	72 large energy-intensive	N.a.	Increased priority to energy and environment	Public sponsored audit	N.a.
PFE, 2005-	Long term agreements	117 energy-intensive	Electricity saving	N.a.	Tax discount	N.a.
SEA-seminars, 2006	Seminars, information	N.a.	N.a.	Increased awareness, low implement-tation	N.a.	N.a.
Project Highland, 2003-2008	Energy audits	340 small and medium-sized	Energy (including electricity) saving	Barriers and driving forces, interviews	Public sponsored audit	Electricity savings , total energy savings
Sparkraft, 2000-2003	Energy audits	Mainly service sector	N.a.	N.a.	Public sponsored audit	N.a.
Oskarshamn, 2000-2001	Energy audits	9 largest in-dustries in Oskarshamn	N.a.	Barriers and driving forces	Public sponsored audit	Electricity saving 48%, total energy saving 40%
Elost,	Energy audits	7	N.a.	N.a.	Public sponsored audit	Electricity saving 58%
Energieffektiva Västra Göta-land, -2005	Energy audits	9	N.a.	N.a.	Public sponsored audit	Total energy saving 16%
Sustainable municipalities, 2004-2006	Energy audits	40	N.a.	N.a.	Public sponsored audit	Electricity saving 20-60%, total energy saving 30-38%

As stated previously, the Swedish government's support towards energy end-use efficiency for the non-energy-intensive and small- and medium-sized manufacturing industry as a whole, has been scarce. Thollander et al. (2007) have compiled the Swedish programs which have taken place since the beginning of the 1990s. These are outlined in Table 2.

In Thollander et al., 2007, the authors also evaluated the first part of the largest Swedish industrial energy program, in terms of participating companies, project Highland. The model of the evaluated program, up until the audit was conducted, is outlined in Figure 1.

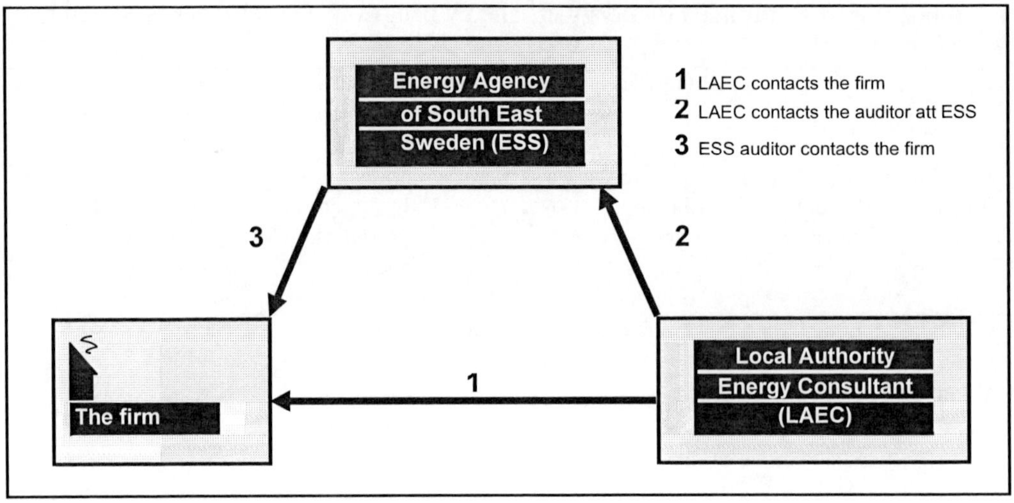

Figure 1. The beginning of an energy audit in project Highland (Thollander et al., 2007).

POLICY DISCUSSION

In sectors where technical risk such as the risk of production disruptions and the cost of production disruption/hassle/inconvenience are considered to be barriers of major importance, there is a need for thorough studies regarding energy efficiency issues before energy efficiency measures can be justified. In the foundry industry, one such means and which have shown to be successful is the MIND method, where optimization of different investment options are possible (Thollander et al., 2008). Yet another such means is other types of investment decision support such as manufacturing simulation, e.g. Solding and Thollander (2006), which have been shown to lower the use in the production processes by up to 10 %.

Another type of barrier is the possible poor performance of purchased equipment considered to be a barrier of somewhat importance in the energy-intensive foundry industry (6th) while the barrier showed to be of less importance in the studies comprising non-energy-intensive manufacturers and SMEs (11th and 20th respectively). In sectors where this barrier is of greater importance, plausible policy measures could be practical research on different energy efficiency measures and examples of successfully implemented energy efficiency measures in different industries.

Lack of time and other priorities was considered to be a major barrier in all studies except for the foundries. As regards SMEs and non-energy-intensive manufacturers this clearly indicates a need for simplicity when it comes to adopting energy end-use efficiency policies. While larger energy-intensive companies have resources to join for example an LTA program such as the Swedish PFE, non-energy-intensive manufacturers and industrial SMEs in particular do not have the time to invest in such activity; neither do they have much time to apply for different investment funds.

One policy means which has been used in e.g. the Netherlands, and which is also proposed in the ESD is that of investment funds. The fact that lack of access to capital was considered to be a barrier of major importance in all but one study indicates that this might be a sound policy adoption. However, the use of the latter in the Dutch industry revealed that the free-rider effect using such a measure was at least 85 % (Farla and Block, 1995). Among German SMEs, using a subsidy for energy audits did not prove to be successful. Swedish experience from the 1970s indicates similar results - only about 10 % of the Swedish companies used investment funds (Persson, 1990). Interestingly, results from the presented barriers studies seem to indicate why – energy is not prioritized among non-energy-intensive manufacturers and SMEs in particular. This means that even though investment funding is offered as a policy measure it will most likely not be shown to be successful in companies where lack of time and other priorities is a major barrier, i.e. in non-energy-intensive industries. On the contrary, such measure would most likely be a success in, e.g. the Swedish foundry industry, where lack of time and other priorities is not a high ranked barrier. These companies will most likely take the time to fill out forms and apply for such funds. Addressing policy measures toward the former, an LTA may be sound approach. Results from the Swedish pulp- and paper industry regarding driving forces indicate that the current LTA, the Swedish PFE, is much appreciated (Thollander and Ottosson, 2008). As regards non-energy-intensive manufacturers and SMEs in particular, a plausible policy measure, and which have been shown to be successful in both Sweden (Thollander et al., 2007) and internationally (Anderson and Newell, 2004) is to offer energy audits free of charge. Interview results from the evaluation of the Swedish project Highland indicate that such an approach is also useful for lowering barriers such as lack of budget funding and other priorities for capital investments – barriers which have shown to be of major importance in several of the Swedish studies on barriers to energy efficiency. Results from Germany also indicate that energy audits, lowers barriers to energy efficiency (Schleich, 2004). The question then remains who are to carry out such audits toward the non-energy-intensive sectors and SMEs. Using actors which are considered trustworthy is of importance which indicates that sector organizations, in cases where they are considered trustworthy, would be a sound approach. However, the regional Swedish energy agencies could also be used. One important thing however, needs to be added regarding the issue –both in the ESD as well as in the reports written in regard to the Swedish NEEAP, on behalf of the directive, it has been stated to be of major importance that the authorities are acting as role models models (EC; 2006, Neij, 2007).

Conclusion

In conclusion, the model presented in Figure 1 in this chapter may be a highly useful approach towards non-energy-intensive manufacturers and SMEs in particular, not only and which have been stated in the evaluation, from a public cost point of view, but also from the perspective of the results from the available Swedish studies regarding barriers to energy efficiency as well as the fact that such a model signals interest from the authorities regards the energy end-use efficiency issue. Moreover, using that model, the local authority energy consultants are able to gain more experience in the area and thus be able to provide the industrial sectors with more thorough support. However, the conductor of the audit does not necessarily have to be the public – using energy consultants would in the long run build up a greater energy audit competence in Sweden. It is left out to the Swedish authorities to design the NEEAP and later adopt the proposed policy measures. Based on the results presented in this chapter, a policy approach towards non-energy-intensive manufacturers and industrial SMEs should primarily include the providing of energy audits free of charge and involve the local authority energy consultants. After the provision of energy audits, LTAs, investment funds and perhaps, when investment decision support have been studied more thoroughly from a policy, perspective also include such means as well.

ACKNOWLEDGMENTS

The work has been carried out under the auspices of the Energy System Program, which is financed by the Swedish Foundation for Strategic Research and the Swedish Energy Agency. I kindly thank Mats Söderström and Jenny Palm, Linköping University, for valuable comments on earlier drafts of this paper. The usual disclaimer holds.

REFERENCES

Anderson, S.T., Newell, R.G., 2004. Information programs for technology adoption: the case of energy-efficiency audits. *Resource and Energy Economics* 26 (1), 27-50.

Bertoldi, P., 1999. *The use of Long Term Agreements to improve energy efficiency in the Industrial sector: Overview of the European experience and proposals for a common framework.* In Proceedings of the 1999 SAVE conference "For An Energy Efficient Millenium", Session III, 1-10.

EC (European Commission), 2006. Directive 2006/32/EC of the European Parliament and of the Council of 5 April 2006 on energy end-use efficiency and energy services and repealing Council Directive 93/76/ EEC, Brussels.

EEC (Energy Efficiency Committee), 2008. *Ett energieffektivare Sverige* [A more energy efficient Sweden]. SOU 2008:25. Stockholm.

Farla, J.C.M., Blok, K., 1995. Energy conservation investment behaviour of firms: analysis of investments in energy efficiency in the Netherlands in the 1980s. In: *Proceedings of the 1995 ACEEE summer study on energy efficiency in industry,* Washington D.C.

Gruber, E., Brand, M., 1991. Promoting energy conservation in small and medium-sized companies. *Energy Policy* 19, 279–287.

Harris, J., Andersson, J., & Shafron, W., 2000. Investment in energy efficiency: a survey of Australian firms. *Energy Policy* 28 (12), 867-876.

Neij, L., 2007. Barriärer och styrmedel för en effektivare energianvändning [Barriers and policies for an effective energy use]. Retrieved 1 September, 2008, from the Swedish Government's homepage: http://www.sou.gov.se/Energieffektiv/pdf/Barri%C3%A4 rer%20och%20styrmedel%20f%C3%B6r%20en%20effektivare%20energianv%C3%A4ndning.pdf

Ottosson, C., & Peterson, K. (2007). First results from the Swedish LTA programme for energy efficiency in industry. Paper presented at the 2007 European Council for an Energy-Efficient Economy (ECEEE) summer study *"Saving energy – just do it",* 1517-1525.

Persson, A., 1990. Nyttiga lärdomar. Erfarenheter av industriell energirådgivning [*Important lessons. Experience from industrial energy consulting*]. Statens industriverk. Stockholm.

Rohdin, P., Thollander, P., & Solding, P., 2007. Barriers to and drivers for energy efficiency in the Swedish foundry industry. *Energy Policy* 35 (1), 672–677.

Rohdin, P., Thollander, P., 2006. Barriers to and driving forces for energy efficiency in the non-energy intensive manufacturing industry in Sweden. *Energy* 31 (12), 1836–1844.

Schleich, J., & Gruber, E., 2008. Beyond case studies: Barriers to energy efficiency in commerce and the services sector. *Energy Economics* 30 (2), 449–464.

Schleich, J., 2004. Do energy audits help reduce barriers to energy efficiency? An empirical analysis for Germany, *International Journal of Energy Technology and Policy* 2 (3), 226–239.

Solding, P., Thollander, P., 2006. Increased energy efficiency in a Swedish iron foundry through use of discrete event simulation. *Paper presented at the 2006 Winter Simulation Conference* (WSC), 1971-1975.

Sorrell, S., Schleich, J., Scott, S., O'Malley, E., Trace, F., Boede, E., Ostertag, K. & Radgen, P., 2000. *Reducing Barriers to Energy Efficiency in Public and Private Organizations.* Retrieved May 7, 2008, from: http://www.sussex.ac.uk/Units/spru /publications/reports/ barriers/final.html

Thollander, P., Mardan, N., & Karlsson, M., 2008. *Optimization as investment decision support in a Swedish medium-sized iron foundry – a move beyond traditional energy auditing.* Submitted for Journal Publication

Thollander, P., Ottosson, M., 2008. An energy efficient Swedish pulp and paper industry – exploring barriers to and driving forces for cost-effective energy efficiency investments. *Energy Efficiency* 1 (1), 21-34.

Thollander, P., Danestig, M., & Rohdin, P., 2007. Energy policies for increased industrial energy efficiency: Evaluation of a local energy programme for manufacturing SMEs. *Energy Policy* 35 (11), 5774–5783.

Uggla, U., Avasoo, D., 2001. *EKO-Energi - Successful voluntary agreements on energy efficiency and environmental control in Swedish industry.* Paper presented at the 2001 European Council for an Energy-Efficient Economy (ECEEE) summer study "Further than ever from Kyoto? *Rethinking energy efficiency can get us there"*, 252-260.

INDEX

D

E

F

G

H

I

J

K

L

M

S

T

U